from Windfall to Curse?

Jonathan Di John

from
Windfall
to Curse?

OIL AND INDUSTRIALIZATION IN VENEZUELA,
1920 TO THE PRESENT

THE PENNSYLVANIA STATE UNIVERSITY PRESS
UNIVERSITY PARK, PENNSYLVANIA

Library of Congress Cataloging-in-Publication Data

DiJohn, Jonathan.
From windfall to curse? : oil and industrialization in Venezuela, 1920 to the present / Jonathan Di John.
    p.   cm.
Includes bibliographical references and index.
Summary: "Examines the political economy of growth in Venezuela since the discovery of oil in 1920"—Provided by publisher.
ISBN 978-0-271-03553-6 (cloth : alk. paper)
1. Venezuela—Economic conditions—1918–1958.
2. Venezuela—Economic conditions—1958–.
3. Venezuela—Economic policy.
4. Petroleum industry and trade—Venezuela.
5. Industrialization—Venezuela.
I. Title.

HC237.D55 2009
338.2′728209870904—dc22
                        2009016316

Copyright © 2009 The Pennsylvania State University
All rights reserved
Printed in the United States of America
Published by The Pennsylvania State University Press,
University Park, PA 16802-1003

The Pennsylvania State University Press is a member of the Association of American University Presses.

It is the policy of The Pennsylvania State University Press to use acid-free paper. Publications on uncoated stock satisfy the minimum requirements of American National Standard for Information Sciences—Permanence of Paper for Printed Library Material, ANSI Z39.48–1992.

CONTENTS

Preface and Acknowledgments vii
List of Abbreviations xv

PART ONE: INTRODUCTION

1 Accounting for Growth and Decline in Venezuela 3
2 Trends and Cycles in the Venezuelan Economy 15

PART TWO: A CRITICAL SURVEY OF THE "RESOURCE CURSE" LITERATURE

3 Economic Explanations of the Growth Collapse in Venezuela 35
4 Political Economy Explanations of the Growth Collapse in Venezuela 77
5 Economic Liberalization, Political Instability, and State Capacity in Venezuela 108

PART THREE: AN ALTERNATIVE POLITICAL ECONOMY OF VENEZUELAN GROWTH AND DECLINE

6 Toward a New Political Economy of Late Industrialization 133
7 Periodization of Industrialization Stages and Strategies in Venezuela 169
8 The Structure of and Changes in Political Settlements in Venezuela 186
9 A New View on the Political Economy of Growth in Venezuela 226

PART FOUR: BEYOND THE VENEZUELAN CASE

10 The Political Economy of Growth in Malaysia and Venezuela 271
11 Conclusion: Rethinking the Political Economy of Growth 285

References 303
Index 327

PREFACE AND ACKNOWLEDGMENTS

Understanding why the wealth of less developed nations increases over time is one of the oldest concerns of political economy and development economics. Reigning theories of economic growth postulate that there are linear tendencies in the growth process. Depending on the assumptions, growth theories predict either divergence or convergence of the income per capita of less developed counties trying to catch up with advanced industrial countries. Reality paints a different picture of the growth process. The fastest-growing countries are never the countries with the highest per capita incomes, but always a small subset of lower-income countries. Economic theories of growth are thus of limited use in explaining the wide and persistent divergence of performance among late developing countries. Furthermore, growth theories cannot explain why some countries sustain rapid growth for long periods only to fade into long periods of stagnation and even downward spiral and disintegration. A major challenge of political economy is to explain why economic growth varies so widely within countries over time.

The Venezuelan experience since the discovery of oil in the 1920s provides one such interesting example. Venezuela was among the fastest-growing economies in Latin America in the period 1920–80. However, two important trends mark the period 1968–2005. First, from the mid-1960s the non-oil economy, and particularly the manufacturing sector, experienced a dramatic decline in non-oil and manufacturing productivity growth. Second, output growth in the manufacturing sector (and the non-oil economy more generally) collapsed in the period 1980–2003. Accompanying this decline was the virtual collapse of once vibrant political parties and the rise of a more volatile and unstable form of antiparty politics and governance in the first decade of the 2000s.

More perplexing is the fact that worsening economic performance coincided with many favorable initial conditions such as high levels of human and physical capital investment, a competitive democratic system since 1958, and plentiful resources to finance growth. Was this worsening economic performance due to policy errors or inappropriate institutions, or did oil windfalls themselves become a "curse" by crowding out the development of non-oil sectors such as manufacturing, or were there other factors that explain the slowdown in growth? This book attempts to answer these questions.

Explaining what factors contribute to growth (such as high levels of investment, secure property rights, macroeconomic stability) does not tell us how or why such factors are achieved. This is where role of the state becomes relevant, since the state is the set of institutions that is responsible for the creating and implementing the regulatory structure of the economy, which determines the incentives of economic actors. One of the oldest (and most ideologically charged) debates in the political economy of development is identifying what types of state intervention are most likely to promote sustained economic growth.

Advocates of laissez-faire suggest that state intervention distorts markets, stifles competition, and generates corruption, all of which create obstacles to potentially growth-enhancing investments. In recent times, this view has dominated as manifested in the adoption of widespread economic liberalization, particularly within Latin America. The outcome of such reforms in Venezuela as well as in most of the region, however, has been disappointing. Economic liberalization has failed to revive investment and growth to the levels of 1950s and 1960s, when more interventionist policies were followed. While the state in Venezuela and many other less developed economies have failed to generate and implement growth-enhancing regulatory structures, market liberalization does not eliminate the market failures and risks of late development that justified state intervention in the first place. The failure of economic liberalization to deliver positive developmental outcomes is not confined to the world of less developed countries. For instance, advocates of laissez-faire saw little need to regulate financial markets because it was thought that market competition would weed out poorly performing investments and thus generate socially productive resource allocation. The world financial and economic crises that emerged in 2008 suggest that such confidence in lightly regulated financial markets was unfounded.

In contrast, developmental state theorists have pointed to the benefits that state intervention, and particularly industrial policy, can have in promoting learning and technology acquisition, and in socializing the risks infant industries face in the context of trying to catch up with more experienced and productive firms in advanced industrial countries. Advocates of more dirigiste strategies will point to the effectiveness of various types of industrial strategies in East Asia. And indeed, the Venezuelan state attempted to implement state-led industrial policy from the late 1950s. While the potential benefits of state intervention may be accepted, developmental state theorists do not explain why most countries that tried policies similar to those of more successful industrial policy states have been far less successful.

Preface and Acknowledgments..ix

The historical evidence suggests that there are a variety of models of state intervention that can promote economic development. The broader question is why leaders in economies undergoing economic decline do not alter the regulatory structure that approximates one of the various production strategies that could enhance growth. The analytical narrative of the Venezuelan growth process since the 1920s presented in this book also attempts to answer this broad question.

Part One summarizes the main argument and introduces the main economic trends in the Venezuelan economy. Part Two provides an extensive critical examination of arguments have dominated discourse on Venezuela's long-run economic decline in the post-1968 period. The most important is the so-called resource curse, which is the idea that oil abundance harms the prospects of economic growth. There is a vast "resource curse" literature in economics, political science, and political economy which attempts to explain the negative effects oil windfalls and busts can have on the structure of the economy and on patterns of governance.

The main problem with these arguments is that they cannot explain why oil abundance has been compatible with cycles of growth and stagnation in Venezuela over the period 1920–2005. Comparative evidence presented in this book suggests that Venezuela has not been an "exceptional" case in Latin America. Problems of growth slowdown, external debt, and capital flight (which featured in post-1980 Venezuela) were a feature of most countries in the region. This suggests that understanding Venezuela's growth trajectory may have resonance for other middle-income Latin American and oil-abundant economies in other regions.

Part Three offers a new explanation of the long-run cycles of growth and decline that Venezuela has experienced since 1920. It argues that reigning explanations of Venezuela's growth slowdown (as well as the more general literature on state intervention whether from a neoliberal or developmental state perspective) do not examine the extent to which different development strategies and stages of import-substitution affect the economic and political challenges and conflicts that state policymakers face in attempting to implement industrial policies. Detailed historical evidence presented in Part Three suggests that the process of industrialization involved two very different strategies over time. The period from the early 1900s to the late 1960s was a period of "easy" import-substitution strategies, where infant industry production tends to be technologically simple and small-scale and where require major coordination of investment by the state is not essential for rapid growth to occur. Moreover, export growth was not central to the prospects of sustained capital accumulation under this strategy.

The period 1960–73 is characterized by a switch in the development strategy that emphasized more advanced import-substituting industrialization (ISI) in intermediate goods and heavy industry, and in particular a "big push" in natural-resource-based industries in oil refining, steel, aluminum, chemicals, and hydroelectric power. This more advanced ISI strategy dominated the state-led industrial development strategy from the early 1970s on. What distinguishes the advanced ISI/big-push strategy from the "easy ISI" strategies is that infant industrial production tends to be more technologically demanding, large-scale, and longer-gestating and tends to require major coordination of large-scale investments by the state and private sector for effective implementation.

A main contribution of this book is to demonstrate how and why this distinction between these two types of development strategy matters for understanding the growth trajectory of the Venezuelan economy. In early ISI strategies, state-led industrial policy is not very demanding in either economic or political terms. Because scale economies were generally small, and exporting not decisive to firm development at the early stage of ISI, the state was not compelled to be selective in the distribution of protection and subsidies, nor was it necessary that state be able to discipline the recipients of subsidies. In contrast to the early ISI strategies, it is shown that the capacity of the state to coordinate industrial investment, to be selective in the subsidy process, and to discipline the recipients of subsidies becomes relevant for the effective implementation of more advanced ISI/big push industrialization strategies. This is because scale economies are more central to the development of competitive firms, and because many of the investments are interdependent.

More advanced ISI strategies require a polity that is capable of coordinating investments and effectively monitoring the deployment of subsidies to infant industries. This in turn means that large-scale fragmentation of political organizations and the state and major disagreements over policy among contending political parties, factions of capital, and labor unions are likely to negatively affect productivity and output growth, particularly in the manufacturing sector. Given the contingent nature of political institutions and contests, there is no reason to expect that the appropriate institutional structure and politics will emerge to accommodate a country's stage of development and changing technological challenges. The mapping of political strategies and types of development trajectory followed opens up the possibility of explaining growth accelerations as the result of the compatibility between development strategy and political settlements and, analogously, growth slowdowns as the result of a growing incompatibility, or mismatch between political settlements and development strategies.

The final chapter of Part Three maps the evolution of development strategies and political settlements in the period 1920–2005. In the course of this mapping exercise (which is based on detailed historical, economic, and political literature on Venezuela), it is demonstrated how and why growth was achieved as a result of a broad compatibility between the development strategy and the organization of political competition in the period 1920–68, and how and why output and productivity growth stagnated and eventually collapsed as a result of a growing incompatibility between the development strategy and organization of political competition in the period 1968–2005. It is argued that theories of the "resource curse," as well as more general theories of state intervention (whether of the neoliberal or developmental state variety), will be better able to explain variations and changes in growth within and across less developed economies by incorporating historically specific politics (especially the structure of political organizations and the shifting grounds of legitimate rule) and production (stage of development and development strategy) in a more interactive way than is generally undertaken. The arguments in this book also suggest that an understanding of the variation and change in growth across and within mineral-abundant less developed economies needs fundamental rethinking that moves beyond the simplistic notions of "windfall" and "curse."

There are many who have provided invaluable support and guidance through the often lonely and dark wood of a long-term research project. I would first of all like to thank Gabriel Palma, who taught me many lessons (during my doctoral studies at the University of Cambridge) on the economic and political history of Latin America that formed the basis of much of my thinking on Venezuela. His critical analysis of my work made correcting earlier versions of this book a rewarding experience. His unending encouragement and friendship have been and continue to be an invaluable source of inspiration.

I am indebted to many of my colleagues and friends who have inspired and helped me to think through difficult concepts and ideas. In particular, I want to thank Christopher Cramer, Jonathan Pincus, Dennis Rodgers, and Ha-Joon Chang for invaluable support. John Sender was one of the best mentors an aspiring development economist could have. He really changed the way I viewed development processes and inspired me, as a master's student in Cambridge, to explore the difficult terrain of political economy. His insistence on the importance of critical thinking and the marshaling of supporting evidence has remained with me. I especially want to thank Mushtaq Khan, who has not only been a close friend, but whose thinking and scholarship on the political economy of institutions have profoundly influenced my ideas. James Putzel, a colleague during my six years as a lecturer at the London School of Economics and still a close and dear friend, read many earlier versions of this work, and his constructive critical eye helped me greatly. James instilled in me the importance of examining history and politics, but his incalculable support, encouragement, and intellectual stimulation provided me with the confidence to find my way out of the dark wood. Finally, I want to thank Peter Nolan and John Weeks, who both read earlier versions of this manuscript and provided valuable comments.

Many thanks to the Crisis States Research Center at the London School of Economics for providing research funding, which permitted me to gather information on the post-1998 period, especially relevant for Chapters 7 and 8.

There are many who made my eight-year residence in Venezuela a rewarding experience both intellectually and personally. Dr. Jaime Torres helped orient me to the country and provided valuable guidance and friendship throughout my first years. This project would never have seen the light of day without the many academics and friends at the Instituto Superiores de Administración (IESA) in Caracas, who provided me with valuable guidance and contacts that helped me explore the magic realism of the Venezuelan political system and economy. Ricardo Hausmann, whom I first met in England, provided the first encouragement for me to study Venezuela and helped me get settled at IESA. His wit, intellect, and passion for economic development are contagious. I also want to thank Asdrúbal Baptista, Gustavo Garcia, Rosa Amelia González, José Malavé, Gustavo Márquez, Ramón Piñango, and the late Janet Kelly for all their help. I particularly am indebted to Carmen Portela, who always believed that my exploration of Venezuelan economic history was worthwhile and who provided support and friendship throughout my time in Venezuela and with whom I share a wonderful bond through our daughter, Iria.

I also would like to thank Francisco Rodríguez, who shared his impressive command of economic theory and historical knowledge of Venezuela, and who provided very insightful comments on earlier drafts. I would equally like to thank Javier Corrales, who also read earlier drafts and offered very useful comments, particularly on the evolution of the Venezuelan political system.

I am also grateful to several people to those who agreed to extensive interviews that contributed substantially to my knowledge of important economic and political trends in Venezuela. I would especially like to thank Alberto Cudemus, José Giacopinni Zárraga, Marcel Granier, Margarita López Maya, Tibisay Lucena, Osmel Manzano, Michael Penfold, Luis Lander, Gumersindo Rodríguez, and Miguel Rodríguez.

My wife, Anne, has been a constant inspiration. Her gentle grace and unconditional support made me see the light in the dark wood, and I am forever grateful. The completion of this work has coincided with the arrival of our daughter, Sophia, who has provided both joy and inspiration in the final editing of the manuscript.

I owe the deepest gratitude to my brother David and my sister Joanne for all their moral support and friendship. I am most indebted to my parents, Rose and John, who have never given up on me and whose affections and support always enabled me to restore my tranquillity, belief, and perseverance. I dedicate this work to them.

ABBREVIATIONS

AD	Acción Democrática (Venezuelan Social-Democratic Party)
ALCASA	Aluminio del Caroní (state aluminum company)
BAUXIVEN	Bauxita de Venezuela (state bauxite company)
BCV	Banco Central de Venezuela (Central Bank of Venezuela)
CANTV	Compañía Anónima Nacional Teléfonos de Venezuela (state telephone company)
CENDES	Centro de Estudios del Desarrollo
CEPR	Center for Economic and Policy Research
CNE	Consejo Nacional Electoral (National Electoral Commission)
COPEI	Comité de Organización Política Electoral Independiente (Christian Democratic Party)
COPRE	Comisión Presidencial para la Reforma del Estado (Presidential Commission on State Reform)
CORDIPLAN	Oficina Central de Coordinación y Planificación de la Presidencia de la Repúbica (Central Office for Coordination and Planning)
CRIAP	Comisión para el Estudio de la Reforma Integral de la Administración Pública (Public Administration Reform Commission)
CTV	Condeferación de Trabajadores de Venezuela (Confederation of Venezuelan Workers)
CVG	Corporación Venezolana de Guyana (Venezuelan Guyana Corporation, state holding company)
ECLAC	Economic Commission of Latin America and the Caribbean
FEDECAMERAS	Federación Venezolana de Cámeras y Asociaciones de Comercio y Producción (main Venezuelan business chamber)

FINEXPO	Fondo de Exportaciones (Venezuelan state export credit fund)
FIV	Fondo de Inversiones de Venezuela (Venezuelan investment fund)
GDP	gross domestic product
IADB	Inter-American Development Bank
ICOR	incremental capital output ratio
IESA	Instituto Superiores de Administración (leading Venezuelan business school)
IMF	International Monetary Fund
INE	Instituto Nacional de Estadística (Venezuelan national statistics office)
INTEVEP	Instituto de Tecnología Venezolana para el Petróleo (Venezuelan Petroleum Technology Institute)
IPC	International Poverty Centre
ISI	import-substituting industrialization
ISIC	international standard industrial classification
LDC	less developed country
LOSPP	Ley Orgánica de Salvaguardia del Patrimonio Público (Venezuelan state property protection law)
MBR-200	Movimiento Bolivariano Revolucionario-200 (Bolivarian Revolutionary Movement 200)
MVR	Movimiento Quinta República (Fifth Republic Movement)
NBER	National Bureau of Economic Research
NEP	New Economic Policy (Malaysian state development plan)
NIC	newly industrializing country
NRBI	natural resource-based industrialization
OCEI	Oficina Central de Estadística e Informática (Central Office of Statistics and Information)
OCEPRE	Oficina Central del Presupuesto (National Budget Office)
OECD	Organization for Economic Cooperation and Development

OPEC	Organisation of Petroleum Exporting Countries
PDVSA	Petróleos de Venezuela (state oil company)
PE	public enterprise
PEQUIVEN	Petroquímica de Venezuela (state petrochemical company)
POSCO	Pohang Iron and Steel Company, Ltd. (South Korean state steel company)
PP	Patriotic Pole (alliance of political parties supporting Hugo Chávez)
PSUV	Partido Socialista Unida de Venezuela (United Socialist Party of Venezuela)
RECADI	Régimen de Cambios Diferenciales (Preferential Exchange Regime)
SIDOR	Siderúrgica de Orinoco, S.A. (state steel company)
SOE	state-owned enterprise
TFP	total factor productivity
UMNO	United Malays National Organization
UNCTAD	United Nations Committee on Trade and Development
UNDP	United Nations Development Programme
UNESCO	United Nations Educational, Scientific, and Cultural Organization
UNIDO	United Nations Industrial Development Organization
URD	Unión Republicana Democrática (Democratic Republican Union party)
VENALUM	Industrias Venezolanas de Alumunio (state aluminum company)
WIDER	World Institute for Development Economics Research
WTO	World Trade Organization

PART 1

INTRODUCTION

1

ACCOUNTING FOR GROWTH AND DECLINE IN VENEZUELA

This work examines the political economy of industrial policy and economic growth in Venezuela in the period 1920–2005. Soon after the discovery of oil in Venezuela during the 1920s, the idea of "sowing the oil," that is, diversifying the production and export structure, was an important organizing concept among economic and political elites (Baptista and Mommer 1987). Since the early 1950s, the industrialization process became increasingly state led. The role of the state was marked by a purposeful policy of import substitution that coincided with the transition to democracy in 1958. Moreover, political leaders set Venezuela on a path of a more pronounced state-led, "big-push" natural-resource-based heavy industrialization that focused on the development of state-owned enterprises in steel, aluminum, petrochemicals, oil refining, and hydroelectric power. The Venezuelan state-led, big-push heavy industrialization strategy was not dissimilar in intent, ambition, and scope from state-led industrialization strategies in South Korea, Taiwan, Malaysia, and Brazil, as well as some other oil-exporting developing economies.

Venezuela's growth trends have, however, several distinguishing and perplexing characteristics. On the one hand, Venezuela was among the fastest-growing economies in Latin America from 1920 to 1965, and its manufacturing growth rate was among the most rapid until the mid-1970s. However, from the mid-1960s the non-oil economy, and particularly the manufacturing sector, experienced a dramatic decline in total factor productivity growth and labor productivity growth. Then, output growth in the manufacturing sector (and the non-oil economy more generally) collapsed in the period 1980–2003. The only factor that kept growth rates afloat in the 1970s was the reflationary impact of high government spending as a result of two oil booms. Even so, after 1980 Venezuela's growth was among lowest of all late-developing economies.

A profound political crisis accompanied this decline. From the early 1990s to 2005, the Venezuelan polity was characterized by deteriorating political order, growing instability, and legitimacy crises, characterized by increasingly frequent coup attempts, alarming increases in voter absenteeism, the growing use of corruption scandals as instruments of political competition, and the virtual disappearance of the traditional political parties that were central to a country that was once considered a model of democratic stability. In contrast to the literature that studies crises in authoritarian rule (e.g., Malloy 1977), or in transitions to democracy (e.g., Przeworski 1991), this book examines the tensions and processes of late development within a long-standing democracy.

Such dramatic changes in long-run growth rates are not well understood in mainstream economic theory. This is because growth theory is concerned with linear trajectories among late-developing economies. "Old" neoclassical growth theories (Solow 1956) predicted convergence in income levels across countries, arguing that given the potentially greater returns to capital in less developed areas, late developers would "catch up." "New" theories of endogenous growth (Romer 1986; Lucas 1988) argue that because investments generate greater externalities in the context of larger stocks of human and physical capital, divergence in income levels across the world will persist or even grow. However, neither the old nor the new growth theories predict the actually observed relationship that the fastest-growing countries are *never* the countries with the highest per capita incomes, but always a subset of lower-income countries (Olson [1996] 2000, 57). Thus, economic theories of growth are of limited use in explaining the wide and persistent divergence of performance among late-developing countries. Indeed, neoclassical growth theories, on their own, cannot explain why some countries sustain rapid growth for long periods only to fade into long periods of stagnation and even downward spiral and disintegration.

The dramatic slowdown in growth was paradoxical, since Venezuela seemed to be a likely candidate to maintain its rapid growth. First, in the period 1974–85, Venezuela received an enormous increase in resource availability as a result of oil windfalls. Second, Venezuela maintained relatively high levels of physical and human capital investment in the context of a relatively accountable long-standing democratic polity. The dominant literature on governance and institutions asserts that greater democratic competition is central to checking the potentially ineffective use of discretion and authority by state leaders, thus assuring effective governance (North 1990; Olson 1993; World Bank 1997a; Sen 1999). Third, the Venezuelan state has not had to contend with ethnic, regional, caste, or religious conflicts, which make governance and the maintenance of social order particularly difficult in many parts of the developing world, including such long-standing democracies as India and Sri Lanka (Clague et al. 1997). In the Latin American context, Venezuela also maintained among the least inequitable distributions of income. In sum, Venezuela appeared to possess many favorable "initial conditions" and "social capabilities" (Abramowitz 1986) for rapid catch-up.

Despite massive resource availability in the 1970s and 1980s, the state became increasingly ineffective in channeling oil revenues in productivity-enhancing and growth-enhancing ways. The failure of the state and other actors within the economy to provide policies and institutions to maintain historically high growth rates, especially in comparison with other similarly endowed middle-income economies, such as Malaysia or South Korea, who were following similar types of industrial policies, is therefore interesting in providing insights into the political economy of growth in Venezuela. Was the failure due to policy errors or inappropriate institutions, or did oil windfalls themselves become a "curse" by crowding out the development of non-oil sectors such as manufacturing, or were there other factors that explain the slowdown in growth?

The structure of the book is as follows. Chapter 2 examines the main trends in output growth, productivity growth, and investment in the non-oil sector in general, and the manufacturing sector in particular. There are two main points that emerge from the long-run growth trends in Venezuela. The first is that worsening economic performance since 1968 cannot be attributed to poor "initial conditions." Indeed, I demonstrate that the growth slowdown and later implosion coincided with historically and comparatively high levels of physical and human capital investment in the context of a political economy with relatively equal income distribution and a competitive democratic system. The second main trend uncovered is the reversal in the efficiency of the economy in managing oil abundance. This is done through simple statistical analysis. The period 1920–65 is characterized by a positive correlation of non-oil growth and real oil prices, while the subsequent period 1965–98 is characterized by a significantly negative relationship between these two variables. This underlying pattern suggests a decline in the efficiency of the fiscal linkage in Venezuela and forms an important organizing trend from which to test competing theories about economic growth and industrial policy in twentieth-century Venezuela.

Part Two (Chapters 3–5) critically examines the reigning explanations of the output and productivity growth slowdown in the post-1968 period. Chapters 3 and 4 present a detailed examination of the reigning economic and political economy explanations of decline in Venezuelan non-oil and manufacturing productivity growth and output growth in the post-1968 period and assess the extent to which such explanations are defensible in light of the comparative and historical evidence. These two chapters provide an intellectual history of the political economy of Venezuelan development and bring together a disparate but influential set of ideas that have permeated economic and political discourse in the country for some time. Moreover, many of the explanations treated in these chapters are similar to explanations of economic failure in many other Latin American countries, as well as other developing countries that export oil, and thus have a wider relevance beyond the Venezuelan case.

The first three sections of Chapter 3 explore different versions of the "resource paradox," particularly "Dutch Disease" models, and related open economy macro models. These models focus on the effects of rigid and high wages, and suggest that oil booms likely lead to overambitious and inefficient state spending. One of the main threads of these models is that oil booms produce exchange rate revaluations, which reduce the incentives to invest in manufacturing and generally makes manufacturing production uncompetitive. While economic models of the "resource curse" are consistent with the coexistence of the slowdown in manufacturing growth in Venezuela with the oil booms in the 1970s and early 1980s, the longer-run correlation of oil resource availability and manufacturing investment and growth in twentieth-century Venezuela is not found to be robust. Oil abundance has coincided with both rapid growth (1930–80) and stagnation (1980–2005). Moreover, these models do not explain why output and productivity growth rates slowed down generally in the Venezuelan non-oil economy or why some manufacturing sectors remained more dynamic than other sectors over time. The fourth section examines neoliberal arguments that claim that growing inefficiency of investment was due to an overly interventionist and centralized state that promoted too many inefficient state enterprises and supported too many inefficient infant industries in the private sector. The comparative evidence on growth, however, suggests that the *quality* rather than the *level* of state intervention is what matters for sustained economic development. One common lacuna in the neoliberal explanations is the failure to explain why policy failures in managing investment funds have persisted in Venezuela from the 1970s until the early years of the twenty-first century.

Chapter 4 explores the rentier state models and the theories that have influenced such models, notably the theories of rent-seeking and corruption. The main thrust of this literature claims that oil abundance induces centralized public authority and excessive state interventionism and discretion, which, in turn, causes growth-restricting and productivity-restricting corruption and rent-seeking. The value of this "paradox of plenty" argument is that it attempts to move beyond identifying possible problems with policy and institutional design. The rentier state argument attempts to explain why ineffective policies and institutions are pursued and, more important, persist. However, the proposition that oil abundance induces extraordinary corruption, rent-seeking, and centralized interventionism and that these processes necessarily restrict productivity growth and output growth is not supported by the comparative evidence or the historical evidence in Venezuela. Similar levels/degrees of state centralization and corruption have coincided with cycles of growth *and* stagnation in Venezuela and elsewhere.

The second part of Chapter 4 critically reviews the literature on rent-seeking and corruption. The literature on rent-seeking and corruption aims to demonstrate why centralized public authority and state-led centralized rent deployment generates waste and growth-restricting incentives among economic agents. It is argued that the mainstream literature on rent-seeking fails to explain why countries with similar levels of corruption can produce divergent developmental outcomes over long periods. Mainstream analysts of rent-seeking focus on the costs of the rent-seeking process. The focus on costs is incomplete and misleading. As with any institution or social process such as production, a complete evaluation of the rent-seeking process needs to address not only the inputs or costs of the process (which include both legal and illegal attempts to influence the state), but also the outputs, or outcomes of the rent-seeking process (which include economic rights created, maintained, destroyed, or transferred to create specific economic rents and the policy conditions under which such rights can be maintained).

The outcomes of the rent-seeking process can vary substantially. For instance, two entrepreneurs in the same industry in two different countries may successfully offer the same bribe amount to a government official to obtain subsidized credit, with the first using the credit to reinvest in his or her business and the other using it to purchase dollars to enhance his or her capital flight portfolio. The first case is likely to contribute to a more favorable developmental outcome than the second. To take another example, two entrepreneurs offer a similar bribe to obtain an import license to produce steel. In the first case, the rent associated with the import license is tied to some performance criterion such as productivity growth or export targets. In the second case, the rent associated with the import license is not subject to performance criteria of any sort. The former is likely to contribute to economic development much more than the latter despite the level of corruption being the same in the two cases. In the end, my examination suggests that outcomes associated with the institution of centralized rent deployment in Venezuela (or anywhere else) over time depend on other conditions that are not specified in mainstream models of rent-seeking and corruption.

On methodological grounds, neoclassical models of rent-seeking and corruption (which tend to support large-scale economic liberalization policies), as well as the more dirigiste approaches to state intervention and governance, have focused on defining the "right" institutions and the "right" incentives. These approaches (whether supporting laissez-faire or more statist solutions) suggest that ineffective institutions result from incorrect mental models or knowledge gaps on the part of leaders and decision-makers, which, in turn, generate growth-restricting incentives and policy failures. One main premise of this study is that institutions such as property rights are not only an incentive structure but also simultaneously reflect power relations. As such, it is not possible to separate issues of incentives and efficiency from issues of distribution and equity, nor is it possible to know, a priori, the dynamic efficiency consequences of a given institution. This means that the actual distribution and use of rents requires political analysis, which is neglected in mainstream approaches.

I also establish that policymakers were well aware of the problems of corruption and the inadequacies of the state in imposing performance criteria on the infant industries. This suggests that if policy errors persisted, they were not due to a lack of knowledge, as many mainstream analysts maintain. Finally, I examine in Chapter 5 the results of economic liberalization reforms initiated in 1989. I find, in contrast to what rent-seeking theory would predict, reductions in state-created rents have not led to either noticeable improvements in state capacity, reductions in corruption, or an increase in investment and growth.

Part Three (Chapters 6–9) develops an alternative explanation for explaining why oil abundance has coincided with long periods of both growth and decline in productivity and output in the non-oil and, particularly, manufacturing sector in Venezuela. Chapter 6 presents the core ideas of this explanation of why centralized state rent deployment and management, in general, and industrial policy, in particular, became increasingly more inefficient in

Venezuela over time. The framework incorporates the historical interaction of economic and political processes.

I suggest that the technologies that accompany different development strategies require different levels of selectivity and discipline in the deployment of subsidies and infant industry protection, coordination of investment, and concentrations of economic and political power to be initiated and consolidated. If this is a reasonable proposition, the possibility then arises that the historically specific nature of political settlements may not be compatible with the successful implementation of a given development strategy.[1] This implies that variations and changes in state capacity may be a function of the *type* of economic challenges facing the state *and* the extent to which state-led strategies generate resistance and legitimacy problems among powerful actors excluded from the distributional benefits of those strategies.

I argue that the effective implementation of the easy, small-scale stage of import substitution (which is generally characterized by promoting small-scale, technologically simple technologies) poses fewer economic and political challenges than the more advanced stage of ISI, including big-push industrialization strategies. In particular, the possibility of rapid growth during early ISI strategies depends much less on the state to impose stringent levels of selectivity in the subsidy deployment process, monitor the performance of subsidy recipients, and coordinate investment compared with advanced ISI and big-push industrialization strategies. This is because more advanced ISI and big-push industrialization strategies involve the creation of firms where scale economies, technological upgrading and learning, and export performance are generally more binding constraints for such firms to become competitive and maintain rapid productivity growth. Imposing selectivity and discipline in the subsidy deployment process is as much a *political* challenge as it is an economic one. When a state is selective in, say, granting import licenses for steel production, this means *excluding* many potentially powerful business groups from state patronage. It also means limiting employment creation, which in the context of significant levels of underemployment, is politically difficult to justify. Thus different development strategies and stages of development affect the types of politics that are compatible with growth. The developmental state, the rent-seeking, and the rentier state literature do not take into account the extent to

---

1. Political settlements refer to the balance or distribution of power between contending groups and classes in society. Political settlements manifest themselves in the structure of property rights and entitlements, which give some social actors more distributional advantages than others, and in the regulatory structure of the state.

which different stages of development or development strategies may affect the nature of political conflicts.

In this chapter, the main propositions of the argument are presented. It having been argued on theoretical grounds that early ISI strategies present fewer economic and political challenges than more advanced ISI strategies, it is then imperative to consider the extent to which different political organizations and contests are likely to be compatible with the two broad development strategies and stages that were previously identified. Drawing on the rich political science literature on Venezuela, it is possible to distinguish two broad types of polities that have been salient in Venezuela over the period 1920–2005. In the twentieth century, Venezuela can reasonably be characterized as a *consolidated state* in the sense that it maintained a monopoly over the means of violence and was able to maintain political and social order for most of the period under study. However, the nature of political organization and competition changed substantially in Venezuela over the course of the twentieth century. A consolidated state, as the literature on Venezuela corroborates, has coincided with two very different types of political organization and contestations: a consolidated state with *centralized* political organizations, and a consolidated state with *fragmented* (and increasingly polarized) political organizations (see more on this distinction below).

The first set of propositions considers the growth prospects of polities that are generally characterized by a *consolidated state with centralized political organizations*. In such polities, patronage structures are controlled by the executive in a centralized fashion. The deployment of patronage in such polities can take place under a cohesive military regime, a centralized one-party state, or through a high degree of cooperation between two contending political parties. The Venezuelan polity, I argue, was of this type during most of the period 1920–68. These type polities, I argue, are most likely to promote economic growth through the early stage of ISI and most likely to be able to meet the political and economic challenges of big-push/advanced ISI development strategies.

The second set of propositions concerns the growth prospects of a *consolidated state with fragmented political organizations*, a common polity type in many less developed countries (and the type Venezuela more closely approximated in the period 1968–2005). In such polities, patronage structures are less predictable and coherent because there is either less political party cooperation, or new factions either within the dominant party system or from outside successfully capture part of the state. I argue that polities of this type may be capable of generating relatively rapid growth when attempting to implement early stage ISI strategies, but are much less likely to successfully manage the more difficult economic and political challenges of big-push/advanced ISI development strategies. This section spells out the logic of these propositions and provides brief examples from other less developed countries.

In order to assess the propositions made concerning compatibility and incompatibility, it is necessary to first trace the evolution of development strategies and political settlements. Then it becomes necessary to map or identify periods when compatibilities and incompatibilities between development strategies and political settlements are salient. Once periods of compatibility and incompatibility are identified, it is then possible to examine the extent to which the evidence on economic growth and productivity growth are consistent with the propositions made. The first step of this mapping exercise is undertaken in Chapter 7, where I demonstrate empirically that the development strategy and stage of ISI switched from the easy to the more advanced big-push stage in the period 1960–73. The political and economic implications of this periodization are explored.

Chapter 8 then traces the transformations in the nature of political settlements and state-society relations in Venezuela over the period 1920–2005. Drawing on a vast political science literature, I argue that, in the period 1920–68, the Venezuelan polity can be viewed as a consolidated state with centralized political organizations. This type of polity was maintained under both authoritarian rule (1920–58) and the period of democratic transition and consolidation (1958–68). The chapter then traces how and why the Venezuelan polity transformed from a consolidated state with centralized political organizations to a consolidated state characterized by fragmented political organizations where political contestation becomes progressively more clientelist, populist, and polarized, particularly after 1973. This chapter enables us to map the changing nature of political settlements and political organizations onto the periodization of the development strategies and stages of ISI identified in Chapter 7.

Chapter 9 presents the main results of the historical mapping of development strategies and political settlements presented in the previous two chapters and provides an alterative explanation of Venezuelan growth and productivity trends to those provided by the reigning economic and political economy arguments. I argue that the period 1920–68 is characterized by rapid manufacturing output growth and respectable productivity growth because there was basic compatibility between the development strategy (early ISI strategies) and politics (a consolidated state with centralized political organizations). I further argue that the period 1968–2005 was characterized by rapid declines in non-oil and manufacturing productivity growth and a collapse in non-oil and manufacturing output growth from 1980 to 2003 because there was a basic incompatibility between the development strategy (more advanced ISI strategies and a big-push, natural-resource-based heavy industrialization strategy) and politics (a consolidated state with increasingly fragmented and eventually polarized political organizations and contests).

In the period 1968–2005, the political science literature establishes that the Venezuelan political system became increasingly populist, clientelist, and factionalized. The basic incompatibility I identify is that politics became increasingly more factionalized and accommodating precisely at a time when the development strategy and stage of import substitution required a more unified and exclusionary rent/subsidy deployment pattern. I incorporate the *economic* consequences of the increasingly factionalized political contests between and within political parties attempting to build political clienteles and accommodate urban middle-class groups, emerging family conglomerates, and labor unions competing for access to centrally allocated state resources. Much of this accommodation was closely related to efforts on the part of the dominant political parties to preserve democratic rule, and by implication, prevent a return to the authoritarianism of most of the first half of the twentieth century. The coordination failures of the big-push industrialization strategy were manifested in the low monitoring of state-created rents and subsidies, excessive entry of private sector firms into protected sectors, massive proliferation of public sector employment, and state-owned enterprises in the decentralized public sector.

The chapter provides detailed empirical evidence on macroeconomic coordination failures (especially during the big-push industrialization strategies of the 1970s and 1980s) as well as microeconomic evidence to support the claim that growing political fragmentation and populist clientelist patronage were salient features of the polity, particularly after 1968. I also discuss how the explanation put forward complements or advances recent political science and economic explanations of Venezuela's decline, particularly since 1980. It is argued that the alternative framework developed is more defensible in light of both historical and comparative evidence.

The chapters in Part Three contribute to institutional and political economy approaches to growth theory in general and advance our understanding of the historical political economy of development in twentieth-century Venezuela in particular. Explaining the growth slowdown in Venezuela as a mismatch of economic strategy and historically specific political strategies contributes to recent analytical narratives on economic growth (Rodrik 2003). Analytical growth narratives attempt to "draw the connections between specific country experiences, on the one side, and growth theory and cross country empirics, on the other ... in order to extend our understanding of economic growth using country narratives as a back drop" (3).

Given the diversity of growth experiences among developing countries and the large variation of growth rates within countries over time, it is likely that a collection of growth models will be required to understand economic performance at different stages of development and under different political institutional regimes (Pritchett 2003). In comparing the growth trajectories of Vietnam and the Philippines, Lant Pritchett suggests that the processes of escaping from low-level poverty traps may be fundamentally different from the processes of igniting or sustaining growth in middle-income countries. In middle-income countries, the obstacles to growth, Pritchett argues, often lie not in insufficient human and capital stocks, but in increasing uncertainties concerning institutions (that is, the "rules of the game"). Pritchett posits that the institutional requirements of re-igniting growth in a middle-income country may be significantly more demanding than in low-income contexts. However, the analytical growth narratives explored by Pritchett and other authors in Rodrik 2003 have not explained either why or how transitions in development strategy or stage of development may generate more contentious sets of political conflicts (and therefore, uncertainties in the "rules of the game") and more difficult institutional challenges for both the state and private sector decision-makers.

Part Four (Chapters 10 and 11) takes the analysis beyond the Venezuelan context. Chapter 10 provides a brief comparison with the Malaysian case in order to illustrate the importance of how different political settlements and the nature of threat can affect the efficiency of centralized rent deployment in two mineral-abundant, middle-income countries. The concluding chapter considers the methodological and empirical implications of a comparative, historical, and interactive approach to understanding variations and changes in state capacity and long-run growth. In particular, it addresses the failure of both neoliberal and developmental state theorists to incorporate politics into an understanding of state capacity. It argues that a more fruitful approach to growth lies in exploring institutional performance as a historically specific interactive process between political and economic development strategies. It also addresses the need to focus on bringing production (and particularly the different stages of late development) and politics (in particular the shifting grounds of legitimate rules) into analyses of governance and economic growth.

This book does not offer a universal theory. It is not likely that a general theory of growth will ever emerge precisely because the political institutions and coalitions underlying the state are historically contingent and often fragile. Rather, it is argued that such a historical and integrative political economy approach will enable a more adequate account of growth cycles and explain more adequately why well-intentioned policymakers fail to change inefficient regulatory and institutional arrangements even when they have knowledge about the problem.

Suggestions for further research are also discussed as well as a brief policy proposal for reviving sustained growth in Venezuela. The revival of economic growth in recent years (2004–7) has indeed been impressive. But this recent growth spurt has been the product of a dramatic boom in oil revenues and an economic rebound in the wake of the greatest growth collapse in contemporary Venezuelan history, in the period 1999–2003. This recent growth spurt is not, the evidence suggests, the result of a coherent production strategy; nor has it led to the transformation or diversification of non-oil exports. The historical evidence presented in this book suggests that without a substantial transformation of the non-oil economy, particularly the manufacturing sector, this growth acceleration is likely to be ephemeral and reproduce what many Venezuelan policymakers have tried for decades to overcome, namely the dependence on the vagaries of oil.

2

TRENDS AND CYCLES IN THE VENEZUELAN ECONOMY

This chapter presents some of the main trends in economic output and productivity growth, investment, and socioeconomic indicators in Venezuela from 1920 to 2005.[1] The principal patterns that emerged are as follows. In the period 1920–80, Venezuela was among the fastest-growing economies in Latin America, and its manufacturing growth rate was among the most rapid until the mid-1960s. Over this period, investment rates were also relatively high in the non-oil economy generally and the manufacturing sector in particular. However, two important trends mark the period 1968–2005. First, beginning in the mid-1960s the non-oil economy, and particularly the manufacturing sector, experienced a dramatic decline in total factor productivity growth and labor productivity growth. Second, output growth and investment rates in the manufacturing sector, and the non-oil economy more generally, collapsed in the period 1980–2003.

What remains paradoxical about this slowdown is that it occurred in an economy that possessed many favorable "initial conditions" in terms of resource availability, human capital investment, democratic accountability of leaders, and income distribution. The chapter is organized as follows. Section 2.1 presents

---

1. The period 1998–2005 is not systematically analyzed apart from non-oil and manufacturing growth rates (data on investment trends are covered for the period 1998–2002). There are several reasons for this. First, this period represents a significant transformation in the nature of politics toward much greater polarization and the decline in the relevance of political parties (see Chapter 8). Second, time series data for many of the variables are either not consistent with the prior period, or are unavailable. Third, because the economic collapse in the period 1998–2003 was particularly severe (see table 2.1), including this period will bias downward the long-run indicators of economic performance.

data on trends in non-oil and manufacturing output and productivity growth, and total and non-oil investment rates in the period 1920–2005. Section 2.2 briefly documents the socioeconomic and political impact of long-run stagnation in the period 1980–2003. Section 2.3 examines the statistical relationship between real oil prices and non-oil growth and demonstrates that there has been a dramatic reversal in the efficiency with which state leaders have managed increases in oil prices and resource availability over time. Section 2.4 examines the extent to which the decline in growth has been due to comparatively low levels of investment in human and physical capital.

2.1 Growth and Investment Trends in the Venezuelan Non-oil and Manufacturing Sectors, 1920–2005

For the period 1920–80, Venezuela achieved a growth rate of 6.4 percent per annum in gross domestic product (GDP) between 1920 and 1980, the highest growth rate of any Latin American economy over this period (Maddison 1995, 156–57). Despite maintaining relatively high growth rates in GDP in this period, and despite being the recipient of oil revenue windfalls in the period 1974–85, Venezuela entered one of the worst growth implosions in Latin America in the period 1980–98, which worsened even further in the period 1998–2003. Annual average GDP grew only 1.5 percent over the period 1980–98; this rate collapsed to −2.5 percent in the period 1998–2003. Average annual growth did recover in the period 2004–5 with non-oil GDP growing at 13.4 percent and manufacturing growing at 15.4 percent. Such high growth rates were the result of dramatic increases in oil revenues in the context of an economy that had witnessed four years of significant declines in economic growth. As indicated in table 2.1, the growth performance of the Venezuelan economy also exhibits a marked long-run decline in the rate of growth in the non-oil and manufacturing sectors, with growth in these sectors significantly lower in the period 1980–2003 when compared with the prior period of 1920–80.[2] The importance of presenting

---

2. In an oil export economy, growth accounting is sensitive to the choice of base year and price deflators used. This is because price fluctuations in oil can be large and because oil has been a significant part of Venezuelan economic output (ranging from 20 to 30 percent of GDP in the period 1930–2005). Such price fluctuations can greatly affect relative prices in the economy as a whole and thus radically change the impact a particular base year can have on subsequent calculation of production. See Baptista 1997, 105–21, and F. Rodríguez 2006 for a discussion of the national accounts in Venezuela in the period 1920–98. These problems notwithstanding, *long-run* growth rates show a significant decline since 1980 regardless of base period used.

TABLE 2.1 Average annual growth rates in the non-oil economy of Venezuela, 1920–2003 (%)

	Non-oil GDP	Manufacturing
1920–30	10.2	n.a.
1930–40	2.7	n.a.
1940–50	9.6	6.6
1950–57	9.1	15.0
1957–70	7.1	7.7
1970–80	5.7	9.7
1980–90	1.1	2.8
1990–98	2.7	1.2
1998–03	−3.5	−5.1

SOURCE: Baptista 1997 for period 1920–57; Banco Central de Venezuela, *Informe anual*, various years, for period 1957–2003.

NOTE: All output series in 1984 bolívares.

non-oil growth data cannot be overstated in the Venezuelan case. By focusing on non-oil growth, it is possible to identify that the slowdown in Venezuelan growth was an economy-wide phenomenon and not isolated to the well-known stagnation in oil production since 1965 (Baptista 1997; F. Rodríguez 2006).[3]

Table 2.2 summarizes the comparative growth performance of the Venezuelan economy with other medium-sized middle-income Latin American and East and Southeast Asian economies. Venezuela's growth performance in the period 1965–98, before its severe economic decline in 1998–2003, has been similar to other middle-income Latin American economies, but poor in comparison with the more dynamic middle-income economies. Interestingly, despite the slowdown in manufacturing growth in Venezuela, the long-run growth rate of manufacturing was the highest among the middle-income, medium-sized Latin American economies. That Venezuela's growth performance was similar

---

3. From 1957 to 1970, oil production steadily increased and reached a peak of 3.4 million barrels per day in 1970. By 1975, as a result of sustained multinational disinvestment, production declined dramatically to 2.4 million barrels per day despite price increases (Espinasa and Mommer 1992, 112). Two government policies in the 1960s contributed to this disinvestment. The first was the Ley de Reversion (Reversion Law) in 1960, which decreed that oil-producing facilities would be turned over to the state in 1983. The second was the introduction in 1968 of fiscal reference prices, "an instrument which converted the income tax based on market prices into a tax per barrel, which became, in effect, an excise tax" (112). The exercise tax lowered profitability considerably (M. Rodríguez 2002, 12). Subsequent to 1975, production was restricted, since Venezuela, a founding member of OPEC, adhered to the cartel's policy of limiting the amount of oil produced and exported. Oil production stagnated in the period 1975–95, though investment in exploration increased after nationalization in 1976.

TABLE 2.2 Venezuelan economic performance in comparative perspective, 1965–1998

	Average annual growth rate (%) of GDP and manufacturing (in brackets)				GNP per capita*
	1965–80	1980–90	1990–98	1965–98	1998
Venezuela	3.7 [5.7]	1.1 [4.3]	2.2 [1.5]	2.5 [4.3]	$3,500
Colombia	5.7 [6.4]	3.7 [3.5]	3.9 [−1.1]	4.2 [3.8]	2,600
Chile	1.9 [0.7]	4.2 [3.4]	7.9 [5.7]	4.0 [2.7]	4,810
Argentina	3.4 [2.7]	−0.7 [−0.8]	5.6 [4.3]	1.5 [0.8]	8,970
South Korea	9.9 [16.4]	9.4 [13.0]	6.1 [6.9]	8.8 [13.0]	7,970
Malaysia	7.4 [8.1]	5.3 [8.9]	7.4 [10.8]	6.8 [9.7]	3,600
Thailand	7.3 [11.2]	7.6 [9.5]	5.7 [7.7]	7.0 [9.8]	2,200

Source: World Bank, *World Development Report*, 1991 and 1999/2000; World Bank, *World Development Indicators*, 2000.

* In 1998 U.S. dollars.

to the pattern of medium-sized, middle-income Latin American economies contradicts a view widely held in the literature that Venezuela is an outlier, or an "exceptional" case in the context of twentieth-century Latin American development because of its oil abundance (Karl 1997).

Total investment and non-oil investment also declined dramatically over time. In the period 1990–2002, total investment rates fell from an annual average 27.7 percent of GDP in the 1950s to under 16 percent of GDP; and non-oil investment as a percentage of non-oil GDP declined from an annual average of nearly 35 percent in the 1950s to under 14 percent in the 1990s (see table 2.3). Declines in private sector investment rates accounted for most of the overall decline in total investment. Overall public investment rates declined from 10.8 percent of GDP in the 1970s to 10.6 percent of GDP in the 1980s and to 9.9 percent in the period 1990–98 (with, however, a more substantial decline to 6.6 percent of GDP in the period 1999–2002). Private sector investment rates, however, fell to a much greater extent: from 23.6 percent of GDP in the 1970s to 10.7 in 1980s, only to fall even further to 6.9 percent in the period 1990–98. Although private sector investment rates recovered to 8.5 percent in the period 1999–2002, they still remained well below their average levels compared to the period 1950–80. A similar trend exists for non-oil investment with the small difference that declines in public investment account for slightly more of the overall decline in total non-oil investment rates. That the slowdown in investment is largely a *private* sector phenomenon implies that changes or shortfalls in fiscal

TABLE 2.3 Gross fixed investment rates: Venezuela, 1950–2002 (in current prices)

	Annual averages of all investments (as % of GDP)			Annual averages of non-oil investments (as % of non-oil GDP)		
	total	public	private	total	public	private
1950–60	27.7	10.5	17.2	34.8	16.9	17.8
1960–70	24.2	8.4	15.8	26.1	9.1	17.0
1970–80	34.4	10.8	23.6	36.8	9.6	27.2
1980–90	21.3	10.6	10.7	22.5	8.7	13.8
1990–98*	15.8	9.9	6.9	13.7	5.6	8.1
1999–02	15.1	6.6	8.5	n.a.	n.a.	n.a.

SOURCE: Baptista 1997; Banco Central de Venezuela, *Informe anual*, various years; Banco Central de Venezuela 1992, vol. 1; OCEI, *Encuesta industrial*, various years.

* Non-oil investment data for the period 1990–95 only.

revenue, or the levels of public investment, are not central to the story of stagnation in capital formation in Venezuela in the period 1980–2003.

2.2 The Socioeconomic and Political Consequences of Venezuelan Economic Stagnation

The socioeconomic and political effects of long-run economic stagnation have been dramatic and devastating. Stagnation in growth has negatively affected the demand for labor. First, there was a dramatic decline in average real wages, which in 1995 had already fallen below the levels attained in 1950 (Baptista 1997, table IV-1). Second, there were important increases in unemployment rates, especially since the mid-1980s. In the period 1965–83, unemployment rates fell steadily and averaged 7.4 percent. From 1984 to 1989, annual average unemployment rates rose to 10.5 percent. By 1999 they had risen to 11.5 percent.[4] By 2002, average annual unemployment rates rose again, to 14.4 percent (F. Rodríguez 2003, table 1), and reached a peak of 18.4 percent in 2003 in the wake of national strikes and political instability (see section 8.2). By the end of 2005, however, unemployment declined to 11.8 percent as a result of the boom in oil prices in 2004–5 (Weisbrot and Sandoval 2007, table 4).

4. Unemployment data are from corresponding editions of OCEI, *Encuesta de hogares* and *Encuesta de empleo*, and from Baptista 1997, table I.4.

A third important consequence of the post-1980 decline has been the increases in levels of urban informal employment.[5] In the period 1975–80, the rate of informal employment of the nonagricultural labor force averaged 32 percent. From 1981 to 1990, this rate increased to 39.5 percent, which signified an important decline in viable accumulation possibilities in the formal industrial and service sectors. If the problem had been simply overregulation, then it would have been reasonable to expect the rapid and profound economic liberalization policies launched in 1989 to have improved employment generation opportunities in the formal sector.[6] They did not. The rate of informal employment averaged 44.5 percent in the period 1991–95, 48.5 percent in the period 1994–95, and 50.1 percent in the period 1998–2002 (F. Rodríguez 2003, table 1). By 2005, however, informal unemployment had dropped to 41.1 percent as a result of the oil boom in 2004–5 (Weisbrot and Sandoval 2007, table 4).

Fourth, prolonged stagnation has led to a significant increase in the incidence of poverty.[7] According to Gustavo Márquez and Carola Alvarez (1996), 20 percent of Venezuelan households were below the poverty line in 1980–81. The share of poor households increased to 36 percent in 1985–86; 42 percent in 1989–90; and 51 percent in 1994–95. When non-oil growth collapsed in the period 1998–2003, poverty rates rose to a peak of 62 percent by the end of 2003 (Weisbrot, Sandoval, and Rosnick 2007, table 1). Poverty rates did decline significantly to 45 percent in 2005 in the wake of the oil boom, expansionary fiscal policy, and increased social spending on the poor (1–2; also see Weisbrot and Sandoval 2007, 17).

Finally, prolonged economic stagnation, decline in real incomes of the vast majority of the population, and increase in poverty levels contributed to severe crises of legitimacy in the political system. Such declines became particularly explosive given the high standards of living and expectations of a more prosperous future that large sections of the population maintained since the

---

5. Following International Labor Office definitions, informal employment in Venezuela is defined as illegal operations (that is, operations not registered with the state) that employ fewer than five workers. This definition includes own-account workers and domestic service workers. See Márquez and Portela 1991 for analyses and measurement of informal employment in Venezuela. Employment data on Venezuela are calculated from the corresponding editions of OCEI, *Encuesta de hogares* and *Encuesta de empleo*.

6. See Chapter 5 for a discussion of the economic liberalization program.

7. The incidence of poverty measures used in Márquez and Alvarez 1996 is the head-count ratio. The head-count ratio measures the percentage of the population below the poverty line. The head-count ratio does not tell us anything about the poverty gap, which measures the percentage by which the mean income of the poor falls short of the poverty line—or about the distribution of income among the poor.

consolidation of democracy in the early 1960s. In the period 1960–89, political stability and democratic consolidation in Venezuela contrasted with cycles of authoritarianism and regime instability throughout South America. However, the period 1989–2000 saw dramatic episodes of destabilization in the political order. The magnitude of the state of crisis in governability was evident with the decline in the legitimacy of the dominant political parties, two military coups attempts in 1992, and the rise of political outsiders and new forms of populist rule (see Chapters 5 and 8).

2.3 Sowing the Oil? Changes in the Efficiency of the Fiscal Linkage in Venezuela, 1920–1998

Because Venezuela is predominately an oil-exporting economy, the accrual of significant revenues from the oil sector to the government confers an important role on the state in the process of economic development. State economic management and effective planning are particularly important since the main sources of foreign exchange, finance, and taxation are centralized in the state. While oil rents simultaneously relax traditional resource constraints on growth (namely the foreign exchange, fiscal, and savings constraints), the nature of the oil industry as an "enclave" with few dynamic forward or backward linkages can negate further direct growth inducements that accompany other industrial activities.[8] For oil-exporting late developers, the "fiscal" linkage—the process through which the state "taxes, transfers and spends" oil revenues is central to processes of economic diversification and particularly, industrialization.

The discovery of massive oil reserves in the 1920s transformed the resource mobilization capacity of the Venezuelan economy. At the time, Venezuela had been among the poorest and most stagnant agrarian economies in Latin America. The traditional exports, coffee and cocoa, and an agricultural structure based on the latifundia system provided little stimulus to capital accumulation. In 1920, the per capita income in Venezuela was one-half the Latin American average (Maddison 1995, 202).[9] Through an arduous process of bargaining and conflict, the state has been able to capture an increasing share of the oil proceeds

---

8. See Hirschman 1958 and 1981a on the concept of forward, backward, and fiscal linkages and their effects on patterns of economic development. For a discussion of linkages in the context of oil economies, see Gelb and Associates 1988, 14–31.

9. Such comparisons should be treated as rough approximations, given the limited amount of consistent and comparable data for the Latin American economies before 1950.

from powerful multinationals in a country that in the 1950s was the world's second-largest producer of oil and continues today to be one of the largest and most profitable producers with among the highest reserves in the world.[10]

For many political leaders in Venezuela, oil abundance in the 1970s and 1980s provided an opportunity to embark on one of the boldest and most ambitious industrialization and modernization projects in Latin American economic history. This oil windfall represented "a unique opportunity for Venezuela to undertake major structural transformations of the industrial and export structure of the economy in order to reduce the dependence on oil, while simultaneously permitting high growth and the accomplishment of distributive goals" (M. Rodríguez 1991, 238).

The magnitude of the financial resources available to the Venezuelan economy during these oil boom years was indeed significant. One indicator of the size of windfall can be seen in export earnings. In the period 1960–72, the price of Venezuelan oil exports was relatively stable and low, fluctuating between US$2 and $2.80 a barrel. Between 1972 and 1974, the average price of Venezuelan crude oil rose from $2.50 to $10.50 a barrel (Gelb and Associates 1988, 293). This extraordinary increase in oil prices ushered in a prolonged period of historically high earnings from the export of oil, which amounted to $247 billion over the period 1974 to 1985.[11] In comparison, Brazil, an economy with nearly *ten times* the population of Venezuela, exported $194 billion over this period. A second indicator of the windfall is the size of the increased exports relative to the non-mining economy. Gelb (Gelb and Associates 1988, 56–65) calculates the Venezuelan oil windfall as the change in the ratio of resources generated from oil relative to the non-mining economy in the boom years compared with a specified pre-boom period (in this case, 1970–72). The first oil windfall (1974–78) was 10.8 percent of non-mining GDP; while the second windfall (1979–81) represented 8.7 percent of non-mining GDP.[12] The fiscal windfall from the oil boom and nationalization was massive. In the period 1974–78, fiscal revenues totaled 82.4 billion bolívares (in 2001 bolívares), which was nearly as much as the *combined* fiscal income of 84.8 billion bolívares collected in the previous

---

10. The bargaining process over the share of profits from the oil industry and its nationalization in 1976 has received ample attention in the literature (see Tugwell 1975 and Vallenilla 1975). For more recent treatment of management of the oil industry, see Boué 1993 and Manzano 2006.

11. All of the export values in this paragraph are in 2000 U.S. dollars.

12. These windfalls, while large, were below the mean windfall increase of 22.1 percent for all developing-country oil exporters in the sample for the period 1974–78, and well below the mean windfall increase of 23.3 percent for all developing-country oil exporters for the period 1979–81 (Gelb and Associates 1988, 62–65).

fifty years![13] In the period 1979–81, the fiscal revenues totaled 48.9 billion bolívares, which brought the total fiscal take in the period 1974–81 to 131.4 billion bolívares (Gelb and Associates 1988). Oil revenues contributed an average of 68 percent of total fiscal revenues in the same period (calculated from De Krivoy 2002, table 1).

One of the remarkable characteristics of the Venezuelan economy is that its growth slowdown occurred despite increases in resource availability in the period 1974–85. This slowdown provides important insights into the political economy of growth in Venezuela. As an oil economy, the Venezuelan state and societal actors with influence over the state used the fiscal linkage to promote industrialization and non-oil growth much more efficiently between 1920 and 1965 than they did after 1965. One of the most telling statistics concerning the long-run dynamics of the political economy of growth in Venezuela I discovered is the reversal in the relationship of real oil prices and non-oil growth in the two periods, as indicated in table 2.4:

TABLE 2.4 Correlation of real oil export prices and non-oil growth in Venezuela, 1920–1998

Period	Coefficient of correlation
1920–65	0.58
1965–98	−0.44

SOURCE: Baptista 1997; Banco Central de Venezuela, *Informe anual*, various years; IMF, various years; Ministerio de Energía y Minas, *Informe anual*, various years.

NOTE: Changes in the correlation coefficient of less than 5 percent are not significant. Real oil prices refer to annual average Venezuelan export prices per barrel in 2000 U.S. dollars; non-oil growth was calculated in 1984 bolívares.

In the period, 1920–65, the coefficient of correlation between real oil prices and non-oil growth was 0.58. This indicates that there was a significant and very positive relationship between increases in real oil prices and non-oil growth. In this period, the state used the fiscal linkage effectively to "sow the oil." However, during the period 1965–98, there was a dramatic reversal in this relationship as the coefficient of correlation between real oil prices and non-growth turned to *negative* 0.44. A further corroboration of this trend can be seen in the patterns of non-oil total factor productivity growth. According to Francisco Rodríguez (2006), non-oil total factor productivity declined

---

13. Based on data from Obregón and Rodríguez 2001, 30.

from an annual average of 1.10 percent in the period 1950–68 to negative 1.45 percent in the period 1968–84. While non-oil total factor productivity average annual growth improved to 0.31 percent in the period 1984–98, this growth rate was less than one-third the growth rate achieved in the period 1950–68. This implies that the Venezuelan state and the private sector became very inefficient in managing the dramatic increases in oil windfalls during the latter period.[14]

2.4 The Venezuelan Paradox: Growth Slowdown in the Context of Favorable Initial Conditions

The growth slowdown in the Venezuelan non-oil economy and manufacturing sector represents an interesting paradox for both the "good governance" paradigm, which stresses the importance of democracy in making state leaders accountable (World Bank 1997a), and some versions of endogenous growth theory (Lucas 1988), which stress the importance of human capital investment in maintaining growth. The purpose of this section is to document some of the favorable initial conditions that the Venezuelan economy possessed in the area of human capital investments and capabilities by the mid-1960s relative to selected medium-sized middle-income countries in Latin America and East Asia. In particular, it is noteworthy that the Venezuelan growth slowdown and collapse coincided with historically and comparatively high levels of physical and human capital investment in the context of a political economy with among the least unequal income distributions and (as mentioned) a relatively competitive democratic system.

One advantage the economy possessed was a relatively high level of resource mobilization. As indicated in table 2.5, Venezuela had a relatively high level of national savings. High levels of national savings are necessary to finance high levels of investment without incurring external debt. In the period 1960–80, Venezuela had a higher national savings rate than (the more rapidly growing) East Asian and the other Latin American economies in the sample. In the 1980s, Venezuela still maintained the highest savings rate in Latin America and had a rate roughly similar to Thailand's. It is only in the 1990s that national savings rates in Venezuela fall 10 percentage points below the East Asian average, almost

---

14. See section 9.2 for a more detailed discussion of declines in non-oil total factor productivity growth and in manufacturing labor productivity growth in the period 1968–98.

TABLE 2.5 Venezuelan gross national savings in comparative perspective, 1960–1998 (annual averages, as a percentage of GDP)

	1960–70	1970–80	1980–90	1990–98
Venezuela	37.0	37.5	25.3	23.9
Colombia	18.3	19.5	21.0	17.2
Chile	19.8	16.5	20.1	24.8
Argentina	25.3	27.1	21.7	17.7
South Korea	8.6	22.2	31.2	35.0
Malaysia	26.6	30.4	33.4	39.3
Thailand	21.2	22.4	27.6	35.8

SOURCE: World Bank, *World Development Indicators*, various years.

thirty years after the beginning of significant declines in manufacturing labor productivity growth and non-oil total factor productivity growth.[15]

National accounts statistics on gross investment do not include investments in human capital such as health and education, the latter being an important factor for some recent models of endogenous growth (Lucas 1988). In terms of human capital investments and social indicators, Venezuela compared favorably to other middle-income countries. In terms of education spending, Venezuela actually rated highest among the selected countries over the whole period, as can be seen in table 2.6.

A similar pattern emerges with respect to investments in health. As indicated in table 2.7, health expenditure rates and levels in the 1990s were surpassed only

TABLE 2.6 Comparative public expenditure on education, selected years (as a percentage of GDP)

	1960	1965	1970	1980	1985	1993–98
Venezuela	4.0	4.1	5.5	4.4	5.4	5.1
Chile	2.7	2.7	4.5	4.6	4.4	3.3
Colombia	n.a.	n.a.	2.2	1.9	2.8	4.1*
South Korea	3.2	1.8	3.8	3.7	4.8	4.1
Malaysia	4.0	4.1	5.5	6.0	6.6	4.4
Thailand	2.5	3.5	2.9	n.a.	n.a.	5.1

SOURCE: UNESCO, *Statistical Yearbook*, various years; UNDP, *Human Development Report*, 1997, 2001
* Refers to 1995–97 only.

15. Baptista 1997; OCEI, *Encuesta industrial*, various years; Banco Central de Venezuela, *Serie estadística*, various years; Amsden 1997, table 1; Katz 2000, table 1; F. Rodríguez 2006, table 10.

TABLE 2.7 Venezuelan health care investments: A comparative perspective, 1965–1998

	Health expenditure (1990–98)		Physicians (per 1,000 people)		
	% of GDP	$PPP per capita*	1965	1980	1990–98
Venezuela	7.5	426	0.8	0.8	2.4
Chile	3.9	344	0.5	0.8	1.1
Colombia	9.4	594	0.4	0.8	1.1
South Korea	5.6	824	0.4	2.5	n.a.
Malaysia	2.4	180	0.2	0.3	0.5
Thailand	6.2	329	0.1	0.1	0.4

SOURCE: World Bank, *World Development Report*, 1991; World Bank, *World Development Indicators*, 2000.

* PPP = "purchasing power parity."

by South Korea and Colombia, whereas the investments in physicians since 1965 have generally been the highest (apart from South Korea in 1980) among the countries in the sample.

Comparative evidence in health and education indicators (a proxy for human capital) also suggests that Venezuela had favorable "initial conditions" that might have contributed to a sustained industrialization drive similar to what occurred in the East Asian "tigers" in the 1970s and 1980s. As indicated in table 2.8, Venezuela's illiteracy rates steadily declined over the period of growth slowdown from a rate of 48 percent in 1950 to 16 percent in 1998. In comparative terms, illiteracy rates in Venezuela were similar to those of Colombia and Malaysia and considerably lower than in Thailand's over the period 1950–98.

TABLE 2.8 Venezuelan illiteracy rates in comparative perspective, 1950–1998 (as a percentage of population fifteen years old and older)

	1950 Total	1960 Total	1970 Total	1998		
				Total	Male	Female
Venezuela	48	37	25	16	7	9
Chile	20	16	11	7	4	3
Colombia	n.a.	n.a.	22	18	9	9
South Korea	78	29	12	5	1	4
Malaysia	48	32	21	10	3	7
Thailand	62	47	40	27	9	18

SOURCE: UNESCO, *Statistical Yearbook*, various years; World Bank, *World Development Report*, various years.

As indicated in table 2.9, Venezuela's enrollment rates also increased for all levels of education between 1965 and 1998. In comparative terms, Venezuela's primary enrollment rates were similar to all the East Asian economies in both 1965 and 1998 and were higher than Colombia's in both years. In 1965, Venezuela's secondary school enrollment rates were higher than Colombia's and Thailand's and similar to Malaysia's. By 1998, they were still nearly twice as high as Thailand's and similar to those of Colombia and Malaysia. In 1965 Venezuela's tertiary enrollment rates were higher than those of all the other countries in the sample, and in 1998 higher than those of all the other countries in the sample except South Korea's.[16]

TABLE 2.9 Venezuelan school enrollment rates in comparative perspective, 1965 and 1998 (percentage of relevant age group enrolled in education)

	1965			1998		
	Primary	Secondary	Tertiary	Primary	Secondary	Tertiary
Venezuela	94	27	7	106	54	27
Chile	124	34	6	102	74	18
Colombia	84	17	3	104	56	14
South Korea	101	35	6	104	87	37
Malaysia	90	28	2	102	57	7
Thailand	78	14	2	87	28	16

SOURCE: World Bank, *World Development Report*, various years.

NOTE: Enrollment rates are measured as the ratio of the number of people enrolled in primary education over the number of children in the primary school age bracket (7–12) in the population. Therefore, percentages of greater than 100 are possible if, for example, there is full enrollment among children (aged 7–12) and there are older people also enrolled in primary education.

Note that there is no conclusive evidence to suggest that education levels across countries explain differential growth performance; nor is there evidence that significant increases in education coverage have resulted in growth increases in countries over time (Pritchett 2001). This is because growth depends on other factors, including an institutional structure that provides the incentives for growth-enhancing activities and penalizes those activities which do

16. The enrollment indicators in table 2.9 do not, of course, reveal anything about the *quality* of the education provided. See Di John 2007 for an examination of why public education quality and student performance on international tests in Latin America lagged behind those in East Asia and Eastern Europe in the period 1980–2000.

not contribute to growth or social welfare. Education levels are one of these factors but not necessarily the most important.[17]

As indicated in table 2.10, the period 1960–98 saw a dramatic improvement in the basic health indicators in Venezuela. Life expectancy increased from a total of 57 years in 1960 to 70 years for males and 76 for females in 1998. Infant mortality rates dropped from 65 per thousand live births to 21 in 1998. In comparative terms, Venezuelans had a higher life expectancy than all the three East Asian economies in 1960 and 1975 and a similar rate to Chile's throughout the whole period. With respect to infant mortality rates, Venezuela achieved substantial improvements over the period 1965–98. Infant mortality rates declined from 65 per 1,000 live births in 1995 to 21 per 1,000 live births in 1998. By 1998, however, Malaysia, South Korea, and Chile had a 50 percent lower infant mortality rate than Venezuela's, though Colombia and Thailand had *higher* infant mortality rates than all countries in the sample.

TABLE 2.10 Venezuelan health indicators in comparative perspective, 1960–1998

	Life expectancy				Infant mortality rates (per 1,000 live births)		
	1960 (all)	1975 (all)	1998 (males)	1998 (females)	1965	1989	1998
Venezuela	57	65	70	76	65	35	21
Chile	56	63	72	78	101	19	10
Colombia	n.a.	n.a.	67	73	86	38	23
South Korea	53	61	69	76	62	23	9
Malaysia	52	59	70	75	55	22	8
Thailand	49	58	66	72	88	28	29

SOURCE: World Bank, *World Development Report,* various years.

Venezuela also appeared to be in a particularly favorable position in terms of assimilating the more technically challenging endeavors in the heavy industrial sectors that were promoted in the 1970s and 1980s. As indicated in table 2.11, Venezuela did not appear to have any shortage of skilled manpower in relative terms. In 1960, Venezuela had more scientists per million than all the

17. The best evidence of this is that the initial level of education was similar in the 1970s in East Asia and Latin America; yet East Asian economies grew substantially more rapidly in the subsequent three decades. To take another example, the socialist economies in the Soviet bloc and in Cuba have among the most advanced education systems for developing countries; yet growth rates have been slow in the post-1970 period because of other disincentives to investment such as the absence of market competition and private property rights.

other countries in the sample. In the period 1981–95, there were more scientists and engineers working in research and development in Venezuela than in Colombia, Malaysia, and Thailand, despite the much higher rates of manufacturing growth in the last two in the period 1965–98.

TABLE 2.11 Scientific and technical labor skills in Venezuela: A comparative perspective

	Scientists (per million in the 1960s for the years indicated)		Scientists and engineers in R&D (per 100,000; annual averages 1981–95)*
Venezuela	16.5	(1964)	208
Chile	15.2	(1962)	364
Colombia	n.a.		39
South Korea	7.1	(1969)	2,636
Malaysia	12.2	(1966)	87
Thailand	1.6	(1964)	173

SOURCE: Chang 1998, table 2; UNESCO, *Statistical Yearbook*, various years; UNDP, *Human Development Report*, 1997, 2001.
* Latest year with data on all selected countries.

The availability of skilled labor power was also enhanced through immigration. In the period 1940–60, Venezuela received among the highest levels of European immigration relative to the country's population in the Americas, the majority of whom were classified as educated and/or skilled (Mörner and Sims 1985). In the 1970s and early 1980s, it was only Venezuela among Latin American countries that received substantial numbers of highly skilled immigrants from the Southern Cone countries—namely, Argentina, Uruguay, and Chile (Chen, Picuoet, and Urquijo 1983).

Finally, Venezuela has been far from being the most unequal country in terms of income distribution in the period 1970 to the mid-1990s. Tables 2.12 and 2.13 compare income distribution across several Latin American and East Asian economies in 1970 and the mid-1990s.[18] In 1970, the ratio of the income of the highest quintile to the lowest 40 percent in Venezuela was 4.0, which was

18. Cross-country comparisons of income distribution should be treated with caution. This is because there are serous problems of comparability in the methodology, sampling, and the coverage of the household surveys that are used to generate income distribution data (Székely and Hilgert 1999). Serious problems include large differences in sample sizes of the survey and the interpretation of data. For instance, differences in the ability of country surveys to capture the information on the richest sectors of society and to accurately capture their sources of income, namely profits or capital income, may bias the cross-country rankings of income distribution.

the lowest ratio in Latin America and lower than in Malaysia and the Philippines. By 1996, Venezuela still maintained the least unequal income distribution in Latin America and a Gini coefficient similar to that of Malaysia, which was growing much more rapidly in period 1970–98.

While there is no conclusive evidence that greater income inequality causes slower growth (Deininger and Squire 1998), there is a substantial literature that suggests that greater egalitarianism enhances the prospects of constructing developmental states. For instance, the role of land reform and the role of promoting a more equal distribution of income more generally has been considered by many to be essential to contributing to political legitimacy and stability and

TABLE 2.12 A comparison of indicators of income inequality, 1970

	Income share of lowest 20%	Income share of lowest 40%	Ratio of top quintile's share to that of the bottom 40%
South Korea	7.3	19.6	2.2
Thailand	5.1	15.2	3.3
Malaysia	4.0	11.7	4.8
Philippines	3.6	11.7	4.6
Colombia	4.7	11.1	5.2
Chile	4.3	13.1	4.0
Venezuela	3.6	13.0	4.0
Mexico	2.8	7.9	8.1
Brazil	2.0	7.0	9.6

SOURCE: Calculated from Deininger and Squire 1996.

TABLE 2.13 A comparative view of income inequality in the 1990s

		Income share				
Country	Year of survey	of top 10%	of top 20%	of lowest 40%	of lowest 20%	Gini index
Venezuela	1996	37.0	53.1	8.4	3.7	48.8
Chile	1994	46.1	61.0	6.6	3.5	56.5
Colombia	1996	46.1	60.9	6.6	3.0	57.1
Mexico	1995	42.8	58.2	7.2	3.6	53.7
Brazil	1996	47.6	63.8	5.5	2.5	60.0
Malaysia	1995	37.9	53.8	8.3	4.5	48.5
Philippines	1997	36.6	52.3	8.8	5.4	46.2
Thailand	1998	32.4	48.4	9.8	6.4	41.4
South Korea	1993	24.3	39.3	12.9	7.5	31.6

SOURCE: World Bank, *World Development Report*, 1999; World Bank, *World Development Indicators*, 2000.

hence the rise of developmental states in Scandinavia (Blomström and Meller 1991) and East Asia (Putzel 1992; Kohli 1999). In the literature on "the macroeconomics of populism," greater income inequality is thought to lead to unsustainable levels of consumption spending on the part of politicians who attempt to win short-term support by directing subsidies toward disadvantaged groups (Dornbusch and Edwards 1990). Finally, Stanley Engerman and Kenneth Sokoloff (1997) argue that one of the reasons why Latin America fell behind the United States and Canada in the nineteenth century was the greater concentrations of economic wealth and political power among elites in Latin America. This concentration of wealth originated in differences in factor endowments and colonial policies in the two regions. Latin America's concentration of wealth derived from the development of large-scale plantation agriculture producing cash crops, while much of North America's early agricultural development was based on a more egalitarian structure of small-scale family farms. Such concentrations of wealth and power supposedly created incentives for elites to keep access to state resources limited to upper-income groups. As a result, the development of human capital was narrow, which limited the prospects of innovation and technological change in Latin America. Whatever the merit of these models, they all would have predicted a much more efficient fiscal linkage and certainly higher *relative* growth rate than Venezuela achieved in the period 1980–2005.

2.5 Conclusion

This chapter has made four main points in relation to the literature concerning the growth process in Venezuela. First, in the period 1920–80, non-oil and manufacturing growth was rapid. Venezuela fared well in comparison with most of the Latin American economies in this period. Second, the period 1965–2005 is marked by two important trends: the non-oil economy, and particularly the manufacturing sector, experienced a dramatic decline in total factor productivity growth and labor productivity growth in the period 1968–98, and output growth and investment rates in the manufacturing sector and the non-oil economy more generally collapsed in the period 1980–2005. Third, there was a dramatic reversal in the relationship between non-oil growth and real oil prices within the period 1920–98. In the period 1920–65, there was a strong and significant positive correlation, while, in the period 1965–98, there was a strong and significant negative correlation. This underlying pattern indicates an important decline in the dynamic efficiency of the fiscal linkage in Venezuela and a reversal in the ability of the government to manage oil revenues in ways that promote growth. Finally, it was demonstrated that the growth slowdown occurred despite many favorable "initial" conditions in the economic and political realm.

The growth implosion in Venezuela represents a challenge to many of the growth theories and institutional theories examining state capacity. The basic theme in both of these paradigms is that growth requires certain preconditions—such as a developed bureaucracy, high investment, human capital, political representation, and so on. The Venezuelan case is an important corrective to the "prerequisite" view of development that reigns in the dominant paradigm of "good governance" (World Bank 1997a), and in new growth theory (Romer 1994). Two basic features of the Venezuelan growth experience in the period 1920–2005 challenge the prerequisite view. First, the period with the most rapid rates of output and productivity growth (1920–58) were marked by authoritarian rule and relatively low levels of human capital formation (see Chapters 8 and 9 for details). Second, the growing stagnation in productivity growth and output growth (1968–2005) were marked by much higher levels of human capital formation and the existence of competitive party democracy (see Chapter 8). The fact that Venezuela began its decline in productivity growth rates by the late 1960s despite the existence of a relatively favorable human and physical capital stock (and despite the existence of one of the more stable competitive democracies among middle-income countries anywhere in the world) suggests that the prerequisite view of development cannot explain growth trajectories. More generally, growth theories are linear models that predict either catching up or stagnation. The Venezuelan case (and others in Latin America and elsewhere, where long-run rapid growth has evolved into long-run stagnation) provides an opportunity to examine the *dynamics of growth path changes* that feature prominently in reality, but are not adequately addressed in neoclassical growth theory.

PART 2

A CRITICAL SURVEY OF THE "RESOURCE CURSE" LITERATURE

3

ECONOMIC EXPLANATIONS OF THE GROWTH COLLAPSE IN VENEZUELA

The relationship between mineral and fuel wealth and economic development has been the subject of intense debates over the past century. Central to the effect mineral and fuel wealth has on long-run economic growth is the stimulus it can provide to industrial activities. One of the main lessons of world economic history of the past two centuries is that sustained economic growth is achieved with sustained and successful industrialization (Kaldor 1967; Chenery 1979). The idea that commodity exports generate domestic demand for manufacturers has long been emphasized by development economists (Lewis 1954). Similarly, the "staple thesis" demonstrated that growth in backward areas commonly began through the initial stimuli that primary product exports brought in terms of attracting capital and labor and inducing a more diversified production structure (Innis 1930; Watkins 1963).[1] Natural resource rents, to the extent they are appropriated by state governments, can relax common resource constraints to growth—namely the savings, foreign exchange, and fiscal constraints (Gelb and Associates 1988, 17–18).

Despite the historically positive association of abundant natural resources and industrial growth in many now advanced countries, the literature covering less developed countries in the twentieth century has largely drawn the opposite conclusion. Natural resources, for most poor countries, are deemed to be more of a curse than a blessing.[2] Structuralist theory, dependency theory, and

---

1. Findlay and Lundahl (1999) find that there is a positive correlation between natural resource abundance and economic growth in the period 1870–1914.
2. Sachs and Warner (1995) find, in the period 1971–89, that mineral exporters, on average,

some Marxist theories of imperialism (Prebisch 1950; Furtado 1970; Baran and Sweezy 1966; Amin 1976) all argue that specialization in the export of natural resources does not lead to sustained industrialization. For structuralists, primary products are subject to declining terms of trade and destabilizing price volatility. For dependency theorists, natural resources are unlikely to stimulate growth, particularly if foreign multinationals dominate resource extraction and are allowed to repatriate profits. Marxists, such as Paul Baran, argue that governments in poor economies are dominated by local elites (the so-called comprador bourgeoisie) whose interests are allied not with national development but with foreign multinationals.[3] Some authors have argued that the dependent nature of Venezuela's insertion into the world economy, and in particular the lack of national technological and industrial capacity, lies behind the failure of the Venezuelan economy to sustain dynamic manufacturing development (Maza Zavala 1974). The "enclave" nature of mineral exports, it was argued by dependency theorists and structuralists, also meant that the leading export sector had few linkages with the rest of the economy, lessening the stimulus resource booms have on investment in non-mineral sectors.

The theoretical and empirical weakness of dependency theory, at least as it applies to Latin American historiography, has received thorough treatment (Palma 1978; Weeks 1985). The variation and change of industrial growth within and across poor economies in Latin America make sweeping generalizations about the "development of underdevelopment" untenable. Moreover, the emphasis on the consequences of Latin America "being behind" largely ignores the origins of Latin American underdevelopment, that is, "how (resource-rich) Latin America fell behind" (Haber 1997).

Within debates over the role of natural resources on economic development, the role of mineral- and fuel-exporting activities has received special attention. One of the reasons that oil-exporting nations have been considered special cases revolves around some peculiar features of petroleum production. These peculiar features include the fact that such activities are conducted in enclaves, are large-scale and capital-intensive (usually with close links to multinationals),

---

grew more slowly than non-mineral exporters. However, Lederman and Maloney (2007), using the Sachs and Warner data, find that there is no robust evidence to suggest that resource abundance negatively affects growth.

3. While dependency theories focus on declining terms of trade and the stranglehold of "constraining" political alliances at the international level, more contemporary "resource curse" arguments focus more on the role resource abundance plays in altering the internal price and production structure of the economy and the role macroeconomic policy can play in terms of trade shocks.

and pay high wages. The well-above-average profit margins of oil production are characteristic of the low extraction costs and predominance of ground rent in the value of such production. The peculiar characteristics of mineral and fuel rents have largely been viewed to have negative consequences for the macro economy. Keynes (1930), for instance, argued that in the sixteenth century discoveries of precious metals in America and the subsequent inflow of income had an adverse effect on Spain's domestic industries by raising wages above competitive levels.

There is an important body of literature that examines the effects of oil on Venezuelan economic development. The main purpose of this chapter is to examine the extent to which reigning economic explanations of the slowdown in Venezuelan manufacturing growth are defensible. Sections 3.1 and 3.2 examine different economic versions of "resource curse" arguments, particularly Dutch Disease models. A common theme in these models is that oil booms produce exchange rate revaluations, which supposedly reduce the incentives to invest in manufacturing and generally make manufacturing production uncompetitive. While economic models of the "resource curse" are consistent with the coexistence of the slowdown in manufacturing growth in Venezuela with the oil booms in the 1970s and early 1980s, the longer-run correlation of oil resource availability and manufacturing investment and growth in twentieth-century Venezuela proves to be inconclusive. Moreover, these models do not address why output and productivity growth rates slowed down generally in the Venezuelan non-oil economy, or why some manufacturing sectors remained more dynamic than others over time.

Section 3.3 examines the argument that poor economic performance in the period 1973–98 was due to the myopia of state decision-makers. The main problem identified here is that the government turned oil revenues into large-scale, long-gestating industrial and infrastructural investments only to find that the increase in oil revenues was not permanent, as they had assumed. The resulting fiscal deficits were, according to this argument, a principal factor behind the growing inflation and macroeconomic instability that has characterized the Venezuelan economy since the mid-1970s. While this feature of Venezuelan policymaking captures an important part of the problem of increasingly poor economic performance, it does not explain why short-run macroeconomic crises necessarily prevented a long-run transformation of the economy or why other resource-rich countries (such as Malaysia) that have experienced substantial macroeconomic crises were able to transform the structure of their economies in ways that contributed to rapid long-run growth.

Section 3.4 focuses on explanations that place emphasis on the failures in policy and institutional design, particularly neoliberal arguments that emphasize the importance of an overly centralized, interventionist state. First, I will examine arguments that claim that poor industrial performance was due to the rise of public enterprises within manufacturing and the increasing inefficiency of these investments. Second, I will assess arguments that blame the slowdown on protectionist policies of import substitution, and the supposedly negative effects of concentrated industrial structure those policies generated. While these arguments capture some aspects of the slowdown in manufacturing and non-oil growth, the main problem identified is their failure to explain why policies and institutions took the course they did, or why policies and institutions were not changed over time. Finally, it will be shown that these arguments are difficult to sustain in light of Venezuelan economic performance over time and in comparative perspective.

3.1 Economic Theories of the Resource Curse: The Dutch Disease Model

3.1.1 Early Models

In the Venezuelan context, José Antonio Mayorbe ([1944] 1990) was one of the earliest to argue that an expanding oil industry generated an expensive exchange rate in the Venezuelan economy. This in turn rendered nascent industries uncompetitive and retarded industrialization in Venezuela (see also Rangel [1968] 1990 and Hausmann 1990, 23–54). The common theme in these arguments is that natural resource booms have adverse effects on the structure of the economy. The question of structural change in oil-exporting economies has received considerable attention in the economics literature, particularly after the discovery of North Sea oil and its impact on the industrial structure of such economies as the Netherlands and the United Kingdom. The concern was focused on the subsequent deindustrialization in output and employment that took place in these two economies following their resource booms, a phenomenon that has come to be called "the Dutch Disease."

The logic of the simple Dutch Disease theories, as Ricardo Hausmann has it (1990, 23–59) can be described as follows. In an economy in full employment equilibrium, a permanent increase in the inflow of external funds results in a change in relative prices in favor of nontraded goods (services and construction) and against non-oil traded goods (manufacturing and agriculture) leading to the crowding out of non-oil tradables by nontradables. That is, an appreciation of the exchange rate leads to a decline in the competitiveness, and hence production and employment, of the traded-goods sector. The mechanism through which this change takes place follows directly from the model's assumptions of full employment equilibrium and static technology. With these assumptions, the external funds (from an oil boom) can be translated into real domestic expenditure only if the flow of imports increases. However, since nontraded goods cannot be imported easily (or only at prohibitive costs), a relative contraction of the traded-goods sector is inevitable, otherwise the resources needed to enhance the growth of the nontraded sector would not be available.

Thus, the model predicts that oil booms inevitably lead to deindustrialization.[4] It is important to note that even without the restrictive assumptions of full employment oil booms can induce more investment in nontraded investments and thus discourage manufacturing investment. This is because the price of nontraded goods rises relative to the price of *non-oil* traded goods as a result of exchange-rate appreciation. A second mechanism through which manufacturing can become less competitive in this model is through the increase in manufacturing wages that result from increases in aggregate demand for labor that oil booms can generate. In the short run, when productivity levels are fixed, unit labor costs in manufacturing rise, which can, in the absence of compensating policies, lead to a loss in manufacturing competitiveness.

The characterization of "deindustrialization" as a "disease" stems from the unique growth-enhancing characteristics the manufacturing sector can potentially embody. The primacy of industrialization advocated in academic and policy circles in the postwar period developed as a result of numerous studies examining the causes of backwardness in developing economies (Prebisch 1950). Primary products (as well as services) were not believed to possess the "external dynamic economies" (Young 1928) observed in manufacturing industry, where faster growth apparently led to increasing productivity manifested ultimately in the dynamic specialization of employment (Verdoorn 1949; Kaldor 1966).[5]

The potential dynamism that manufacturing can generate opens up an important role for policy in affecting the growth outcomes of oil booms. In the simple Dutch Disease model, technology is assumed to be a given, which

---

4. Dutch Disease models are summarized in Corden and Neary 1982 and Neary and van Wijnbergen 1986.

5. While the service sector has increased its role as a driver of growth in recent years, there is still robust evidence that the manufacturing sector is central to the prospects of output and employment growth and technological development in most less developed countries (Dasgupta and Singh 2006).

means that additional foreign exchange is not of particular relevance to economic growth. However, when a late-developing country faces a technological gap, additional export revenues, if channeled by an appropriate industrial policy, can play an important part in closing that gap, since the additional foreign exchange can accelerate the process of importing advanced technology and the machines that embody them. Additionally, if the industrialization strategy promotes "learning," additional revenues can theoretically accelerate the growth process. For instance, during the boom, the government could promote industry by channeling resources to that sector through protection, subsidies, or financial incentives. This can serve to modernize the manufacturing capital stock, which in turn can improve productivity.[6]

This means that the structural change against non-oil tradables, such as manufacturing, is not inevitable; rather, the outcomes of resource booms depend on state policy responses (Neary and van Wijnbergen 1986; Gelb and Associates 1988). Peter J. Neary and Sweder van Wijnbergen (1986, 10–11) note: "In so far as one general conclusion can be drawn [from our collection of empirical studies] it is that a country's economic performance following a resource boom depends to a considerable extent on the policies followed by its government.... [E]ven small economies have considerable influence over their own economic performance." As will become clear, Venezuela did attempt to protect its industry but that most of its infant industries failed to develop.

If Dutch Disease models are to provide a convincing explanation of the slowdown in Venezuelan manufacturing growth, then it is necessary to establish that growing oil revenues over time have led to a decline in both the relative share of manufacturing in the economy and to a decline in manufacturing investment. In terms of the first factor, these models have been used to explain why Venezuela, in 1973, on the eve of the first large oil boom (1973), had an economic structure that was not only more dependent on oil exports than in 1960 but whose manufacturing share in *non-oil* GDP (17.5 percent) was significantly (greater than two standard deviations) *below* the Chenery norms for its income level and population size (Gelb and Associates 1988, 88). Given Venezuela's income and population size in 1973, manufacturing should have accounted for 27 percent of non-oil GDP.

The oil booms have indeed coincided with a "premature tertiarization" of the Venezuelan economy,[7] as indicated in table 3.1. By 1950, services already

---

6. Moreover, oil boom revenues can be used to invest in physical infrastructure, which is central to maintaining rapid growth (Easterly and Servén 2003).

7. "Premature tertiarization" (or alternatively "premature deindustrialization"), a relatively

accounted for 52 percent of non-oil GDP in Venezuela. The effect of the oil boom in 1973 further accelerated the structural change in favor of services. The relative share of services reached 60 percent by 1973. The share of services thereafter remained relatively high (59 percent in 1981) as a result of the second oil boom in 1979–80. Interestingly, the share of services continued to increase throughout the 1990s despite continued *declines* in per capita oil revenues through the 1990s (see table 3.1).

While the share of the non-oil GDP of services increased from 1950, the share of non-manufacturing industrial sectors ("Other industry" in table 3.1), including mining, construction, electricity, and water, declined by similar proportions. The share of the non-manufacturing industrial sectors declined from 24 percent of non-oil GDP in 1960 to 17 percent by 1981 and fell further to 10 percent by 1998. The share of manufacturing, however, did *not* experience a corresponding contraction after 1973. In 1950 the share of manufacturing was relatively low at 8 percent of non-oil GDP. This can be explained by the maintenance of a liberal trade policy in a context of rapidly rising oil revenues in the period 1930–50 (see sections 7.1 and 8.1). However, the share of manufacturing in non-oil GDP steadily *increased* to 16 percent in 1973 and rose further to 18 percent by 1981 despite two oil booms. Moreover, the share of manufacturing rose slightly to 20 percent of non-oil GDP by 1998 despite both oil booms and busts throughout the 1990s.

TABLE 3.1 Structural change in the composition of the non-oil national product of Venezuela, 1950–1998 (at 1984 prices and expressed as a percentage of non-oil GDP)

	Agriculture	Manufacturing	Other industry*	Services
1950	8	8	22	52
1960	6	15	24	55
1973	7	16	17	60
1981	6	18	17	59
1990	6	20	10	64
1995	6	19	14	61
1998	6	20	10	64

SOURCE: Banco Central de Venezuela, *Serie estadística,* various years.

* Refers to construction, mining, electricity, and water sectors.

---

high share of services in output terms *before* the development of an internationally competitive industrial base, is a common structural feature of all Latin American economies, not just the oil-exporting economies (Palma 2002).

There are two principal reasons for the increases in the share of manufacturing in non-oil GDP after 1950. The first was that the government protected domestic industry through tariff and import quotas from 1958 until 1989. Such protection meant that domestic firms were sheltered from the effects of whatever exchange-rate appreciation occurred because of the oil booms (see Chapter 7 for discussion on protectionist policies).

The second concerns the evolution of exchange rates. Table 3.2 provides data on the evolution of real exchange rates in Venezuela in the period 1950–98. The evidence suggests that there were, in fact, very *few* episodes of rapid and sustained appreciation of the exchange rate. The 1950s saw little movement in the real exchange rate. In 1961, a balance of payments crisis induced the government to *devalue* the bolívar by 25 percent.[8] In the context of very low inflation (less than 2 percent per year) throughout the 1960s, the large-scale devaluation translated into a real devaluation of nearly 24 percent. During the oil boom of the 1974–78, the effective exchange rate depreciated. This was achieved largely through the maintenance of price controls and satisfying excess demand, where needed, through imports (Gelb and Associates 1988, 82). The only appreciation occurred in the period 1979–84. In the period 1979–81, the local currency, the bolívar, appreciated 8 percent above its 1974–78 level and subsequently appreciated in 1982–83 by a dramatic 27 percent above the 1974–78 level.[9] After a large-scale devaluation in 1984, there is little evidence of episodes of massive appreciations of the real value of the bolívar. In fact, compared to the level of

TABLE 3.2 Real exchange-rate movements in Venezuela, 1950–1998 (average trade-weighted, real effective exchange rate)

(1950–52=100)	1952–61 102.3	1961–70 76.5		
(1970–72=100)	1974–78 92.9	1979–81 100.1	1982–83 119.8	1984 86.7
(1984–85=100)	1986 81.8	1988–94 56.8	1995–96 69.9	1996–98 76.6

SOURCE: Banco Central de Venezuela, *Informe anual,* various years, and *Serie estadística,* various years; IMF, *International Financial Statistics,* various years; Gelb and Associates 1988, table 6-1.
NOTE: An increase in the index indicates an appreciation of the real exchange rate.

8. See Hausmann 1990, 265–87, for a discussion of exchange-rate policy during the balance of payments crisis of the period 1958–62.
9. See M. Rodríguez 1991 for a discussion of why it was ill advised to maintain an appreciating (and increasingly overvalued) exchange rate in this period.

1984–85, the bolívar generally depreciated over the period 1988–98. This latter period of depreciation was, however associated with low rates of manufacturing growth and (as we shall see) declining real investment in the manufacturing sector. This is the opposite of what the Dutch Disease model predicts.

In terms of the second factor, there is *no* evidence that oil booms in Venezuela have been associated with declines in manufacturing investment. In fact, the experience of Venezuela shows that rapid growth of oil revenues is normally associated with high investment in economic activity in traded goods, and in particular the manufacturing sector. Moreover, falling oil revenues are associated with slower growth and investment in manufacturing industry. In the Venezuelan case, if we compare manufacturing investment rates in figure 1 with the evolution of oil revenues in figure 2, we see there is a broadly *positive* correlation between oil revenues and manufacturing investment over the period 1960–98. Even where oil booms are associated with downturns in industrial activity, as in Venezuela in the period 1979–82, this was due more to restrictive fiscal and monetary policy and the maintenance of a highly overvalued currency and less to the oil boom per se (M. Rodríguez 1991).

Again, state policy had a decisive impact on the use of oil windfalls. In the case of manufacturing investment, public sector manufacturing investment in natural-resource-based industries (hydroelectric power, oil refining, petrochemicals, steel, and aluminum) was important in maintaining the high levels of total manufacturing investment during the oil booms of the 1970s. For example, while the public manufacturing investment share (excluding oil refining) accounted for 24 percent of total manufacturing investment in the period 1968–71, the public share rises to 41 percent in the period 1972–80 (see Chapter 7 for more detailed evidence).

In sum, there is no evidence to suggest that the period of the slowdown and collapse of growth (1980–2003) coincided with a decline in the share of the manufacturing sector in the non-oil economy. If anything, the share of manufacturing increased slightly over this period. Thus, the decline in the growth rate of the entire non-oil economy cannot plausibly be due to a decline in the relative share of manufacturing. Moreover, declines in manufacturing growth coincided with declines in real oil exports over the same period. Second, Dutch Disease models, even when accounting for the importance of industrial policy and investment policies, cannot explain why the growth rate of manufacturing *itself* slowed down dramatically after 1980. While it can be argued that the poor growth of the manufacturing sector could be due to inadequate policies (a theme examined in section 3.4), the stress on policy failures cannot explain

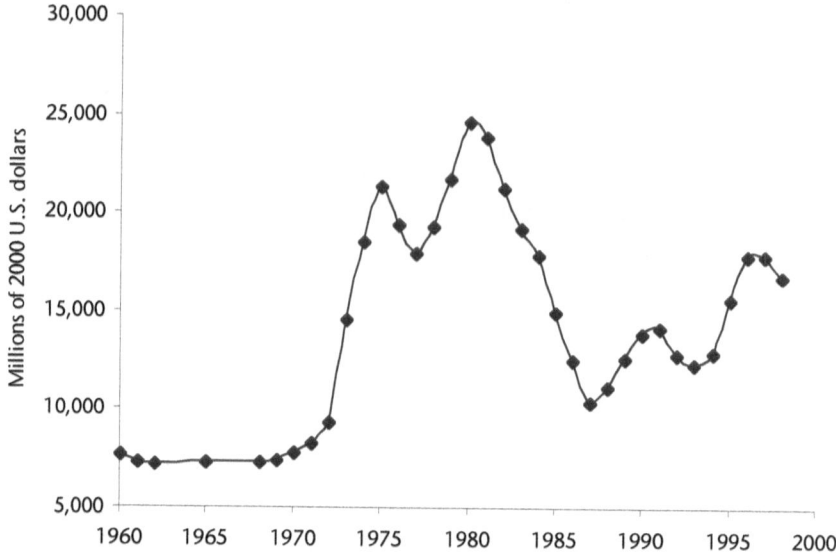

FIGURE 1. Venezuela: Real oil exports, 1960–1998. The data points represent a three-year moving average of the sum of the oil export revenues deposited in the Banco Central, the value of oil sector salaries, and the purchases of domestic products by oil companies operating in Venezuela. This sum is deflated by U.S. producer prices and converted into 2000 U.S. dollars. Ministerio de Minas y Energía, *Petroleo y otros daos estadísticas*, various years; Banco Central de Venezuela, *Serie estadística*, various years; IMF, *International Financial Statistics*, various years.

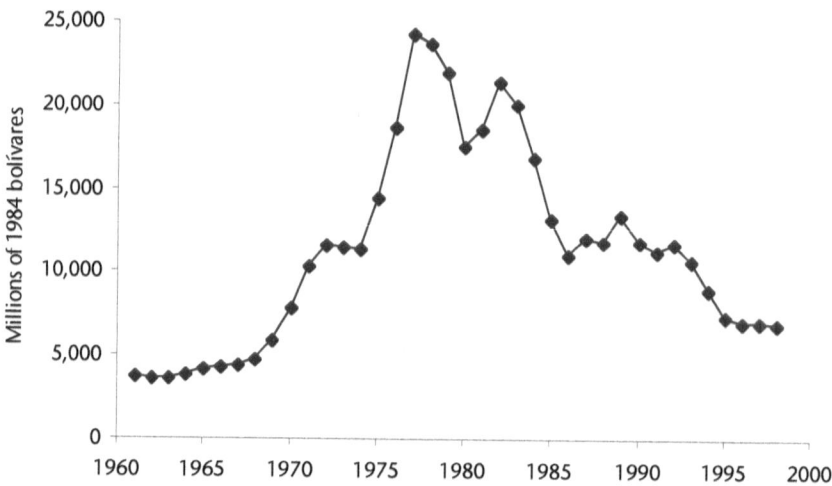

FIGURE 2. Real manufacturing investment in Venezuela, 1960–1998. The data represent a three-year moving average. OCEI, *Encuestra industrial*, various years; Banco Central de Venezuela, *Serie estadística*, various years.

Economic Explanations of the Growth Collapse 45

why governments in countries with stagnant long-run growth persistently fail to take corrective action (see section 4.5).

3.1.2 More Recent Models

Two recent studies have presented a more nuanced view of the problem of the resource curse. The first, by Francisco Rodríguez and Jeffrey Sachs (1999), attempts to explain the role natural resource abundance has played in the Venezuelan growth experience. They posit that there is an association between lower growth rates and higher levels of income. Resource-abundant countries grow more slowly *precisely* because they have an unusually high level of income (278). The basic intuition is that even though the economy has the same *steady state* as an economy without these resources, its abundance allows it to enjoy abnormally high levels of consumption for an extended period. However, in the long run, it will not be able to sustain its capital stock on the basis of its foreign exchange earnings, since those foreign exchange earnings tend to zero over time. Thus declining output marks the process of adjustment toward the steady state (282).

There are several assumptions that underpin the model. The first restriction generating overshooting is the lack of access to international markets (284). Since it is assumed that domestic agents cannot save internationally, the only way to save resources is to invest domestically. An oil boom thus induces an investment level that overshoots its steady state. If the natural resource–abundant economy were able to sell its natural resource internationally and permanently consume the interest on the assets derived from that, it would avoid the overshooting result. Why these economies would decide to do otherwise is an interesting question, and not addressed in the model.

The second restriction of the model that generates overshooting is the assumption of a balanced current account. The economy can avoid overshooting by selling resources in international markets, depositing the revenues at a fixed interest rate, and consuming the interest on these deposits (295). Although the authors argue that the assumption of a balanced current account is consistent with the behavior of Venezuela's current account during the 1972 to 1993 period (during which the current account surplus averaged 1.65 percent of GDP), a fully satisfactory account of the overshooting phenomenon should provide an account of why such a restriction seems to hold in practice.[10]

---

10. In fact, there was a significant accumulation of external public liabilities and external private assets in the period 1972–93. The Venezuelan public and publicly guaranteed external debt went

This model provides a valuable contribution to understanding the growth collapse in Venezuela. However, it is subject to some important shortcomings. Hausmann (2003) points to two of these shortcomings: first, the large and increasing volume of oil revenues makes arguments about exhaustion less compelling as a determinant of current trends; second, the assumption of a home bias in investment implies that residents must be investing at rates of return below those of the world economy, but decide to keep their savings at home anyway. While limits on borrowing can be explained as the consequence of contract enforcement or sovereign risk, it is much harder to argue that overinvestment is caused by limits to the outward mobility of capital. Moreover, there is scant evidence that profit rates were unusually low during the boom years in Venezuela (see section 3.2).

A third shortcoming is that the model assumes the long-run trends of the Venezuelan economy can be explained as a result of the natural response of an economy on its balanced growth path to an increase in the availability of resource rents, *given a certain level of productivity.* There is no attempt to explain *why* productivity growth stagnated in Venezuela. The authors acknowledge that Venezuela's growth downturn owes as much to a lack of productivity growth as to its natural resource abundance (296). "What is really striking is that a constant rate of productivity growth is not an unreasonable assumption for Venezuela. If Venezuela were able to achieve rates of productivity growth similar to those undergone by the strongest-performing developing countries, the downward tendency in the growth rate could be reversed" (296). The Rodríguez and Sachs model fails to explain why productivity growth stagnated in Venezuela from the late 1960s to 2005.

A second recent work on the problem of the "resource curse" is by Ricardo Hausmann and Francisco Rodríguez (2006), who show that the decline in capital accumulation and growth in Venezuela can be accounted for by the country's inability to develop an alternative export industry. Nearly complete oil specialization is, they argue, a necessary ingredient of an explanation of the Venezuelan economic collapse (10).

There was some non-oil export growth in Venezuela in the period 1981–2002, but such growth was generally modest and was concentrated in mineral-based sectors:

---

from US$1.41 billion to US$26.9 billion (from 4 to 47 percent of GDP) from 1972 to 1993, whereas privately held external assets (excluding interest) increased by US$38.4 billion during the same period (M. Rodríguez 1991, 302). To understand the reason overshooting was generated, one must effectively understand the reasons leading the state to take actions that ultimately resulted in offsetting private decisions to invest resources internationally.

Per capita real non-oil exports (measured in 2000 US$) grew by 42% in the period 1982–2002. Their share of total exports grew from 7.1% to 19.7% of total exports, but mainly due to decline in oil exports. The annual real growth rate of per capita exports, at 2.01%, is the third lowest in the group of 10 oil exporters that suffered important collapses in oil exports in the 1980s and 1990s. Three-fifths of that growth has been in sectors such as iron ore and steel, chemicals and non-ferrous metals that heavily rely on the economy's comparative advantage in petroleum and energy. Although non-energy intensive non-oil exports have grown at a satisfactory rate of 5.2% a year, this is partly due to fact that it was an incredibly small sector, providing only $39 per capita in export revenue in 1982. This growth is also surprisingly weak if one views it in the light of the considerable real exchange rate depreciation that occurred between the early eighties and late nineties (a nearly 50% fall). (11)

Had there existed an alternative export sector in Venezuela in 1980, the authors argue, the growth of that sector would have played a stabilizing role in that country's reaction to falling oil revenues. In its absence, the domestic economy had to react to adverse oil shocks by contractions in domestic production. This process must continue until either the fall in oil revenues is halted[11] or the real exchange rate falls sufficiently to make the production of non-oil tradables competitive (Hausmann and F. Rodríguez 2006, 11). The authors show that of the ten oil-exporting developing countries that experienced significant export collapses in the period 1981–2002 (Mexico, Oman, Bahrain, Indonesia, Saudi Arabia, Trinidad and Tobago, Venezuela, Ecuador, Algeria, and Nigeria), only two of these countries (Mexico and Indonesia) were able to develop a significantly strong growth of their non-oil exports (especially in manufacturing) to compensate for the decline in oil exports and generate an overall positive economic growth.

Hausmann and Rodríguez provide an important contribution to the resource curse argument by identifying that the degree of export diversification matters for the growth prospects *among* oil-exporting late developers. And indeed they argue that policies that increase productivity in the non-oil tradable sector would have improved the growth prospects of the Venezuelan economy (12). However, their argument does not explain why non-oil tradable (and especially manufacturing) productivity was so poor and as a result, why non-oil export sectors did not emerge in a more dynamic and diversified manner in

---

11. In this perspective, the increase in economic growth in 2004 and 2005 owes much to the boom in oil exports.

Venezuela. This is because their model assumes a given path of productivity growth among oil exporters. A more complete understanding of the growth collapse in Venezuela must not only identify the failure of export diversification, but also explain the mechanisms underlying poor productivity growth in the Venezuelan non-oil tradable sectors.

3.2 Dutch Disease Models: The Wage Rigidity Argument

Not all Dutch Disease models are purely economic in their logic. In the discourse addressing worsening economic performance in Venezuela, there is an influential political economy version of Dutch Disease models that attempts to explain *why* governments do not implement policies to counteract the lack of competitiveness the manufacturing sector faces in light of the "expensive" exchange rates (and accompanying "expensive" wage costs) that oil abundance can generate. The main argument of this wage rigidity model is that politically dominant *urban* workers, the middle-class, and infant industry owners benefit from and thus lobby for the maintenance of "expensive" or overvalued exchange rates. These models predict that in highly urbanized countries implementing protectionist policies, there are few lobby groups that would favor devaluation, which is central to the prospects of making the non-oil tradable sectors (agriculture and manufacturing) competitive.

These models also focus on the "expensive" wage levels of the oil-determined exchange rate to explain poor manufacturing performance in Venezuela (Márquez 1987; Hausmann 1990).[12] Unit labor costs are deemed too "expensive" for the level of Venezuelan productivity at any given moment. James Mahon (1992), Márquez (1987), and Hausmann (1990, 342) further argue that the high rates of urbanization and few exporter interests imply that there are few interest groups favoring devaluation, which such authors view as crucial to removing the anti-export bias in protectionist import-substitution policies.[13]

Devaluation (when accompanied by a less than proportional increase in nominal domestic wage rates) implies a reduction in real wages, which according to these theories would imply a reduction in urban middle- and working-class incomes. Such devaluations are deemed to be politically infeasible, given

---

12. These models are part of the "macroeconomics of populism" approach (Sachs 1985; Dornbusch and Edwards 1990). See also Baptista 1995 for a discussion of the limits that an oil-determined exchange rate places on the export possibilities of the manufacturing sector.

13. See also Sachs 1985.

the powerful "voice" urban interests have had in the Venezuelan polity since 1958. "Expensive" wages tend to be downwardly rigid in these models. These macroeconomic theories are a version of the "urban bias" scenario, which posits that high levels of urbanization foster the entrenchment of powerful rent-seeking interest groups that, once entrenched, oppose devaluation *and* reductions in effective rates of protection of urban-based industries.[14]

These theories attempt to identify the political and economic mechanisms of policy failure. The greater emphasis on the political economy of exchange rate and wage policies distinguishes these models from standard Dutch Disease models. However, the focus on the exchange rate policy as the source of the growing slowdown in Venezuelan manufacturing is subject to some important shortcomings. First of all, even if we assume that exchange rate levels and policy is central to manufacturing performance, comparative and historical evidence suggests that it is not clear that the degree of urbanization explains the extent to which policymakers will maintain competitive devaluations. For instance, many countries with largely *rural* populations (in sub-Saharan Africa and South Asia) have not been immune to urban rent-seeking interest groups resisting restructuring, nor have growth rates in such economies necessarily been very impressive (Bardhan 1984 on India; Sandbrook 1986 on sub-Saharan Africa).

Within Venezuela, resistance to devaluation has not always or even primarily been determined by the degree of urbanization. In 1934, when 75 percent of the population was rural (Baptista 1997, 29), the Venezuelan government did not decide to follow the U.S. devaluation with respect to gold. As a result, the bolívar appreciated from Bs.5.20. to Bs.3.06. to the dollar. The dominant interests at the time were military leaders and merchant classes interested in maximizing the amount of dollars retained through oil exports (Rangel [1968] 1990, 207–27).[15] The coffee and cocoa sectors, more numerous in terms of employment,

14. These open economy macroeconomic models follow standard trade theory in that protection represents a "distortion" from the optimum "free trade" policy, which would align an economy with its comparative advantage. The resistance to the dismantling of tariffs is due to excess over normal profits that protected industries appropriate. Resistance to reductions in effective rates of protection may occur not only through opposing trade liberalization but also by resisting increases in tariffs of intermediate and capital goods industries, which provide inputs into consumer goods industries. As Hirschman notes: "The numerous studies which have been devoted to the topic of effective protection have paid little attention to this important mechanism, presumably because they were rooted in the desire to denounce the evils and influences of ISI" (1971, 110 n. 28).

15. The decision to revalue the bolívar in the early 1930s did not occur before vigorous and heated debates between influential economists and bankers. Adriani ([1931] 1990) argued that devaluation was necessary for Venezuela as a part of constructing a more dynamic agricultural

were not powerful enough to resist the revaluation of the bolívar, a policy change that, in the context of liberal trade policies, was to contribute to the dramatic decline in those sectors in relative and absolute terms (Baptista and Mommer 1987, 27–31).

In 1961, facing a severe balance of payments crisis, the Betancourt administration devalued the bolívar from 3.30 bolívares per U.S. dollar to 4.50, a real devaluation of 36 percent that was maintained until the mid-1970s. This long-run real devaluation was, however, achieved in an economy with an urbanization rate of 70 percent, which was significantly higher than the 26 percent urbanization rate in the mid-1930s, an era that, as we mentioned, experienced significant revaluations of the bolívar. Moreover, the rapid real devaluation of the bolívar in the period 1982–99 as a result of balance of payments and capital flight crises occurred in a period where the degree of urbanization remained unchanged. This suggests that the there may be more important social and political conflicts than those between the tradable and nontradable sectors or between urban and rural interests more generally.

Second, the focus on an overvalued exchange rate as a cause of inefficient import-substituting or inward-oriented industrial polices misleadingly assumes macroeconomic "distortions" (exchange rates) are necessarily correlated with microeconomic policies (industrial policy, degree of openness, and so on). Strictly speaking, overvalued exchange rates, by definition, result in unsustainably large trade deficits. Trade protection, per se, is indeterminate with respect to long-run economic growth. The distinction between micro reforms and macro reforms is commonly lost in many studies that associate activist industrial policies with macroeconomic instability (Díaz-Alejandro 1975, 115–16; Rodrik 1996, 13–17).[16] Moreover, the implicit assumption that greater "outward orientation" or "openness" in trade policy generates faster economic growth is *not* established either on methodological or empirical grounds (Lucas 1988; Pritchett 1996). Exports are more plausibly seen as the result (and not the cause) of rapid investment, output and productivity growth, and internal competition

---

sector. Vicente Lecuna, president of the principal commercial bank (Banco de Venezuela), argued in favor of a strong bolívar. Lecuna argued that because the oil companies paid salaries and made local purchases in local currency, a revaluation would generate an increase in the number of dollars to be spent to acquire the bolívares needed to make payments in the country (Astorga 2000, 230 n. 17).

16. As Rodrik (1996, 114) points out, the confusion between macroeconomic policies and microeconomic policies often reveals itself in empirical studies, such as in Dollar's (1992) "real exchange rate distortion" index used in World Bank 1993, that uncover a negative relationship between measures of exchange rate distortion and economic growth, and then attribute the effect to the lack of openness in terms of trade protection.

and industrial policies. Surely, Brazilian football players are exported because they score lots of goals; they do not score lots of goals because they are exported.

A third shortcoming of open economy models is that they do not incorporate the potential role of high wages in providing productivity-enhancing incentives. It is recognized in the theory of efficiency wages that firms often pay wages in excess of opportunity costs in order to motivate workers and managers (Shapiro and Stiglitz 1984). Higher wages may also increase the applicant pool, which provide employees a greater chance of finding employees whose skill sets match the needs of the firm more than those of an "average" worker.

A fourth important shortcoming in these models is the failure to systematically compare real wage growth to productivity growth in the manufacturing sector. Real wages raise production costs only when they rise faster than productivity. Table 3.3 provides data on average real wage growth and average labor productivity growth for the manufacturing sector as a whole in the period 1950–98. The evidence suggests that manufacturing real wages did not always grow faster than manufacturing labor productivity. In the period 1950–60, average annual manufacturing labor productivity grew at 9.9 percent, while average annual real wages grew at 5.8 percent. In the period 1960–73, real wages grew 4.1 percent, while labor productivity grew at 3.1 percent—not a particularly large difference. Moreover, manufacturing output growth was still relatively rapid in this period. It was only during the oil boom years, 1973–81, that annual manufacturing real wages grew substantially faster than annual manufacturing productivity growth (3.9 percent and 1.5 percent respectively). However, this was *not* when the growth slowdown in manufacturing was most dramatic. The years 1980–98 represented the slowest growth in manufacturing output over the period 1940–98. However, the period 1981–88 saw real wages and

TABLE 3.3 Venezuela: Evolution of real wages and labor productivity growth in the manufacturing sector, 1950–1998

	Average annual real wage growth (%)	Average annual labor productivity growth (%)
1950–60	5.8	9.6
1960–73	4.1	3.1
1973–81	3.9	1.5
1981–88	1.2	1.2
1988–98	−5.5	0.7

SOURCE: OCEI, *Encuesta industrial*, various years; Banco Central de Venezuela, *Serie estadística*, various years.

productivity growing at the *same* annual rate. The following period, 1988–98, saw real wages actually falling at annual rate of 5.5 percent while average annual labor productivity grew at a modest 0.7 percent. In this latter period, the fall in unit wage costs, which is the difference between the change in productivity levels and the change in real wages, was 6.2 percent per year, which is the greatest fall registered for any of the periods considered.

If high real wages were responsible for the slowdown in manufacturing growth, as the open economy models suggest, then the record fall in unit wage costs in the 1990s should have been associated with increased profitability and increased growth in manufacturing. The average annual growth in manufacturing in the period 1990–98 was below 2.0 percent, which was the *lowest* growth rate for any period since 1950. At least at the aggregate manufacturing level, higher manufacturing real wages are associated with faster manufacturing growth and manufacturing slowdown with a manufacturing wage collapse, a result that does not support the wage rigidity argument. From a late development perspective, what is of more concern is the decline in manufacturing productivity growth rates themselves. This point, which I will return to, is not assessed in such models, since they focus on macro equilibrium issues and not issues of productivity growth.

A more detailed assessment of the wage rigidity argument requires an examination of Venezuelan manufacturing wage rates and productivity performance in comparative perspective and disaggregated into subsectors. The problem of an "overvalued" or "expensive" exchange rate should manifest itself in higher than average wage costs in Venezuela than elsewhere. Is it the case that comparative manufacturing wage costs or an excessive wage share (and hence low profitability) explains the comparative performance of the Venezuelan manufacturing sector? Moreover, is it the case that the evolution of certain subsectors fit the open economy model better than others?

Tables 3.4a, 3.4b, and 3.4c provide evidence on wage levels relative to the U.S. level, productivity levels relative to the U.S. level, and profitability (measured as one minus the percentage share of wages in value-added—a proxy for the operating surplus) in Norway, Venezuela, Colombia, Chile, South Korea, and Malaysia for nine manufacturing sectors of varying degrees of capital intensity over the period 1970–97 (or the latest year in which comparative data are available). Following data on the capital intensity of different manufacturing sectors in the United States during the 1980s provided by Alfred Chandler and Takashi Hikino (1997, 46–50), I have divided the nine manufacturing sectors into three categories: labor-intensive sectors (textiles, clothing, and

TABLE 3.4a Relative wages, productivity, and profitability in labor-intensive manufacturing sectors: Venezuela in comparative and historical perspective, 1975–1995 (wages and productivity relative to U.S. level)

	Textiles						Clothing						Footwear					
	1970	1975	1980	1985	1990	1995	1970	1975	1980	1985	1990	1996	1970	1975	1980	1985	1990	1996
A. Wages (annual wage per employee)																		
Norway	0.58	1.00	1.09	0.81	1.27	1.29	0.63	1.48	1.35	0.88	1.50	1.56	0.62	1.11	1.23	0.84	1.34	1.43
Venezuela	0.45	0.58	0.64	0.54	0.21	0.11	0.37	0.76	0.68	0.54	0.20	0.13	0.36	0.43	0.64	0.46	0.16	0.13
Colombia	0.18	0.16	0.23	0.18	0.11	0.18	0.13	0.15	0.19	0.16	0.10	0.17	0.12	0.12	0.20	0.18	0.11	0.16
Chile	0.24	0.14	0.37	0.19	0.19	0.34	0.26	0.22	0.45	0.23	0.22	0.89	0.30	0.17	0.38	0.22	0.20	0.40
S. Korea	0.07	0.10	0.21	0.19	0.43	0.62	0.09	0.13	0.31	0.23	0.49	0.74	0.10	0.12	0.24	0.24	0.48	0.69
Malaysia	0.07	0.10	0.15	0.16	0.15	0.23	0.07	0.13	0.15	0.16	0.16	0.24	0.10	0.14	0.17	0.17	0.14	0.25
B. Productivity (value-added per employee)																		
Norway	0.61	0.77	0.77	0.50	1.01	0.79	0.57	0.83	0.88	0.55	0.88	0.82	0.52	0.77	0.85	0.50	0.78	0.75
Venezuela	0.61	0.82	0.67	0.26	0.27	0.26	0.61	0.68	0.67	0.48	0.18	0.36	0.29	0.41	0.51	0.39	0.16	0.20
Colombia	0.34	0.30	0.48	0.40	0.37	0.36	0.20	0.17	0.29	0.20	0.15	0.21	0.15	0.16	0.25	0.24	0.19	0.27
Chile	0.46	0.34	0.57	0.38	0.30	0.38	0.49	0.33	0.72	0.43	0.30	0.55	0.55	0.41	0.83	0.35	0.31	0.40
S. Korea	0.12	0.20	0.29	0.28	0.46	0.80	0.12	0.14	0.28	0.23	0.47	0.84	0.14	0.15	0.24	0.27	0.60	0.81
Malaysia	0.10	0.16	0.21	0.18	0.19	0.31	0.10	0.12	0.16	0.13	0.14	0.18	0.09	0.22	0.17	0.10	0.12	0.22
C. Profitability (1 minus the share of wages in value-added, %)																		
Norway	47.8	32.6	31.7	30.7	30.4	32.5	41.0	27.2	25.2	29.2	28.0	27.2	36.0	24.4	35.8	26.3	26.9	22.5
Venezuela	59.8	62.8	54.3	60.3	66.4	82.2	66.9	55.5	50.8	50.7	65.3	85.7	32.2	45.2	44.4	48.7	57.1	74.1
Colombia	70.6	73.5	77.2	77.9	86.5	79.9	66.9	64.4	67.1	64.8	70.3	69.8	56.5	61.7	56.7	67.0	75.4	75.5
Chile	77.8	77.8	68.8	75.8	72.5	63.5	70.2	73.5	69.0	75.8	69.2	37.4	70.9	78.1	40.7	72.2	72.9	59.3
S. Korea	64.7	71.4	64.7	66.8	60.0	68.1	60.9	60.5	60.1	56.6	55.8	66.2	56.2	54.6	62.2	60.7	65.9	65.3
Malaysia	57.6	65.4	63.7	58.0	65.8	70.1	59.1	54.0	53.8	43.8	48.9	49.3	40.2	66.5	66.5	57.8	50.0	53.3

SOURCE: UNIDO, *Handbook of Industrial Statistics*, 1988; UNIDO, *International Yearbook of Industrial Statistics*, various years.

TABLE 3.4b Relative wages, productivity, and profitability in intermediate capital-intensive manufacturing sectors: Venezuela in comparative and historical perspective, 1975–1997 (wages and productivity relative to U.S. level)

	Nonelectrical machinery						Electrical machinery						Transport equipment					
	1970	1975	1980	1985	1990	1997	1970	1975	1980	1985	1990	1996	1970	1975	1980	1985	1990	1997
A. Wages (annual wage per employee)																		
Norway	0.51	0.94	1.03	0.69	1.15	1.01	0.56	0.94	1.07	0.70	1.12	1.13	0.46	0.79	0.79	0.49	0.86	0.89
Venezuela	0.22	0.35	0.37	0.31	0.12	0.08	0.32	0.40	0.46	0.35	0.15	0.12	0.35	0.40	0.47	0.32	0.13	0.06
Colombia	0.13	0.10	0.13	0.10	0.06	0.11	0.12	0.10	0.16	0.14	0.09	0.14	0.11	0.09	0.13	0.12	0.08	0.12
Chile	0.21	0.12	0.42	0.30	0.24	0.42	0.26	0.15	0.48	0.30	0.24	0.35	0.21	0.12	0.25	0.15	0.12	0.24
S. Korea	0.06	0.09	0.18	0.16	0.34	0.48	0.06	0.08	0.16	0.14	0.32	0.45	0.08	0.11	0.18	0.16	0.39	0.40
Malaysia	0.08	0.10	0.13	0.14	0.12	0.17	0.11	0.10	0.11	0.12	0.10	0.17	0.09	0.09	0.13	0.13	0.11	0.15
B. Productivity (value-added per employee)																		
Norway	0.46	0.72	0.72	0.49	0.69	0.57	0.52	0.77	0.68	0.45	0.63	0.53	0.41	0.54	0.52	0.28	0.54	0.42
Venezuela	0.50	0.41	0.45	0.38	0.18	0.18	0.50	0.52	0.49	0.41	0.20	0.17	0.50	0.57	0.58	0.35	0.15	0.36
Colombia	0.21	0.16	0.20	0.17	0.11	0.14	0.25	0.21	0.34	0.26	0.20	0.16	0.22	0.29	0.25	0.20	0.22	0.19
Chile	0.28	0.24	0.37	0.18	0.19	0.33	0.58	0.53	0.55	0.45	0.44	0.41	0.33	0.26	0.54	0.27	0.22	0.35
S. Korea	0.07	0.11	0.20	0.22	0.48	0.60	0.12	0.14	0.19	0.24	0.46	0.68	0.14	0.16	0.23	0.24	0.52	0.66
Malaysia	0.08	0.16	0.18	0.17	0.19	0.23	0.22	0.18	0.16	0.17	0.12	0.17	0.13	0.13	0.19	0.18	0.24	0.24
C. Profitability (1 minus the share of wages in value-added, %)																		
Norway	42.7	38.0	27.5	34.9	28.8	36.2	43.0	50.2	30.8	55.7	30.8	40.4	39.0	49.5	23.2	19.5	30.1	28.2
Venezuela	76.4	59.5	73.5	63.0	72.2	84.4	66.5	61.5	58.9	61.8	70.9	80.6	62.1	64.0	58.2	58.7	63.0	94.2
Colombia	68.1	71.5	77.0	73.0	76.9	73.2	73.6	76.3	79.0	76.3	82.6	75.9	74.8	83.5	72.7	72.6	84.6	79.3
Chile	62.0	75.0	25.5	24.6	46.3	54.0	76.5	85.7	61.2	71.0	78.9	76.4	65.3	77.5	76.4	74.4	76.5	76.4
S. Korea	59.8	60.2	65.7	66.5	69.3	71.0	69.7	72.6	63.7	74.1	72.9	81.2	67.5	63.7	60.3	69.3	67.8	79.3
Malaysia	54.1	66.8	73.2	62.4	72.5	72.8	74.3	71.9	68.7	68.4	69.2	73.4	62.3	60.1	66.0	67.1	81.2	78.1

SOURCE: UNIDO, *Handbook of Industrial Statistics*, 1988; UNIDO, *International Yearbook of Industrial Statistics*, various years.

TABLE 3.4C Relative wages, productivity, and profitability in capital-intensive manufacturing sectors: Venezuela in comparative and historical perspective, 1975–1996 (wages and productivity relative to U.S. level)

	Iron and steel						Nonferrous metals						Industrial chemicals					
	1970	1975	1980	1985	1990	1996	1970	1975	1980	1985	1990	1996	1970	1975	1980	1985	1990	1995
A. Wages (annual wage per employee)																		
Norway	0.48	0.80	0.80	0.58	0.96	0.95	0.54	0.93	1.51	0.75	1.20	1.27	0.47	0.86	0.88	0.56	0.97	0.93
Venezuela	0.43	0.38	0.37	0.42	0.23	0.25	0.39	0.46	0.46	0.41	0.20	0.20	0.25	0.45	0.52	0.36	0.16	0.18
Colombia	0.12	0.10	0.14	0.15	0.08	0.15	0.13	0.11	0.13	0.17	0.09	0.16	0.16	0.14	0.20	0.17	0.11	0.16
Chile	0.32	0.13	0.27	0.26	0.22	0.41	0.35	0.15	0.42	0.37	0.30	0.63	0.24	0.12	0.39	0.23	0.22	0.33
S. Korea	0.08	0.10	0.31	0.18	0.46	0.51	0.08	0.10	0.17	0.20	0.48	0.53	0.09	0.12	0.20	0.17	0.36	0.52
Malaysia	0.10	0.11	0.13	0.13	0.13	0.17	0.14	0.18	0.16	0.12	0.13	0.20	n.a.	n.a.	0.15	0.12	0.19	n.a.
B. Productivity (value-added per employee)																		
Norway	0.55	0.90	0.70	0.52	0.61	0.52	0.79	0.83	1.11	0.92	1.12	0.99	0.40	0.49	0.56	0.43	0.54	1.49
Venezuela	0.56	0.49	0.41	0.63	0.24	0.44	0.59	0.63	0.47	0.98	0.74	0.72	0.32	0.32	0.38	0.45	0.20	0.39
Colombia	0.22	0.20	0.37	0.41	0.39	0.34	0.25	0.16	0.29	0.45	0.36	0.25	0.23	0.24	0.31	0.26	0.17	0.25
Chile	0.68	0.51	0.73	0.66	0.47	0.56	5.70	0.16	2.79	3.13	3.00	0.81	0.23	0.26	0.36	0.36	0.37	0.41
S. Korea	0.14	0.32	0.81	0.54	0.91	1.13	0.10	0.14	0.26	0.10	0.52	0.27	0.15	0.18	0.29	0.32	0.44	0.57
Malaysia	0.20	0.20	0.21	0.22	0.27	0.60	0.17	0.14	0.25	0.09	0.18	0.27	n.a.	n.a.	0.20	0.21	n.a.	n.a.
C. Profitability (1 minus the share of wages in value-added, %)																		
Norway	51.7	54.1	45.0	41.4	31.0	31.9	68.0	50.9	71.5	61.0	54.5	55.1	63.3	56.4	57.8	62.8	62.6	88.5
Venezuela	57.0	60.0	37.1	64.8	58.7	79.0	69.4	68.8	74.8	80.0	88.7	90.2	74.8	64.4	80.1	77.2	83.3	91.4
Colombia	68.6	71.9	79.1	80.9	90.7	82.6	75.2	69.3	88.9	81.9	89.3	77.8	77.9	85.1	87.2	81.7	86.9	88.0
Chile	74.0	87.0	79.4	79.1	79.3	72.5	97.1	97.6	95.9	94.3	95.8	99.1	67.7	88.5	89.9	81.6	87.7	85.0
S. Korea	70.8	84.3	78.0	82.6	77.7	82.7	64.3	73.3	71.5	67.9	61.2	77.1	82.3	83.4	77.6	84.7	82.8	82.9
Malaysia	71.4	71.4	66.7	69.0	79.0	89.0	62.3	42.7	70.4	71.7	71.0	74.1	n.a.	n.a.	91.6	n.a.	n.a.	n.a.

SOURCE: UNIDO, *Handbook of Industrial Statistics*, 1988; UNIDO, *International Yearbook of Industrial Statistics*, various years.

footwear); intermediate-capital intensive sectors (nonelectrical machinery, electrical machinery, and transport equipment); and capital-intensive-sectors (iron and steel, nonferrous metals, and industrial chemicals). The comparative and disaggregated evidence in these tables illuminate some important tendencies in the Venezuelan manufacturing sector in the period 1970–97 but do not provide much support for the wage rigidity argument. Consider the evolution of Venezuelan wage levels, profitability, and productivity growth in the labor-intensive sectors in the period 1970–96 (table 3.4a). The period 1970–85 does indeed indicate that Venezuela was a relatively high wage economy. Venezuelan wages in all three sectors were higher than in all of the developing countries in the sample in this period and indeed closer to the level of Norwegian wages than to the levels of the East Asian economies. For instance, in the textile sector in 1980, Venezuelan wages were 64 percent of the U.S. level, whereas Chilean, Colombian, South Korean, and Malaysian textile wages were respectively 37 percent, 23 percent, 21 percent, and 15 percent of the U.S. level. As an advanced economy, Norwegian wages were highest, at 109 percent of the U.S. level in 1980.

After 1985, the wage picture in the labor-intensive sectors dramatically reverses. In 1990 and 1996 (or 1995 in the case of textiles), wages collapse in all three sectors in Venezuela, while either increasing or nearly maintaining their levels in all of the other countries. For instance, in the footwear sector, Venezuelan wages dropped from 46 percent of the U.S. level in 1985 to 16 percent in 1990 and 13 percent in 1996! Venezuela was rapidly becoming a low-wage economy. The only country in the sample to experience a decline was Colombia, though the drop in the wage level there was much less pronounced. Nevertheless, by 1996 the Venezuelan wage level (in all three sectors) was the *lowest* of all the countries compared.

Despite the decline in wage levels, the average annual growth rates in the period 1988–98 for textiles, clothing, and footwear in Venezuela (−0.4 percent, −8.4 percent, and 0.3 percent respectively) were either declining or stagnant, as can be seen in table 3.5. Moreover, the period 1974–88, when wage levels were higher in each of the three sectors, the average annual growth rates for textiles, clothing, and footwear (1.7 percent, 4.6 percent, and 9.1 percent respectively) were generally much higher.

The evidence on profitability in the labor-intensive sector does not support the wage rigidity arguments either. An estimate of the operating surplus (1 minus wages expressed as a percentage of value-added) is used as a proxy for profitability. As can be seen in table 3.4a, the problem of industrial stagnation in Venezuela has not been one of low profitability in the labor-intensive sectors

TABLE 3.5 Growth rates in selected Venezuelan manufacturing sectors, 1974–1998 (average annual growth in gross output, %)

	1974–88	1988–98
All manufacturing	5.1%	−1.8%
Labor-intensive sectors		
Textiles	1.7	−0.4
Clothing	4.6	−8.4
Footwear	9.1	0.3
Intermediate capital intensive sectors		
Nonelectrical machinery	5.5	−4.1
Electrical machinery	8.5	−5.4
Transport equipment	2.5	2.4
Capital-intensive sectors		
Industrial chemicals	10.6	3.2
Iron and steel	4.0	−4.6
Nonferrous metals	8.6	1.8

SOURCE: OCEI, *Encuesta industrial,* various years; Banco Central de Venezuela, *Series estadísticas en los ultimos cincuenta años,* various years.

over time. The Venezuelan wage share was not abnormally high (which might reduce incentives for employers to invest in labor-intensive manufacturing production). Moreover, the Venezuelan wage share was actually lower than in many countries, both resource-rich (such as Norway and Malaysia) and resource-poor (such as South Korea) over most of the period 1970–96. Let us consider the textile sector. In 1975, the operating surplus in Venezuela was 62.8 percent, which was nearly *double* the operating surplus of 32.6 percent in Norway; the operating surplus in Malaysia was similar to Venezuela's at 65.4 percent. The Colombian, Chilean, and South Korean textile sectors averaged operating surpluses slightly higher at rates of 73.5 percent, 77.8 percent, and 71.4 percent respectively. These comparative figures are similar for 1980 and 1985. However, the operating surplus in Venezuela increases to 66.4 in 1990 and to 82.2 percent in 1996. In 1990, Venezuela's profitability rates were more than double Norway's and were higher than in South Korea and Malaysia, though lower than in Colombia and Chile. In 1996, the profitability rates were higher in Venezuela than in *all* other countries in the comparison. In contrast to the predictions of the wage rigidity argument, increasing industrial stagnation in the textile sector coincided with *increases* in profitability in Venezuela over time.[17] In comparative

---

17. The use of operating surplus as a proxy for profitability is likely to underestimate the degree to which Venezuelan manufacturing firms were more profitable than their counterparts in

perspective, there is little evidence that wage laborers appropriated more of the surplus generated in the textile sector than in the other countries. These trends in profitability are similar for the other two labor-intensive sectors: clothing and footwear (see table 3.4a).

In late developers, catching up with more productive incumbent producers requires rapid reductions in production costs relative to "best practice" producers (Amsden 2001). In this perspective, the central problem of Venezuelan manufacturing in the labor-intensive sectors is not excessively high wages or relatively low profitability, but rather the inability of these sectors to sustain rapid *productivity* growth (measured by growth in value-added per employee).[18] Let us first compare Norway and Venezuela in the labor-intensive sectors. In Norway lower profitability levels and higher wage levels were no obstacle to the investments necessary to generate more rapid productivity growth in the period 1970–95 (as can be seen in table 3.4a). For example, in the textile sector in 1970, both countries' average labor productivity was 61 percent of the U.S. level. By 1980, Norwegian productivity reached 77 percent of the U.S. level and Venezuelan productivity reached 67 percent. However, over the next fifteen years, Venezuelan productivity collapsed to 26 percent of the U.S. level in 1996, but Norway continued to improve slightly, reaching 79 percent of the U.S. level. The same pattern follows in the other two labor-intensive sectors: in clothing, Venezuelan productivity fell from 61 percent of the U.S. level in 1970 to 36 percent in 1996, but in Norway, productivity increased from 57 percent of the U.S. level to 82 percent over the same period; in footwear, Venezuelan productivity fell from 29 percent of the U.S. level in 1970 to 20 percent in 1996, while in Norway, productivity rose from 52 percent of the U.S. level in 1970 to 75 percent in 1996.

In contrast to the Venezuelan pattern, the general path in the other latecomers is a sustained increase in relative productivity levels, with South Korea

---

the other countries in tables 3.4a–c. Considering that Venezuela had one of the *lowest* shares of income tax collection as a share of GDP in Latin America in the period 1990–2003 (Moreno and F. Rodríguez 2005, 63–65) and given that Latin America had a relatively low share of income tax collection compared to Malaysia and South Korea over the period 1985–2002 (Di John 2006), the after-tax profitability of Venezuelan manufacturing firms in 1990 and 1996/7 is likely to be *even higher* relative to manufacturing firms the other countries in tables 3.4a–c.

18. The measurement of labor productivity levels can be affected by trade policies and industrial policies. For instance, protectionist policies that raise tariffs and industrial policies that limit entry into sectors may limit competitive pressures, which can, in turn, affect the prices producers charge. In this case, higher productivity levels can be the product of higher prices charged. Nevertheless, given the sufficiently long time frame under analysis, from 1970 to 1997, and given that 1997 was in an era of relatively low tariffs (and thus a relatively undistorted measure of productivity) in all countries, the data in tables 3.4a–c allow for comparing the effectiveness of infant industry policies in the selected countries over the entire period.

experiencing the fastest catch-up and Colombia and Malaysia the slowest. This difference occurs despite similar levels of profitability across the latecomers in the sample. Venezuela ends up with the lowest productivity levels in textiles and footwear in 1996 and in clothing outperforms only Colombia and Malaysia. The decline in productivity in the Venezuelan labor-intensive sectors requires an explanation that the wage rigidity models do not provide.[19]

Turning to the intermediate capital-intensive sectors (table 3.4b), the historical and comparative pattern for wages and profitability in Venezuela that emerges is generally similar to the evolution of labor-intensive sectors. In the period 1970–85 in all three sectors Venezuela was a relatively high wage economy relative to the latecomers in the sample. For instance, in the electrical machinery sector in 1975, Venezuelan wages were 40 percent of the U.S. level, while Chilean, Colombian, South Korean, and Malaysian electrical machinery wages were respectively 15 percent, 10 percent, 8 percent, and 10 percent of the U.S. level. As expected, Norwegian wages were highest at 94 percent of the U.S. level in 1975.

The wage picture here again reverses after 1985 in all three sectors. In 1990 and 1997 (1996 in the case of electrical machinery), wages collapse in all three sectors in Venezuela, while either increasing or nearly maintaining their levels in all of the other countries in 1990 and 1997. The transport sector is representative of the wage trend: Venezuelan wages dropped from 32 percent of the U.S. level in 1985 to 13 percent in 1990 and 6 percent in 1997! In Norway, the transport sector wage level was 49 percent of the U.S. level in 1985, and increased to 86 percent in 1990 and 89 percent in 1996. The transport sector wage level relative to the U.S. level in 1985 in Chile, Colombia, South Korea, and Malaysia was 15 percent, 12 percent, 16 percent, and 13 percent respectively. By 1997, the wage level increased in South Korea, Chile, and Malaysia to 40 percent, 24 percent, and 15 percent of the U.S. level respectively. The only country in the sample to experience wage stagnation was Colombia, as wages remained at 12 percent of the U.S. level in 1997. Once again, the Venezuelan wage levels in all three sectors were the *lowest* of all the countries by 1997.

The declines in the wage levels were, contrary to the wage rigidity logic, associated with *declines* in the growth rates of all the Venezuelan intermediate

---

19. Moreover, wage rigidity models cannot explain the differential productivity performance within the labor-intensive sectors in Venezuela, despite similar wage and profitability rates in the three sectors. By 1996, the clothing sector had fallen least, to 36 percent of the U.S. level, while the textile and footwear sectors had fallen further to 26 percent and 20 percent of the U.S. level respectively.

capital-intensive sectors in the period 1988–98 compared with the period 1974–88. As can be seen in table 3.5, the average annual growth rate in the period 1988–98 for nonelectrical machinery, electrical machinery, and transport equipment in Venezuela were −4.1 percent, −5.4 percent, and 2.4 percent respectively. In the previous period (1974–88), when wage levels were considerably higher in each of the three sectors, the average annual growth rates for nonelectrical machinery, electrical machinery, and transport equipment (5.5 percent, 8.5 percent, and 2.5 percent respectively) were much *higher* in two of the sectors and the same in the case of transport. In *no* sector were *falling* wages associated with *higher* growth rates.

The evidence on profitability in the intermediate capital-intensive sector again does not support the prediction of wage rigidity models. Comparatively low profitability (that is, a comparatively high wage share) has not been a factor behind the slowdown in growth or the comparatively poor productivity performance of these sectors in Venezuela. Consider the nonelectrical machinery sector—which is again close to the general pattern for the intermediate capital-intensive sectors. In 1975, the operating surplus in Venezuela was 59.5 percent, which was much *higher* than the operating surplus of 38.0 percent in Norway; the operating surplus in South Korea was similar to Venezuela's at 60.2 percent. The Colombian, Chilean, and Malaysian sectors averaged operating surpluses a bit higher at rates of 71.5 percent, 75.0 percent, and 66.8 percent respectively. These comparative figures are similar for 1980 and 1985. However in 1990, the operating surplus in Venezuela increased to 72.2, only to increase further to 84.4 percent in 1997. In 1990 and 1997, Venezuela's profitability rates were more than double Norway's, and were higher than in all other latecomers in the sample. In contrast to the predictions of wage rigidity argument, increasing industrial stagnation in the textile sector coincided with *increases* in profitability in Venezuela over time.

The productivity performance in the Venezuelan intermediate capital-intensive sector was relatively poor despite falling wage levels, similar levels of profitability to the other latecomers in the sample, and a considerably higher profitability level compared to Norway, for the period 1970–97. The nonelectrical machinery sector in Venezuela experienced the worst productivity declines: in 1970 labor productivity was 50 percent of the U.S. level, declining to 38 percent in 1985 and collapsing to 18 percent of the U.S. level by 1997. In comparison, the productivity levels in Norway were 57 percent, and only Colombia had a lower productivity level at 14 percent. The electrical machinery productivity pattern is nearly identical.

The transport equipment sector experienced a less dramatic decline in relative productivity performance: in 1970, labor productivity was 50 percent of the U.S. level, increasing to 58 percent in 1985 and falling to 36 percent of the U.S. level by 1997. In comparison, the productivity levels in Norway were only slightly higher at 42 percent, and in Chile only slightly lower at 35 percent. Among the other latecomers, South Korea, which began the period with a productivity level of only 14 percent of the U.S. level in 1970, had reached a level of productivity of nearly twice Venezuela's at 66 percent of the U.S. level in 1997, a reflection of the effectiveness of Korea's auto industrial policy (Chang 1994, 91–129). Only Malaysia and Colombia reached lower levels of productivity by 1996. Nevertheless, only in Venezuela had all three sectors fallen further behind the U.S. productivity level over the period.

Finally, the historical and comparative patterns in wages and profitability for the Venezuelan capital-intensive sectors (table 3.4c) in the period 1970–96 follow a similar trajectory to the two less capital-intensive sectors just examined. Higher wage levels in the period 1970–85 turn into a wage collapse in 1990 and 1996. In all three sectors, Venezuelan wages become the lowest of all countries in the comparison except Colombia's by the end of the period.

The declines in the wage levels were, again contrary to the wage rigidity logic, associated with *declines* in the growth rates of all the capital-intensive sectors in the period 1988–98, compared with the period 1974–88. As can be seen in table 3.5, the average annual growth rate in the period 1988–98 for industrial chemicals, iron and steel, and nonferrous metals (mainly aluminum) in Venezuela were 3.2 percent, –4.6 percent, and 1.8 percent respectively. In the previous period 1974–88, when wage levels were considerably higher in each of the three sectors, the average annual growth rates for industrial chemicals, iron and steel, and nonferrous metals were much higher at 10.6 percent, 4.0 percent, and 8.6 percent respectively. In no sector were lower wages associated with higher growth rates.

The trends in profitability in the capital-intensive sectors were similar to those in the other less capital-intensive sectors. Comparatively low profitability (that is, a comparatively high wage share) has not been a major factor behind the slowdown in growth or the comparatively poor productivity performance of these sectors in Venezuela. Consider the nonferrous metals sector, which is broadly representative of all three capital-intensive sectors.[20] In 1980, the

---

20. The industrial chemicals sector had a similar profitability pattern to nonferrous metals over the whole period. The steel sector is the one sector with somewhat lower profitability levels than the other late developers in the period 1970–90 though they remained above Norway's profitability levels, and were rising to levels similar to the other late developers by 1997.

operating surplus in the Venezuelan nonferrous metals sector was 74.8 percent, which was slightly *higher* than the operating surplus of 71.5 percent in Norway; the operating surpluses in South Korea and Malaysia were similar to Venezuela's at 71.5 percent and 70.4 percent respectively. Only Chile and Colombia had considerably higher levels of profitability at 95.9 percent and 88.9 percent respectively. These comparative figures were similar for 1985. However in 1990, the operating surplus in Venezuela increased to 80.0 percent, only to increase further to 90.2 percent in 1996. In 1990 and 1996, Venezuela's profitability rates were nearly double Norway's, and were higher than in all other latecomer countries except Chile. In contrast to the predictions of wage rigidity argument, increasing industrial stagnation in the nonferrous metal sector coincided with increases in profitability in Venezuela over time.

The distinguishing aspect of the Venezuelan capital-intensive sectors is that their long-run productivity performance is superior to that of the less capital-intensive sectors despite the fact that all three types of sectors experienced wage level collapses and similar trends in profitability. The wage rigidity argument cannot explain this differential performance across Venezuelan sectors. Indeed, the capital-intensive sectors are the only ones "catching up" to the U.S. level over the whole period (although all three sectors do "fall behind" between 1985 and 1996, their fall is much less than that of the less capital-intensive sectors).

Moreover, the productivity performance of these sectors is considerably better in *comparative* terms, particularly vis-à-vis the Latin American economies and Malaysia. By 1996, Venezuelan productivity levels relative to those of the United States in the iron and steel sector, the nonferrous metals sector, and industrial chemicals sector (for 1995) were 44 percent, 72 percent, and 39 percent respectively (of the intermediate capital-intensive sectors, only the transport sector ended up with a productivity level near these). In comparative terms, the productivity level of the Venezuelan nonferrous sector in 1996 was considerably higher than the productivity levels of the Colombian, South Korean, and Malaysian nonferrous sectors, where levels reached only 25 percent, 27 percent, and 27 percent of the U.S. level respectively, but below Chile's level of 81 percent and Norway's level of 99 percent. By 1996, the productivity of Venezuela's iron and steel sector was higher than Colombia's (34 percent), and not far behind Norway's (52 percent), Chile's (56 percent), and Malaysia's (60 percent). All the countries in the sample were well behind South Korea, where the level reached 113 percent of the U.S. level.[21] The productivity level of the Venezuelan

---

21. Korea's preeminent steel performance owes largely to the development of POSCO, a state-owned enterprise, which by the mid-1980s had become the most efficient steel producer in the world (Amsden 1989, 291–318).

industrial chemicals sector in 1995 was higher than the Colombian (25 percent), similar to the Chilean (41 percent), and not massively behind the South Korean (57 percent), though considerably behind the Norwegian (149 percent). Finally, despite the relatively more dynamic performance of the Venezuelan capital-intensive sectors, it is worth noting that the wage rigidity argument cannot explain the comparative evidence on the capital-intensive sectors, since Norway, South Korea, and Chile were catching up to U.S. levels more rapidly than Venezuela, despite the fact that Venezuelan wages were falling more rapidly than wages in those countries. Moreover, the Venezuelan capital-intensive sectors had similar level of profitability to these latecomers and considerably higher profitability than Norway for the period 1970–96.

The examination of the wages, profitability, and productivity across sectors with three different technological characteristics has confirmed the weakness of the wage rigidity argument in several revealing ways. First, the wage rigidity argument cannot explain wage collapse coinciding with growth stagnation. While Venezuelan wages were initially high relative to other latecomers, dramatic declines in wage levels in Venezuela coincided with growing stagnation in all sectors. Second, given the fact that wages collapsed, it would follow that falling wages would benefit labor-intensive sectors more, yet output growth in the labor-intensive and intermediate capital-intensive sectors slowed down as much as in the capital-intensive sectors. Third, there is little evidence that profitability was unusually high or that it was declining in any of the sectors examined. Indeed, all three sector types examined experienced increases in profitability over time. The wage rigidity argument thus cannot explain stagnation in terms of wage rigidity since in the period 1985–97, wages simply weren't rigid.

The wage rigidity argument does not address the determinants of productivity growth. This is unfortunate since the core problem of most of the Venezuelan sectors, particularly since 1985, appears to have been the inability to catch up with the advanced economies by sustaining rapid *productivity* growth as the other more successful late-developing economies in the sample, both resource-rich (Malaysia and Chile) *and* resource-poor (South Korea), have been able to do. In a dynamic context, poor productivity growth is more plausibly the reason behind the lagging dynamic competitiveness of most Venezuelan manufacturing sectors. Moreover, the capacity of the capital-intensive industries to maintain a more competitive productivity performance over the whole period is not addressed.

In order to explain why productivity growth in Venezuela increasingly declined, it is necessary to examine why the level and efficiency of manufacturing investment *allocation* was relatively poor, particularly after 1968. Dutch Disease models (which examine resource allocation at a given moment in time and which assume that technology is given) are not designed to address this issue.[22] Stagnant and declining Venezuelan industrial performance does not seem to be the result of low investment in (and "crowding out" of) the industrial sector during and following oil boom periods. Rather, it has at least as much to do with the type and productivity of industrial *investment*, which has been relatively poor, a situation created, at least in part, by the ineffective or inefficient economic policies of the Venezuelan state.

The issue of investment effectiveness and allocation is also relevant because some countries such as South Korea and Malaysia were achieving more rapid long-run growth and catch-up with similar investment levels to Venezuela's in the 1970s and maintained rapid catch-up with even higher rates of investment in the 1980s and 1990s.[23] Therefore, it is necessary to incorporate other factors than those considered by Dutch Disease models in order to explain divergent manufacturing performance *across* mineral resource rich economies and to explain cycles of manufacturing performance within Venezuela over time. While such models do indicate that "policy matters," there is no attempt to explain why decision-makers in some polities choose and enforce growth-enhancing policies in the face of mineral booms and others do not.

3.3 Oil Revenue Instability and Economic Planning—
The Curse of Short-Termism?

A potential danger of economic management in oil-exporting economies is the tension between long-run planning and short-run instability in export revenues. There clearly are dangers involved in oil bonanzas. First, oil booms generate large and sudden inflows of capital, which may create absorptive capacity problems, as well as increasing resources at the discretion of public officials, which creates opportunities for wasteful spending and corruption. Second, oil inflows

---

22. With respect to investment levels, the main trend to bear in mind is the dramatic *reduction* in non-oil investment as a proportion of non-oil GDP after 1980 (table 2.3). These declines are mirrored in the data on manufacturing investment (see figure 2). The average annual growth of manufacturing investment increased by 11.5 percent in the period 1960–78 but declined by 5.5 percent from 1978 to 1998.

23. The Venezuelan investment rate as a share of GDP averaged 34 percent in the 1970s while the Korean and Malaysian rates averaged 28.1 and 25.5 in the same period. The Korean and Malaysian investment rates as a share of GDP in the 1980s were 30.1 percent and 30.5 percent respectively and increased to 36.3 percent and 34.1 percent respectively in the period 1990–98.

are "free" resources that do not need to be repaid. In this sense, oil windfalls can lead to even more reckless investment behavior and speculation than large inflows due to foreign borrowing, which needs to be repaid and thus may reduce the proclivity of political leaders to invest resources in a reckless manner.

Many economists argue that Venezuela's poor economic performance in the period 1973–98 was caused by short-sighted state policymaking. Hausmann (1989) argues that the principal problem revolves around the government quickly turning oil boom export revenues into long-gestating industrial and infrastructural investments only to find that the increase was not permanent, as was often assumed by policymakers, but transitory.[24] The resulting fiscal deficits were, according to this argument, a principal factor behind the growing inflation and uncertainty that had characterized the Venezuelan economy since the mid-1970s. While average annual inflation rates in Venezuela were low at 1.1 percent in the 1960s and 6.6 percent in the 1970s, they rose to 23 percent in the 1980s and 50.1 percent in the period 1990–98 (see table 9.3). The aim of this section is to assess whether myopia in Venezuela was *particularly* acute or whether it is justifiable to argue that other variables need to be included to explain increasing stagnation in Venezuela.

A comparison of Venezuela with Malaysia in the 1970s and 1980s is instructive. Venezuela and Malaysia share some basic similarities, which make the comparison useful (see Chapter 10). Both countries have middle-income economies; both are rich in mineral resources; both have nearly identical populations; and both implemented big-push industrialization drives beginning in the mid-1970s and had roughly similar levels of income distribution in 1970s and 1990s.

The relevant issue is whether the process of converting the oil bonanza into public spending on consumption and investment led to higher fiscal deficits and, in turn, whether fiscal deficits were a critical factor in explaining the differences in growth rates between the two economies. As table 3.6 indicates, consolidated public fiscal deficits as a proportion of GDP were in fact significantly higher in Malaysia than in Venezuela in most of the period 1970–98. This is consistent with observations that Malaysian leaders severely overestimated the resources available to undertake large-scale, heavy industrial investments (World Bank 1989, 5). In the period 1990–98 (for which only *central* government public

---

24. The standard policy recommendation that follows from this analysis is the introduction of an oil stabilization fund that can prevent overspending in times of oil booms. The idea of such a fund is that revenues above some conservative target oil price are channeled into offshore accounts held by the government until the domestic economy can absorb them productively. In the event of oil export revenues below some specified target, the offshore fund can be drawn on to smooth expenditure declines.

deficits are presented), the average annual central government fiscal accounts for Venezuela (−1.4 percent of GDP) and Malaysia (1.1 percent of GDP) were both essentially in balance.

What is also missing from Hausmann's analysis is a comparative account of the *private* sector surplus in this period. While I could not obtain reliable data on private sector savings in Malaysia, it is possible to make a reasonable assumption that the problem of generating an investable surplus in Venezuela was the result of declines in the private sector savings rate. Consider table 3.7. Given that the Venezuelan fiscal balance was generally far *more* positive than Malaysia's over the period 1970–88, the significant declines in national savings in Venezuela relative to Malaysia must have been due to much lower *private* sector savings in Venezuela.[25] Venezuelan national savings declines from an annual

TABLE 3.6 Consolidated public sector budgets in Venezuela and Malaysia, 1970–1998 (as a percentage of GDP)

Years	Venezuela	Malaysia
1970–73	−0.8	−7.5
1974–75	15.0	−7.3
1976–80	1.9	−10.2
1981–83	−0.8	−19.2
1984–85	7.3	−9.0
1986–87	−6.7	−9.5
1988	−8.6	−2.0
1989	−1.1	−2.9
1990–98	−1.4	1.1

SOURCE: World Bank 1989, table 2.3; Ariff 1991, table 2.13; M. Rodríguez 1991, table 7.13, table 7.23, and table 7.34; IMF, *International Financial Statistics*, various years.

NOTE: The period 1990–98 refers only to consolidated central government public sector accounts.

TABLE 3.7 Gross national savings, Malaysia and Venezuela, 1960–1998 (annual averages as a percentage of GDP)

	1960–70	1970–80	1980–90	1990–98
Venezuela	37.0	37.5	25.3	23.9
Malaysia	26.6	30.4	33.4	39.3

SOURCE: World Bank, *World Development Indicators*, various years.

25. Given the high levels of capital flight in Venezuela in the period 1974–85 (table 9.2), the problem may not have been so much a decline in private savings per se, but rather a decline in private savings that remained in the country.

average of 37.5 percent in the 1970s to 25.3 percent in the 1980s and drops further to 23.9 percent in the period 1990–98. At the same time, the Malaysian national savings rate increases from annual average of 30.4 percent in the 1970s to 33.4 percent in the 1980s.

This decline in national savings in Venezuela relative to Malaysia is mirrored in the decline in investment rates as indicated in table 3.8. Recalling table 2.3, Venezuelan non-oil private investment as a proportion of non-oil GDP also declined from an annual average of 27.2 percent in the 1970s to 13.8 percent in the 1980s to 8.1 percent in the period 1990–98! While declines in gross fixed investment in Venezuela may have been caused by increased risk factors or higher levels of uncertainty, these trends do not seem to be the direct result of relatively high fiscal deficits. The framework Hausmann employs cannot explain why declines in domestic savings and investment in the Malaysian private sector did not occur despite the fact that fiscal deficits in Malaysia were higher than or similar to fiscal deficits in Venezuela.

Although there are debates about the size of the fiscal deficit in Venezuela in the late 1970s and early 1980s,[26] none of the estimates approaches the size of consolidated public deficits that were incurred in Malaysia in the 1980s. Neither the size of fiscal deficits nor the extent to which there were miscalculations regarding the future availability of budget resources for long-run investment seems to convincingly explain *relatively* poor growth performance in Venezuela in the 1980s and 1990s, at least in comparison with Malaysia. More generally, there are many examples of economies that have for long periods maintained macroeconomic stability and equilibrium without developing the institutional context for promoting dynamic industrial performance (e.g., India, Bangladesh, and the Philippines).

There are indeed many journalistic accounts that document increases in risky, unviable, and even reckless investments that occurred in the 1970s as a

TABLE 3.8 Gross fixed investment, Malaysia and Venezuela, 1960–1998 (annual averages as a percentage of GDP)

	1960–70	1970–80	1980–90	1990–98
Venezuela	24.2	34.4	21.3	16.5
Malaysia	19.2	25.5	30.5	34.1

SOURCE: World Bank, *World Development Indicators*, various years.

26. See section 3.4.1 for details.

result of the oil boom resources that flowed into the country like rain in a tropical downpour. The 1970s were filled with "delusions of grandeur" in Venezuelan investment projects (Naím and Piñango 1984). At one point there were plans to build a *glass* tunnel connecting the mainland to Margarita Island, a major tourist attraction, which would have been the longest in the world. However, the *crisis* that engulfed Venezuela in the 1980s turned into prolonged *stagnation*, whereas the intermittent macroeconomic crises that Malaysia experienced in the 1980s and in 1997 did not prevent a longer-run *structural transformation* of the Malaysian economy. As such, purely macroeconomic arguments that focus on "overambitious" investment and the short-sightedness of its leaders do not adequately explain the failure of the Venezuelan economy to make significant structural transformations.

3.4 Getting the Microeconomics Wrong? Neoliberal Critiques of State Intervention in Venezuela

The Dutch Disease models discussed in the first two sections fail to provide an adequate examination of the allocation of investment and its effect on productivity growth in the manufacturing sector in Venezuela. This section considers two influential neoliberal arguments that Venezuela's poor economic performance after 1968 was the result of an increasingly ineffective allocation of investment. The first identifies "excessive" public investment as the culprit; the second, state-led industrial policies that generated excessive levels of firm concentration and low levels of inter-firm competition. These arguments were central to the intellectual support for economic liberalization in 1989 (see Chapter 5).

3.4.1 Inefficient Public Investment as a Cause of Economic Slowdown in Venezuelan Manufacturing

Some analysts have attributed the slowdown in manufacturing growth in Venezuela to increases in the scale and scope of public enterprise manufacturing investment and the inefficiencies surrounding such investment in the period 1974–88. There was indeed a considerable increase in public manufacturing investment and production in this period. In 1970, state-owned enterprises accounted for 5 percent of manufacturing value-added (excluding oil refining). This share rose to 8 percent in 1980 and reached 18 percent by 1986 (see Chapter 7 for details).

Some argue that the Venezuelan economy was unable to *productively absorb* the ambitious natural resource-based big-push industrial strategy initiated by the Pérez administration in the period 1974–79 (Hausmann 1981, 418–24; Baptista 1995, xxix). These authors point out that Venezuelan workers and managers simply lacked the expertise to handle new, complex tasks in petrochemicals, aluminum, and steel production, which contributed to absorptive capacity problems.

Others focus on the growing capital intensity of public manufacturing investment and production (Gelb and Associates 1988, 316–19; Hausmann 1989, 17–18; Baptista 1995, 151). For instance, Hausmann (1989, 17–18) argues that, in the period 1968–85, the incremental capital output ratio (ICOR) for the total non-oil economy was 6.7, which is relatively high, and was mainly due to increases in public manufacturing investment. According to Hausmann, in the period 1968–85, ICORs for the non-oil private sector were 4.9, while ICORs for the non-oil public sector were 10.8 (10.0 in manufacturing and 18.4 in electricity and water). For Hausmann, "public investment in private goods has been a major cause of the inefficient use of capital resources" (17).

Still others posit that public enterprises were inherently inefficient because public managers were subject to "soft budget constraints," that is, the perpetual bailout of loss-making firms (Torres 1993).[27] For Gerver Torres, both the losses suffered by public enterprises and state efforts to covers those losses were responsible for the waste of resources. The validity of this argument depends on the extent to which there were losses and the extent to which such losses and state bailouts necessarily signal dynamic inefficiencies.

In most studies, aggregate levels of profitability are used to measure the performance of public enterprises.[28] These studies indicate that the public non-oil sector in Venezuela generally lost money. Nelson Segarra (1985) calculates that, for the twenty-one largest non-oil, non-financial public enterprises, annual net losses before taxes averaged nearly 5 billion bolívares (US$1.2 billion), which amounted to 7.6 percent of non-oil mining GDP in the period 1978–82.

Torres (1993, 56–58) also provides aggregate evidence on profitability for *all* non-financial, non-oil public enterprises. For the period 1988–92, public enterprises received larger transfers from the central government than they paid in

27. On the notion of the "soft budget constraints," see Kornai 1980.
28. In Venezuela, evidence of poor performance by public enterprises has been conducted at the aggregate, as opposed to the firm, level (and there are no studies comparing public and private firms of similar scale). This is due to the difficulty of collecting reliable data on input prices and state subsidies for public enterprises; although the problem is significant for collecting data on private firms as well, given widespread tax evasion and the general secrecy within which many private firms operate.

income taxes to the treasury. Moreover, Torres indicates that this negative balance did not include capital injections, the assumption of their external debt by the central government, or transfers from the Venezuelan Investment Fund (which was initially set up by the government in the early 1970s to sterilize oil boom revenues offshore). In 1983, the Venezuelan external debt was above $30 billion, of which over 70 percent was owed by public enterprises. While the central government assumed most of the debt of the public enterprises in 1983, by 1986, nearly half of the restructured external debt of $21.04 billion was still owed by eleven public enterprises (56). Finally, the balance sheet presented does not include, as Torres indicates, the massive equity injections of the Venezuelan Investment Fund to the public enterprises in Guyana in the hydroelectric, steel, and aluminum sectors. From the mid-1970s until 1992, the Venezuelan Investment Fund had accumulated total assets of approximately US$6 billion, 75 percent of which were invested in public enterprises. If these additional variables are included, the aggregate balance sheet of the public enterprises turns a significantly darker shade of red.

Let us consider each of the three previous arguments in turn. The emphasis placed on the inability of the Venezuelan economy to absorb high levels of investment from the 1970s onward is suspect when viewed in comparative perspective. First, Venezuelan investment rates, while high, were not higher than East Asian late developers such as South Korea and Malaysia also undertaking state-led industrialization drives in the 1970s and 1980s. In the 1970s and 1980s, Venezuela's annual average investment rate was 34.4 percent and 21.3 percent respectively while South Korea's and Malaysia's annual average investment rates were 28.1 percent and 30.1 percent respectively and 25.5 percent and 30.5 percent respectively. Moreover, education, health, and other human capital investment indicators in Venezuela compared favorably to East Asian economies (see tables 2.6–2.11) undertaking *similar or higher* levels of investment. Finally, the absorptive capacity argument does not explain why Venezuelan firms have not been able to learn how to acquire and adapt technology over the longer run.

The argument concerning slowdown as the result of rising ICORs in the public sector is also subject to criticism. While very high ICORs of the public sector have surely contributed to poor long-run growth in Venezuela, is the implicit comparison between public and private sector performance valid? First, Hausmann compares the aggregate *non-oil* private sector with the *industrial* public sector. Since disaggregated data on the private manufacturing sector is not available, this comparison is not valid. Second, comparing public sector firm investment (which on average is subject to much higher-scale economies than private sector firms) is inappropriate. Only when constant returns to scale hold for *all* industries, do direct comparisons of productivity, however measured, have any meaning. There is no evidence that convincingly suggests that public enterprises in developing countries have a lower level of technical efficiency than private firms operating at the same scale of operation (Millward 1988). Finally, the emphasis on rising capital-output ratios in the public manufacturing sector is not plausibly the main factor behind the growth slowdown. This is because much of the growth slowdown was due to the collapse of non-oil private sector investment in the 1980s and 1990s and declines in non-oil public investment over the same period (see table 2.3).

The argument that public enterprises (because they are subject to "softer" budget constraints) are inherently less efficient than private firms has some important shortcomings. First, there is nothing in economic theory which predicts that public enterprises cannot in principle produce any good as efficiently as a private enterprise (Milgrom and Roberts 1990). Moreover, it is generally recognized that regulatory structures (i.e., competition policy) are more central to enterprise performance than ownership per se (Stiglitz 1994, 171–96). Second, isolating public ownership as a source of slowdown is itself inadequate since there is no empirically established relationship between the relative share of public ownership and economic performance (Cook and Kirkpatrick 1988; Millward 1988). Third, the hardness of the budget constraint depends on historically specific political economy of state regulation. Clearly, public enterprise performance varies substantially across countries (Chang and Singh 1992). Fourth, the focus on public ownership would not be able to explain why both the public *and* the private sectors are more efficient in South Korea or Taiwan than in Venezuela.

The neoliberal critique of public ownership is also unable to explain why some public firms *within* Venezuela, such as the oil company, Petroleum of Venezuela (PDVSA), has been more efficient than others, such as the steel company (SIDOR). In 1998, PDVSA ranked first among all developing-country conglomerates in terms of total assets and total foreign assets and third in terms of total sales (UNCTAD 2000, table III.9).[29]

29. William Ascher, who extensively documents the ways in which governments waste natural resources in developing countries, argues that PDVSA was, in comparative terms, better managed than counterparts in Mexico, Peru, Indonesia, and Nigeria over the period 1975 to the mid-1990s (Ascher 1999, 221–23). Thus, it would be difficult to argue that the relatively poor growth performance of Venezuela, at least compared to these countries, was due to the relatively poor performance of the economy's leading firm. As a corollary, it is not plausible to argue that the nationalization of the oil industry in 1976 was, on its own, responsible for Venezuela's comparatively poor economic performance.

The fact that public enterprises lose money does not necessarily tell us that they are not performing efficiently. Public enterprises may generate positive externalities in the form of more jobs, higher aggregate demand, higher demands for infant industries, lower input costs for private industries, and so on (Chang and Singh 1992). Even if a public enterprise producing solely for export is performing poorly in terms of current profitability, this does not necessarily indicate that such firms will not reduce their costs over time if they themselves can "learn by doing."[30] The main point is that low profitability and even financial losses do not adequately reflect social contributions, or social profitability (Kaldor 1980). For many years, state-subsidized infant industries such as Toyota (in autos) or Airbus (in aerospace) were incurring huge financial losses. This did not necessarily make their long-run rates of return unviable, as history has dramatically proven in each case. Even if losses were the principal problem, focusing exclusively on accounting analysis does not help explain why the problem has not been *corrected*.

Apart from these theoretical considerations, there is no evidence that Venezuelan public enterprises showed greater losses than public enterprises in other late developers who were growing more rapidly than Venezuela. According to data provided by Robert Short (1984, 168–74), nearly every country surveyed, less developed and industrialized, posted aggregate deficits in the 1970s; indeed, the average deficit for developing-country public enterprises was 3.9 percent of GDP. In the period 1974–77, the Venezuelan overall deficit for public enterprises was 5.2 percent of GDP, whereas in Korea and Taiwan, the overall deficits of public enterprises were 5.4 percent and 7.3 percent respectively. Aggregate losses of public enterprises in Venezuela are hardly unique and hence not a distinguishing factor contributing to poor economic performance.

Moreover, there is evidence that were significant *declines* in deficits of public enterprises in the period 1989–2004 (Moreno and F. Rodríguez 2005, 29–31).[31]

---

30. As Hirschman (1967) notes: "If project planners had known in advance all the difficulties and troubles that were lying in store for a project, they probably never would have touched it, because a gloomy view would have been taken of the country's ability to overcome these difficulties by calling into play political, administrative, or technical creativity" (12). However, the entrepreneurial and technical skills that are obtained in solving difficult problems that emerge in the course of an ambitious project can provide a valuable contribution to the development process (9–34). Such ambitious projects, while not necessarily a financial success in the short or medium run, can, in the long run, contribute more to development than a project that is more successful in financial terms, but turns out to be less technologically demanding and, as such, does not elicit such creative problem-solving skills (27).

31. The average annual consolidated public sector financial balance as a percentage of GDP was −2.7 percent in 1962–73, −1.6 percent in 1974–83, −3.0 percent in 1983–88, −0.4 percent in 1989–98, and −0.1 percent in 1998–2004.

To a great extent this has been a result of the privatization of three key state-owned enterprises: CANTV (the national telephone company), VIASA (the national airline) in 1991, and SIDOR (the steel-producing company) in 1997 (29). As a result, transfers to the public enterprise sector were negligible in the period 1989–2004 (29). These substantial declines in transfers to large public enterprises, while "hardening the budget constraint," has not coincided with increases in growth rates, rather it has coincided with the sustained decline in growth in the period 1989–98, and the collapse of growth rates in the period 1998–2003.

3.4.2 Protectionist Policies and Industrial Concentration as a Cause of Economic Slowdown in Venezuelan Manufacturing

A second common assertion among neoliberal analysts is that state protectionism and excessive regulation stifle the emergence of a competitive private sector. It is common in these analyses to posit that protectionism generates excessive industrial concentration and that such concentration is associated with the decline in competitive pressure facing oligopolistic and/or monopolistic firms. Moisés Naím (1993) argues that in Venezuela "profound structural changes were urgently needed to alleviate problems caused by the highly concentrated, oligopolistic industrial structure, low overall productivity growth and significant obstacles to non-oil exports that had been cultivated over many years of government mismanagement" (41).[32] Miguel Purroy (1982, 238–41) and Terry Lynn Karl (1997, 113) argue that industrial concentration led to a lack of competition, which in turn, caused inefficiencies in the manufacturing sector.[33] Finally, some authors point out that state protectionism generated a purely "inward-looking" strategy that, by the mid-1970s, became dysfunctional to the prospects of capital accumulation because of the "exhaustion" of import substitution (Hausmann 1981; Hausmann and Márquez 1990). The evidence of "exhaustion" of import substitution provided is that the participation of imports in household final consumption of goods had declined from 47.4 percent in 1950 to 11.2 percent by 1976 (Hausmann 1981, 347).

---

32. More generally, analysts have criticized state regulations promoting import-substituting industrialization policies on the grounds that policies such as protection, cheap finance, and import licenses (which can create restrictions on competition) have led to patterns of industrial production where oligopolists or monopolists produce at suboptimal scales in small markets and where the rental profits of such investments do not induce firms to worry much about productivity performance (Balassa 1980).

33. See Chapter 7 for discussion of evolution of protectionist policies and for evidence on growing concentration of production in manufacturing.

While there is clear evidence of growing concentration in manufacturing production in Venezuela over time, the comparative and historical evidence (and conventional economic theory) point to several problems with the claim that protectionist policy causes dynamic inefficiency by exhausting the possibilities of ISI and by creating a concentrated industrial sector. First, the identification of the "exhaustion" of the "easy stage" of import substitution does not explain why some countries have managed to *modify* institutions and policies to overcome this common challenge, while other countries have not (Hirschman 1971; Bruton 1998).[34] Second, the comparative evidence, as seen in table 3.9, suggests it is *not* the case that the Venezuelan industrial sector was unusually concentrated compared to countries with much faster rates of industrial growth. Moreover, in the period since 1989, when economic liberalization, including trade liberalization in the Venezuelan manufacturing sector (see Corrales and Cisneros 1999 for details), industrial concentration increased while growth continued to stagnate.[35]

TABLE 3.9 Concentration ratios in manufacturing: Venezuela compared

Economies	Date of assessment	Share of production by top firms (%)
Three-firm concentration ratios		
Japan	1980	56
South Korea	1981	62
Taiwan	1981	49
Four-firm concentration ratios		
Argentina	1984	43
Brazil	1980	51
Chile	1979	50
Colombia	1984	62
Venezuela	1981 (1988)	54 (52)
France	1969	28
United States	1972	40

SOURCE: Amsden 1989, table 7.1; World Bank 1990, table 3.8; World Bank 1993, table 2.1.

34. Such policies could include export subsidies and preferential exchange rates.
35. One of the arguments made by the proponents of economic liberalization is that foreign multinational firms will enhance the productivity of the economy by providing competitive pressure and by introducing more productive production techniques. However, Aitken and Harrison (1999) find no evidence that foreign direct investment (FDI) generates positive externalities in the Venezuela manufacturing sector, throwing doubts on the neoliberal explanation. While the authors find that foreign participation improves the productivity in small plants (of fewer than fifty employees), the introduction of wholly owned foreign firms negatively affects the productivity of wholly domestically owned firms in the same sector.

Theoretically, models of oligopoly are indeterminate in terms of the dynamic efficiency of firms in a given sector. This is because the regulatory structure of an industry (e.g., antitrust laws, the extent to which subsidies and protection are conditional on firm performance) plays an important role in determining the competitive pressures facing firm owners. Second, in contrast to the perfect competition model, a greater number of firms do not necessarily assure a greater intensity of competitive effort (Demsetz 1997, 137–42). In aerospace, for example, the competition between just two firms, Boeing and Airbus, has not hindered rapid innovation and intense competition. In the presence of increasing returns, industrial concentration may be crucial to achieving the scale economies to compete with "best practice" firms (World Bank 1993, 92–102; Chandler and Hikino 1997, 29–34). Moreover, industrial concentration can play the functional role of creating learning rents that compensate for the risk and uncertainty of undertaking investment in the context of imperfect capital markets and the challenges of late development more generally (Amsden and Hikino 1994). The problem of industrial competitiveness in latecomers is not too high a concentration level, but an overdiversified conglomerate structure and below minimum efficient plant size.[36] If the successful late developers are a useful reference, then competitiveness will more likely be improved in Venezuelan manufacturing through a purposeful industrial strategy that ensures minimum efficient plant size and ensures that firms receiving subsidization and protection are subject to explicit performance criteria. Eliminating restrictions on firm entry is unlikely to be compatible with the coordination of investment of large-scale enterprises that natural-resource-based industrialization requires.

Whether industrial concentration promotes efficiency and good governance or rent-seeking and cronyism is largely an empirical issue. The identification of industrial concentration per se does not take us far in explaining poor economic performance. This is because most late developers are characterized by concentration and the dominance of a handful of diversified business groups (Amsden 2001; Chandler and Hikino 1997).

---

36. According to a World Bank study of the Venezuelan industrial sector, fragmentation of production is viewed as a more serious problem than concentration. In the automotive sector, for example, as of 1989, there were fifteen assemblers in the market, which fell from 163,000 units in 1982 to 26,000 units in 1989 (World Bank 1990, 52). International standards of efficiency normally require a minimum of 100,000 units per plant, which indicates the degree to which Venezuelan auto plants are suboptimal in size. Naím (1988) and Naím and Francés (1995) do in fact identify the problem of overdiversification and firm size as an important problem for the competitiveness of the Venezuelan industrial sector.

In sum, two important neoliberal critiques, the first focusing on the negative effects of public enterprise manufacturing investment and the second focusing on the inefficiencies of import-substitution industrialization, have important theoretical and empirical shortcomings. Comparative evidence on growth suggests that the *quality* rather than the *level* of state intervention is what matters for sustained economic development. One common lacuna in the explanations discussed is the failure to explain *why* policy failures in managing investment funds have persisted in Venezuela from the 1970s until the early years of the twenty-first century.

3.5 Conclusion

This chapter critically reviewed some of the influential economic explanations of the increasingly poor growth and productivity performance of the Venezuelan economy in the post-1968 period. The most important critique is directed toward Dutch Disease models, which do not adequately explain the slowdown in non-oil and, particularly, manufacturing growth in Venezuela. The problems with these models can be summarized as follows. First, the oil booms of the 1970s and early 1980s did not generate a "crowding out" of manufacturing investment; rather, manufacturing investment was positively correlated with changes in oil revenues. Second, there is little evidence that the oil booms decreased the share of manufacturing in non-oil GDP over time. Thus, the decline in the growth rate of the entire non-oil economy cannot plausibly be due to decline in the relative share of manufacturing. Third, these models do not explain why productivity growth rates slowed in the Venezuelan non-oil economy in general and in the manufacturing sector in particular or why some manufacturing sectors remained more dynamic than other sectors over time. Fourth, Dutch Disease models do not explain why relatively rapid manufacturing growth coincided with the expansion of the oil industry from the early 1920s to the late 1960s. Finally, while it can be argued that the poor growth of the manufacturing sector could be due to inadequate microeconomic policies, the stress on policy failures generally would be inadequate in explaining why governments in countries with stagnant long-run growth persistently fail to take corrective action. Neither oil windfalls nor the level of state intervention can explain, in isolation, the Venezuelan growth slowdown in light of the historical and comparative evidence.

4

POLITICAL ECONOMY EXPLANATIONS OF THE GROWTH COLLAPSE IN VENEZUELA

Rentier state models attempt to explain *why* state decision-makers in natural-resource-rich economies create and maintain growth-restricting policies (Mahdavy 1970; Karl 1997; Auty and Gelb 2001). These models move beyond economic models of the resource curse, such as Dutch Disease models, by attempting to show that policymaking and institutional formation are endogenous to a type of governance that typically emerges in oil abundant contexts. The rentier state model posits the need to view poor economic performance in the context of resource abundance as an outcome of historically specific institutional arrangements and not the cause of economic decline. These models are part of a growing trend of reviving the "staples thesis"—the notion that natural factors endowments or technology shape the relations of production, or institutional evolution, of a society (Engerman and Sokoloff 1997; IADB 1998, 96–100).

The rentier state model applies many of the concepts and notions of the theory of rent-seeking. In the rentier state model, oil abundance is assumed to generate extraordinarily large degrees of rent-seeking, and these rent-seeking contests are assumed to be uniformly negative in terms of the developmental outcomes they generate. The main idea behind the theory of rent-seeking is that whenever public agents have monopoly power and discretion over the distribution of valuable rights, incentives for corruption increase. This, in turn, leads to a waste of resources in attempts to influence public authorities. Moreover, the need to keep bribes secret reduces the security of property rights, which lowers investment in long-gestating projects. This is because rents appropriated through corruption have no legal basis and are more vulnerable to political challenges and confiscation.

There are several propositions that follow from the rentier state model and rent-seeking theory. First, the existence of a higher level of mineral rents will increase rent-seeking and corruption relative to economies with lower mineral abundance. Second, increases in rent-seeking and corruption will generate lower growth. Third, the basic policy advice deriving from rent-seeking theory is that the elimination of all state-created rents will eliminate wasteful, that is, unproductive rent-seeking activities. This policy advice is meant to support the case for economic liberalization (including trade liberalization, privatization, and financial deregulation). The assumption is that economic liberalization, by reducing state-created rents and curbing centralized state discretionary authority over rent allocation, will reduce the level of rent-seeking and corruption, which, in turn, will increase investment and growth levels.

This chapter assesses the extent to which the rentier state model and theories of rent-seeking and corruption are consistent with Venezuelan economic performance over time and in comparative perspective. Section 4.1 presents the main arguments of the rentier state theory. Section 4.2 presents some important shortcomings of the rentier state model. Section 4.3 analyzes neoliberal theories of rent-seeking and corruption. Section 4.4 explores the empirical evidence on the relationship between mineral resource abundance, corruption, and growth. The evidence suggests that rents created by a centralized state authority do not generate uniform scales of rent-seeking or corruption and that rent-seeking and corruption do not have uniform developmental consequences. The discussion in this section points to the need to incorporate political analyses in order to more adequately explain the economic effects of centralized rent deployment. Section 4.5 attempts to demonstrate the extent to which there has been knowledge of the problems of mineral abundance and corruption in Venezuela. The neoliberal theories view corruption as a problem of institutions that create incentives for corruption but do not examine why such structures are not changed, particularly when there is knowledge of the problems that neoliberal analysts identify.

4.1 The Rentier State Model

In rentier state theory, a rentier state is likely to emerge when mineral and fuel production accounts for at least 10 percent of GDP and where mineral and fuel exports are at least 40 percent of total exports (the World Bank refers to these as mineral economies) (Nankani 1979). In this specification, Venezuela can be considered a rentier state, or what Karl (1997) calls a "petro-state." Rentier state theorists consider mineral-rich economies distinct from other sorts of economies because they generate natural resource rents that emanate from "point" resources rather than from "diffuse" resources such as land under small farms.[1] Point rents, according to Richard Auty (2001, 6), are associated with staples that are relatively capital intensive and thus concentrate ownership. In contrast, where the staple or mineral resource poses smaller investment barriers to entry, as with rice and maize, and some tree crops such as coffee and cocoa, the rents are likely to be more widely dispersed through the population. Point resources tend to be associated with "enclave" industries in the sense that they generate fewer production and consumption linkages and socioeconomic linkages in poor economies than more "diffuse" resources do (Hirschman 1981a; Auty and Gelb 2000, 141). Within this classification of abundant point-resource economies, oil economies are considered to be the prototypical examples of this category (Karl 1997, 17–19).

The main premise of the rentier state model of governance is that the more states generate their revenues from "unearned" income (such as resource rents or international aid), the less state decision-makers need to "earn" their income by levying domestic taxes. Unearned income (through mineral resource rents) can allow elites to purchase security through corrupt patron-client networks, rather than through the establishment of a social contract based on the exchange of public goods financed through domestic taxation. This in turn makes leaders less accountable to individuals and groups within civil society, more prone to engage in and accommodate rent-seeking and corruption, and less able to formulate growth-enhancing policies.[2] Such arrangements, according to this model, can reduce a regime's legitimacy and relative military, administrative, political, and economic authority and power. James Fearon and David Laitin (2003), drawing on rentier state theorists (Chaudhry 1989; Karl 1997), also posit that oil-dependent poor countries are more prone to conflict than non-oil economies because "oil producers tend to have weaker state apparatuses than one would expect given their level of income because rulers have less need for socially intrusive and elaborate bureaucratic system to raise revenues—a political Dutch Disease" (81).

---

1. Economies characterized by large-scale plantations of wheat and cattle—such as Uruguay—also have very concentrated ownership though they are not considered rentier economies in the literature.

2. Moreno and F. Rodríguez (2005, 43–60) provide comparative data on the relatively low levels of domestic taxation in Venezuela compared with most Latin American economies.

Although there are some distinct variations of the rentier state argument, political economy explanations of the "resource curse" generally argue that "because taxation dominates the domestic linkages that 'point' resources generate, the role of the government is increased so that both the probability of policy failure and the magnitude of its effects are increased" (Auty and Gelb 2001, 141). In sum, the rentier state model develops three main explanations of why state capacity is weak in mineral economies. First, oil rents provide a sufficient fiscal base for the state and thus reduce the need to tax citizens. This in turn reduces political bargaining between state and interest groups, which makes governance more arbitrary, paternalistic, and even predatory. Furthermore, the absence of incentives to tax internally weakens the administrative reach of the state, which results in lower levels of state authority, capacity, and legitimacy to intervene in the economy. Second, relatively higher levels of mineral rents generate greater levels of rent-seeking and corruption than in non-mineral economies. And finally, these higher rates of rent-seeking and corruption lead to lower economic growth rates. This section and the next consider the first argument, and sections 4.3 and 4.4 consider the second and third arguments.

One of the more ambitious rentier state theories is that developed by Karl (1997).[3] Karl argues that, in "petro-states," fiscal dependence on oil rents generates an "unfortunate" gap between the extensive jurisdictional role of petro-states (that results from the sudden inflow of resources) and their weak mechanisms of authority. As such, "this gap between jurisdiction and authority ultimately works to the detriment of the state's ability to flexibly adjust to changing conditions" (59). Moreover, "oil rents reduce the need of state leaders to extract resources by other means.... As a result, petro-states lack the capacity to build extensive, penetrating and coherent bureaucracies that could successfully formulate and implement policies" (60).

Karl makes other sweeping claims regarding petro-states. She argues that oil-exporters generate specific types of social classes, organized interests, and patterns of collective action (16). Resource-rich countries supposedly get locked into a rentier development path because vested interest groups learn to appropriate state resources through unproductive rent-seeking, which crowds out productive activity. Karl further notes:

> Dependence on petroleum revenues produces a distinctive type of institutional setting, the petro-state, which encourages the political distribution

---

3. One of the main reasons for focusing on *The Paradox of Plenty* is that it applies the theoretical analysis with a study of the politics of rent-seeking in Venezuela.

of rents. Such a state is characterized by fiscal reliance on petrodollars, which expands state jurisdiction and weakens authority as other extractive capabilities wither. As a result, when faced with competing pressures, state officials become habituated to relying on the progressive substitution of public spending for statecraft, thereby further weakening state capacity. (16)

Karl's model of the petro-state follows Hossein Mahdavy (1970), who first advanced the concept of the rentier state. Mahdavy argues that resource rents make state officials myopic and resource wealth allows a level of welfare and prosperity that lessens the urgency for change and rapid growth (437). Based on an extensive case study of Venezuela in comparative perspective of other oil exporters, Karl concludes that "oil exporters share a similar path-dependent history and structuration of choice.... Where states characterized by overwhelming incentives for rent-seeking are put in place, institutions at the deepest level of political domination will shape whatever regime type of government grafted on to them" (1997, 227).

Several other analysts have employed the rentier state concept to explain weak state authority and capacity, which they see as the cause of Venezuela's economic decline since 1973 (Uslar Pietri 1984; Naím and Piñango 1984; Naím 1993; Gil Yepes 1992; Thorp and Durand 1997). The main thrust of their argument is that *oil largesse* increased rent-seeking or influencing activities and also reduced the willingness of state leaders to assess the economic efficiency consequences of state subsidies.[4] According to Moisés Naím and Ramon Piñango (1984, 552–63), the reigning political culture of "excessive avoidance of conflict" is indispensable in understanding Venezuela in the democratic era. This is an era which is characterized by the lack of bureaucratic continuity, a lack of priorities (563–64), and the uncontrolled proliferation of organizations (552–53). The authors claim that oil allowed state officials and politicians to accommodate as many demands as possible, regardless of their economic efficiency consequences.

Finally, many analysts and policymakers argue that the large influx of petroleum resources, especially in the boom period of 1973–85, created absorptive capacity problems for political and economic institutions. For instance, Juan Pablo Pérez Alfonzo, cofounder of OPEC and minister of oil and mines in

---

4. The expression "hay pa' todo" (there is enough for everyone) became common in Venezuela in the 1970s and 1980s.

Venezuela in the early 1960s, famously referred to oil as the "devil's excrement," and perceived that Venezuela was sinking in it (Pérez Alfonzo 1976). By 1978, this phrase had become common in Venezuela as oil wealth was perceived as the source of growing corruption, criminality, and the arbitrary expansion of state powers. In sum, in the rentier state model, oil abundance is assumed to generate patterns of state intervention and governance that are inimical to economic development.

4.2 Criticisms of the Rentier State Model

There are several assumptions of the rentier state argument as developed by Karl that drive the results. First, it is assumed that a weak state exists at the time of the discovery of "external rents" (Karl 1997, 42). Karl arbitrarily chooses oil booms as the point in which state formation takes place in late-developing oil economies and makes the case that the *timing* of the discovery of mineral rents is decisive for subsequent state capacity developments (40–43). Karl would argue that Norway and Australia already possessed "strong" bureaucracies before their mineral windfalls arrived and thus had countervailing industrial and agricultural interests groups to ensure that resource rents were well managed.

However, the Venezuelan case, which is the focus of Karl's petro-state study, does not corroborate her argument. Karl arbitrarily chooses the period of oil discovery in the 1920s as *the* period of state formation in Venezuela. The history of Venezuela from Independence in 1810 to the 1920s was a period of mineral-resource scarcity in production terms.[5] If resource abundance determines the character of the state, as Karl argues, then Venezuela had ample time and stimulus to develop a "Norwegian" style bureaucracy. This is because the petro-state problem can be avoided when state building has occurred *prior* to the introduction of the oil-exporting activity (Karl 1997, 13).

In fact, from 1810 to the 1920s in Venezuela mineral-resource scarcity coexisted not with state-building but with incessant factional *caudillo* wars (or "guerra de guerillas") and fragmentation of the national territory (Gilmore 1964; Brito Figueroa 1966).[6] The Federal War (1859–63) was among the most

   5. The key term here is "in production terms." With hindsight, we know that Venezuela had large oil reserves. However, no one knew about them in the nineteenth century, nor was there any technological capacity in the country to discover, let alone develop the vast oil and natural gas reserves that existed.
   6. Venezuela's nineteenth century was very unstable, even by Latin American standards (Moreno and F. Rodríguez 2005, 10). One historian calculated that there were 39 national revolutions and

brutal in nineteenth-century Latin America. The negative effects of "roving banditry" (Olson 1993) were reflected in economic performance: per capita income in Venezuela was approximately one-half the Latin American average by 1920 (Thorp 1998, 353). More generally, the failure of central state consolidation in nineteenth-century Latin America was common and certainly *not* associated with any one pattern of resource endowment (Centeno 1997; López Alves 2001).

Second, rulers are assumed to "own" the natural resources. That is, they are assigned the "property rights" over resources. How rulers appropriate and maintain power is not analyzed. By assigning "rights" to leaders, the whole problematic of how to manage "common pool resources" is neglected, when the real problem of common pool resources is, in fact, analyzing the processes through which rights are assigned, enforced, maintained, and changed (Ostrom 1990). In other words, it is assumed that there are no collective actors within the society who can impose some *domestic* conditionality on how those who occupy the state exercise their power.

Rentier state theory does not adequately address the question of the legitimacy of state leaders and the political strategies that may support or undermine the appropriation and use of mineral rents. The political coalitions underpinning a leader's power surely affect how mineral rents are managed. Since the state is a historically specific *agent of coalitions*, the very process of class, group, and coalition formation drives institutional formation and change (Przeworski 1991, 23). Thus, how mineral rents are used is *not prior to*, but is essentially the by-product of political struggles, bargains, and settlements.

Third, leaders are assumed to have *predatory* as opposed to *developmental* aims. The neglect of the political processes through which a leader appropriates power limits our understanding of the motivations of state leaders. A state is not a thing, much less such a thing as "a predator" or "rent-seeking maximizer," but a set of social relations. Why a particular coalition in power will not use oil revenues to diversify production is not addressed. Even if it is (unrealistically) assumed that the leader has absolute power and is thus the "owner" or "residual claimant"[7] in an economy, it does not necessarily follow that leaders will act in predatory ways. According to Mancur Olson (1993), a leader who has a long

---

127 uprisings of different sorts between independence in 1830 and 1903; another calculated that Venezuela enjoyed barely 16 years of peace while suffering 66 years of civil war and insurrection since independence (Caballero 1993, 34–35, cited in Moreno and F. Rodríguez 2005).

7. In the neoclassical theory of the firm, the residual claimant refers to the firm owner (Alchain and Demsetz 1972). The firm owner in this theory is assigned the right to appropriate the residual, that is, the profits, of the firm's team production. According to this theory, private ownership of firms provides the incentives for owners to monitor team production efficiently.

time horizon, what he calls a "stationary bandit," has the incentive to maximize the rate of economic growth, since this will maximize the resources accruing to the state in the long-run. A dictator who does not have to tax citizens to maintain power can still have developmental as opposed to predatory motivations.[8] Predatory behavior on the part of leaders (such as confiscating property) cannot be assumed or simply described, but needs to be explained. Predation will occur as a consequence of the failure to adopt much more lucrative and broad-based legitimacy-enhancing developmental aims. The decision of leaders to purposefully engage in rapacious acts to accumulate capital thus assumes that they have made a *prior* decision that long-run economic development is either undesirable or politically and/or economically infeasible. However, the conditions under which predatory motivations dominate developmental motivations in a mineral-dominant economy are not addressed in the rentier state model.[9]

It is implicitly assumed that leaders in petro-states are *revenue-satisficers* and not revenue-maximizers (Ross 1999). This is *ad hoc* reasoning. It is assumed that, because mineral rents are available, the leaders of powerful interest groups (again, whether in the state or civil society) will, at some arbitrary point, not be interested in demanding policies that promote growth, compared with leaders in mineral-poor countries. From a supply side (or state) perspective, these models never explain why leaders do not want to maximize growth and revenues as, for example, in Olson's stationary bandit model (1993). From a demand (society) side perspective, the rentier state model cannot explain the *failure* of rent-seeking agents *to continue capturing/influencing the state* to the point of imposing the installation of a stationary bandit agent, or even a "democratic state for capitalists." In sum, the rentier state model does not adequately address the *dynamics* of rights appropriation, of rent-seeking, or of power

---

8. The idea that autocratic rulers could benefit more from promoting economic development rather than engaging in predatory behavior is an old one. Niccolò Machiavelli, in 1571, advocated that monarchs "ought accordingly to encourage his subjects by enabling them to pursue their callings, whether mercantile, agricultural, or any other, in security, so that this man shall not be deterred from beautifying his possessions from the apprehension that they may be taken from him, or that another refrain from opening a trade through fear of taxes; and he should provide rewards for those who desire so to employ themselves, and for all who are disposed in any way to add to the greatness of his City or State" (Machiavelli 1992 [1571], 61).

9. It should be kept in mind that developmental outcomes are not only or even mainly the result of a leader's intentions or aims. Developmental outcomes are often the unintended consequence of conflicts and political struggles (Brenner, 1976; Knight, 1992; Tilly, 1990). Also, the distinction between developmental and predatory is not necessarily in binary opposition. Leaders who do not have developmental aims will not necessarily become predatory. I thank Christopher Cramer for bringing these points to my attention.

struggles. The neglect of dynamics is certainly inappropriate for a theory that purports to assess growth prospects associated with institutional formations.

Fourth, rentier state proponents assume that resource abundance causes high levels of rent-seeking and conflict. However, the causation may be the reverse. It is also possible that war, the most violent form of rent-seeking, may *prevent* an economy from becoming more resource abundant in the first place. If state leaders are to appropriate oil revenues, for instance, they need to secure and enforce property rights in the territory where the oil is located. Oil rents, like all rents, themselves require the specification of rights, which do not occur naturally. State leaders need to be able to either extract taxation from multinationals, or what is even more difficult, extract the mineral through public enterprise production. In the Venezuelan case, the ability of a series of Andean caudillos to eliminate regional warlords and consolidate central state power and authority in the period 1900–1920 was central to the prospects of rapid oil development which occurred from 1920 onwards (see Chapter 8).

Fifth, by choosing oil booms as the point in which state formation takes place in late-developing oil economies, Karl's model is subject to selection bias. By definition, most countries that do not have a diversified agricultural and manufacturing base become mineral dependent. In historical terms, almost all countries began as mineral-dominant economies. For instance, the United States, Canada, Norway, Sweden, the Netherlands, Australia, and Malaysia were, in earlier stages of development, more mineral-dominant, less diversified economies. Not only that, Ronald Findlay and Mats Lundhal (1999) have demonstrated the generally growth-enhancing role natural resources played in stimulating capital accumulation and growth throughout the now advanced countries from 1870 to 1914.

Rentier state theorists do not examine the possibility that mineral abundance can be central to the development of manufacturing industry in particular. For instance, Gavin Wright and Jesse Czelusta (2007) examine how and why technological development and collective learning positively affected the development of natural resources in the U.S. economy. They demonstrate how large-scale investments in exploration, transportation, geological knowledge, and the technologies of mineral extraction, refining, and utilization of natural resources contributed to long-run economic growth and industrialization of the United States. Magnus Blomström and Ari Kokko (2007) explore how the development of natural resources led to increasingly high-tech industrial production in Sweden and Finland during the nineteenth and twentieth centuries. The key policy question to ask is why natural resource revenues are used in ways that sustain economic growth and diversification in some countries and not in others (Chenery 1979). Lack of economic diversification and poor economic growth are why economies are mineral dependent. If that is the case, then it makes sense to ask why political conflicts prevented growth in some mineral-dependent economies and not in others.

Sixth, the Venezuelan case does not corroborate the idea that resource abundance either determines actions or policies in some path-dependent way. As discussed in section 4.5, there has been substantial assessment of the dangers of oil wealth and administrative corruption in Venezuela since the 1930s, and actions have been taken to deal with such deficiencies.[10] As important, there have been important *changes* in economic policy and institutional design that counter the idea put forward by Karl (1997, 225–27) that resource abundance *determines* agency and choice. From 1930 to 1957, when the country was mostly under authoritarian rule, Venezuela had a relatively liberal trade policy by Latin American standards (Karlsson 1975). From 1957 to 1973, the state promoted import substitution. From 1973 to 1989, the state inaugurated the development of public industrial enterprises as part of a natural-resource-based big-push industrialization strategy. The 1990s were characterized by significant changes toward economic liberalization in the form of privatization, trade liberalization, and financial deregulation. Then, from 1998 to 2005, the state renationalized several large sectors, such as telecommunications.[11] There is, in sum, no decisive evidence in Venezuela that oil abundance had led to any one type of development or industrial strategy or policy.

Seventh, even after allowing for selection bias, the *historical* relationship between non-oil growth and mineral resource abundance in Venezuela does not support the rentier state argument. Oil abundance has coexisted with periods of rapid growth, stagnation, and growth implosions ever since 1920. Consider the evolution of oil revenues available to the Venezuelan economy in the twentieth century as indicated in table 4.1. Between 1920 and 1980, real retained oil revenues per capita (in 2000 U.S. dollars) grew at 18.5 percent, while the average real per capita level of such revenues was $1,220 in the period 1950–80. Between 1980 and 2000, the respective growth of real retained oil revenues per capita rate fell to –0.6 percent, and the average level of real retained oil revenues fell to $910 per capita. In the first decade of that period, the respective

10. On the history of policy debates concerning the role and dangers of oil for national economic development in the period 1925–80, see Baptista and Mommer 1987. For more recent debates on the desirability of economic liberalization and political reform, see Goodman et al. 1995.

11. See Chapters 5, 7, and 8 for a discussion of the different development strategies initiated in Venezuela over the course of the twentieth century.

TABLE 4.1 Venezuela: Real retained oil exports per capita, 1920–2000

	Mean level (2000 U.S. dollars)	Mean annual growth (%)
1920–80	$1220	18.5
1920–50	870	28.3
1950–80	1551	8.6
1980–2000	910	–0.6
1980–90	1150	–2.4
1990–2000	685	2.3

SOURCE: Hausmann 2001, table 4, 29; Banco Central de Venezuela, *Informe anual*, various years; Baptista 1997.

NOTE: Real retained oil exports are the sum of the value of oil export revenues deposited in the central bank, the value of oil sector salaries, and the value of the purchases of domestic products by oil companies operating in Venezuela. This sum is deflated by U.S. producer prices and converted into 2000 U.S. dollars.

growth rate was –2.4 and the respective level was $1,150 per capita; in the second, the respective growth rate was 2.3 percent and the respective level dropped to $685 per capita.

The data in table 2.1 shows that the two periods of fastest growth in non-oil GDP in Venezuela—the 1920s and from 1950 through 1957—coincide with the greatest increases in oil production and exports, and oil-related fiscal revenues.[12] The most rapid average annual manufacturing growth (15 percent per year) occurs in the period 1950–57. In both periods, Venezuela was under military rule (Juan Vicente Gómez, 1908–35, and Marcos Pérez Jiménez, 1950–58). Moreover, the level of cronyism between foreign oil companies, a small set of powerful family-run economic groups/conglomerates, and the state in both periods was considered to be extraordinarily high even by Venezuelan standards.[13]

As important, the rentier model cannot explain why the *decline* in the growth and the level of per capita retained oil revenues in the period 1980–2000 has not induced any measurable improvement in governance or growth. The logic of the rentier state argument is that increasing resource scarcity should lead to

12. In the Gómez era, oil was discovered in 1920. By 1928, Venezuela became the largest oil exporter in the world. Between 1950 and 1957, during the reign of Pérez Jiménez, Venezuela accumulated more foreign exchange than any other country in the world other than West Germany (which was receiving a significant share of the aid available under the Marshall Plan) (Karl 1997, 96–97).

13. For a discussion of the networks of corruption and cronyism between the oil companies, the Gómez family, and politically connected merchant groups, an era that was known as "dance of the oil concessions," see McBeth 1983 and Coronil 1998, 82–84. For a discussion of the high levels of public spending and the growing cronyism in the awarding of public construction projects in the Pérez Jimenez era, see Moncada 1995.

more effective economic management. This has not happened in Venezuela. Table 2.1 shows that growth rates of manufacturing and non-oil GDP from 1980 through 2003 were substantially below their average respective growth rates from 1920 through 1980.

Finally, Venezuela's long-run growth performance, when placed in the Latin American context, does not support the rentier state argument. If oil were such a curse to Venezuela's economic development, then surely its level of catch-up with more advanced countries would be slower than that of less oil-dependent Latin American countries. This is not the case. In comparative terms, the growth in GDP in Venezuela was impressive by Latin American standards over the period 1920–80. According to Angus Maddison (1995, 156–57), Venezuela achieved a growth rate of 6.4 percent per annum in GDP between 1920 and 1980, which was the fastest of the seven largest Latin American economies (this compares with 5.5 for Brazil, 4.8 for Peru and Mexico, 4.7 for Colombia, 3.4 for Argentina, and 3.3 for Chile). In the period 1980–2000, Venezuela's growth rate slows to 1.6 percent. However, the growth rate of most of the largest Latin American economies, apart from Chile, slows as well (2.8 for Brazil, 1.5 for Peru, 2.1 for Mexico, 3.3 for Colombia, 1.8 for Argentina and 5.5 for Chile).[14] In this perspective, Venezuela is *not* such an exceptional case in Latin America as rentier state proponents claim.

4.3 Theories of Rent-Seeking and Corruption

In Venezuela, several studies suggest that state intervention in general and state-led industrial policies in particular have led to corruption and cronyism, which have hindered economic development (Naím 1993; Naím and Francés 1995; Pérez Perdomo 1995; Capriles 1991; Karl 1997). Moreover, many analysts posit that massive oil inflows generate political struggles over appropriating oil rents which supposedly dominate an oil-rich polity. As Karl (1997) notes: "Political struggles [in petro-states] were always over the domestic distribution of rents.... It was never over the broader issue of whether, when and what rate they should be permitted to overwhelm the economy and state or about what alternative development strategies might be appropriate" (225). Similarly, Richard Auty and Alan Gelb (2000) argue that natural resource rents, particularly

---

14. Moreover, Chile, the fastest-growing large Latin American economy in the period 1980–2000, is also one of the most mineral abundant, with copper exports dominating foreign exchange revenues.

in concentrated form, offer a magnet for political competition. Resource abundance therefore encourages contests for rents that tend to engender factional or autonomous predatory political states. "To stay in power, governments of resource-abundant countries need to find a way to re-distribute rents to favored groups.... They do so at the expense of a coherent economic policy" (2–3).[15] In the rentier state model, oil abundance is assumed to generate growth-restricting state intervention, extraordinarily large degrees of rent-seeking where these rent-seeking contests are assumed to be uniformly negative in terms of the developmental outcomes they generate.

Modern theories of rent-seeking and corruption form a substantial part of the intellectual foundation of the rentier state model. The basic idea behind these models is that there are substantial costs to the workings of an economy when the allocation of resources is channeled primarily through state leaders, who have discretionary authority, rather than through bargains between private economic agents. In oil economies, because most revenues originate in the central government, the level of state discretion in allocating resources and regulating the economy tend to be higher than in most non-oil economies. In the rentier state model, the predominant view is that oil economies are subject to a higher level of rent-seeking and corruption in comparison with non-mineral-abundant economies. This begs the following questions. What is rent-seeking and why is it necessarily harmful for economic development? And are oil economies in fact subject to greater levels of rent-seeking and corruption?

The early rent-seeking models were developed by liberal economists who were trying to demonstrate that state intervention—and in particular, artificially created rents in the form of monopoly grants (Krueger 1974; Posner 1975; Buchanan 1980)—induced agents to "waste" resources on "unproductive activities" in an attempt to capture the rents.[16] The form that such "wasteful" activities can take, according to rent-seeking theorists, include bribery, smuggling, purposefully creating excess capacity in an industry if this means being favored for an import license, and even by investing "excessive" time and money in acquiring the credentials to become a state bureaucrat if state restrictions imply an increase in state rental income and an increase in public wages (Krueger 1974).

---

15. The same authors, in a later article, note: "A poor natural resource endowment tends to engender autonomous benevolent states that place a premium on investment efficiency whereas resource-abundance tends to foster factional and predatory states that relax market constraints and depress investment efficiency" (Auty and Gelb 2001, 127).

16. The theory of rent-seeking argues that state intervention generates inefficiency in ways that are not recognized by standard texts in welfare economics, namely the "costs of creating monopoly" (Krueger 1974; Posner 1975).

Rent-seeking broadly can be interpreted as activities that seek to create, maintain, or change the rights and institutions on which particular rents are based. Rents refer to the "excess incomes," or the "proportion of earnings in excess of the minimum amount needed to attract a worker to accept a particular job or a firm to enter a particular industry" (Milgrom and Roberts 1992, 269). Rents can take many forms, such as higher than competitive rates of return in monopolies, extra income earned from exclusive ownership of a scarce resource (whether a natural resource or specialized knowledge), and extra income from politically organized transfers such as subsidies (Khan 2000a). Since rents specify incomes that are higher than would otherwise have been earned, they create incentives to create and maintain these rents (Khan 2000b). These influencing activities range from bribing and coercion to political lobbying and advertising.

Rent-seeking theory has been instrumental in creating a negative view of politics and of the state. For instance, models of the welfare state pioneered by Arthur Pigou have been replaced by models that view the state as a "predator" (North 1981), a "bandit" (Olson 1993), "populist" (Dornbusch and Edwards 1990), or as having a "grabbing hand" (Shleifer and Vishny 1998). As Margaret Levi (1988) notes, most rent-seeking theorists "are obsessed with demonstrating the negative impact of government on the economy. They view competitive markets as the most socially efficient means to produce goods and services . . . [and] do not treat the effects of government intervention as *variable,* sometimes reducing and sometimes stimulating social waste" (24, quoted in Hutchcroft 2000, 210; emphasis added). In the mainstream view, the availability of rents is the ultimate source of rent-seeking and corruption (Mauro 1998, 11), and some have even postulated an "iron law" of rent-seeking: "Wherever a rent is to be found, a rent seeker will be there trying to get it" (Mueller 1989, 241).[17]

Aaron Tornell and Philip Lane (1999) provide a more nuanced view of rent-seeking dynamics that may occur in oil-abundant developing countries. Their model considers an economy that lacks strong legal-political institutional

17. The use of the term "rent" in the history of economic thought has undergone significant (ideological) changes. The classical political economists described rent as nonproductive revenues gained either through market power or natural scarcity that were generated *within the private sector.* Unproductive activities in the classical tradition refer to resources used as inputs that do not add value to the corresponding outputs produced. For Ricardo, landowners who owned the more fertile/productive units of land (which, by nature, were in fixed supply) appropriated a rent from farmers. For Adam Smith, most service sector activities were considered unproductive, while Marx highlighted the unproductive nature of private financial services (Boss 1990). In contemporary neoclassical political economy (i.e., theories of public choice and rent-seeking), the definition of rent has shifted to involve revenues that are generated through unproductive political activities (that is, attempts to influence the state), which seek to protect, maintain, or change rent-generating rights (Khan 2000a, 7).

infrastructure and is populated by multiple powerful interest groups. In such an economy, powerful groups dynamically interact via the fiscal process that effectively allows open access to the aggregate capital stock. In equilibrium, this leads to slow economic growth and a "voracity effect" by which a shock such as a terms of trade windfall perversely generates a more-than-proportionate increase in fiscal redistribution and reduces growth.

In the Tornell and Lane model, if there are no institutional barriers to discretionary redistribution, an increase in the raw rate of return in the formal sector reduces growth. The intuition is as follows. An increase in the raw rate of return in the formal sector unleashes two conflicting effects: a direct effect that increases the profitability of investment in the formal sector, and a voracity effect that leads each group to attempt to grab a greater share of the national wealth by demanding more transfers. This is reflected in a higher tax in the formal sector, which induces reallocation of capital to the informal sector, where it is safe from taxation. This shift reduces the growth rate in the economy, counteracting the direct positive effect of an increase in the raw rate of return.

The empirical evidence suggests, the authors argue, that Nigeria, Venezuela, and Mexico enjoyed significant oil windfalls but dissipated their revenues (38–40). Their model predicts a more-than-proportional increase in fiscal spending in response to a positive revenue shock. "The poor growth performance of countries experiencing windfalls is suggestive that increases in public capital expenditure were not productively deployed and that appropriated resources were consumed, invested in safe but inefficient activities, or transferred overseas" (40).

There are two problems with this proposition. First, there is little evidence that non-oil taxation was significant in Venezuela either during or after the oil booms in the 1970s and early 1980s. Indeed, María Antonia Moreno and Francisco Rodríguez (2005, 59–72) find that Venezuela has levels of non-oil taxation to non-oil GDP that are far lower than the Latin American tax-to-GDP ratio average. Second, *declines* in oil exports per capita since the early 1980s (see table 4.1) have not coincided with increased growth rates. If the model predicts that the "voracity effect" of oil booms generate declines in growth rates, then declines in relative oil abundance should have led to less "voracious" rent-seeking and thus greater growth. This has not occurred.

The literature on corruption closely parallels the literature on rent-seeking in the sense that it focuses on the costs of state intervention. Corruption is commonly defined as the violation of formal rules governing the allocation of public resources by officials in response to offers of financial gain or political support (Nye [1967] 1989). For most purposes, this standard definition is useful, since most types of corruption of relevance to resource allocation are illegal in most developing countries.

The mainstream view follows Gunnar Myrdal (1968, 937–58), who argued, in anticipation of later rent-seeking theory, that if corruption is allowed, then government officials will purposefully promote state restrictions and rents in order to take an interest in generating bureaucratic obstacles and thus be in a position to demand bribes. The key to each result is the benchmark used to compare the corrupt act in relation to the resource allocation outcome in absence of corruption. Myrdal and later mainstream theorists view corruption, a priori, as having negative consequences since their benchmark is the model of the perfectly competitive market, which does not need any state intervention to operate at full efficiency.

For neoliberal theorists, the root of the negative impact of corruption is the discretionary centralized authority that accompanies state intervention (Rose-Ackerman 1978). For most oil economies, the concentration of resources in the central state is likely to impart a high degree of centralized government control over resource allocation. Discretionary centralized authority is characterized by the power vested in top decision-makers to intervene in the activities and welfare of subordinates, as well as the relative immunity from intervention by others. The very existence of such authority and intervention makes possible its inappropriate use. In the mainstream models, the costs of corruption include the waste of resources in attempts to influence public authorities and reduction in the security of property rights since corrupt transactions need to be kept secret (Shleifer and Vishny 1993).

Mainstream rent-seeking theory's emphasis on the *costs* of state-created rents and centralized discretionary authority is incomplete. As with any activity, it is important to incorporate the potential benefits as well. There are important positive roles that discretionary centralized authority and state-created rents can play in inducing economic agents to take risks and experiment. In a Schumpeterian framework, the protection of rents by the state may be an essential component of promoting innovation and technical progress. This is why the patent system has been central to technological development in capitalist economies (Chang 1994, 30). The classic problem of Schumpeterian rents is that there is no way to determine how long a rent should be protected to ensure optimal long-run growth (Khan 2000a, 45).

Related to Schumpeterian rents are learning rents. In developing countries, learning rather than innovation drives growth (Amsden and Hikino 1994). In the context of late development, investments (especially in high-productivity sectors where scale and gestation periods are large) entail significant *risks*. First, the existence of learning-by-doing (Arrow 1962) implies that the now more advanced countries have a "first-mover advantage" of accumulated knowledge relative to producers in late developers. Second, the more efficient institutional and infrastructural environment (in the form of more secure property rights, well functioning education systems, well run ports, and so on) in advanced countries represents an implicit subsidy to production relative to less developed economies (Abramowitz 1986). As well, the greater insecurity of property rights in poor economies compared with advanced economies raises the transaction costs of conducting business (North 1990) and thus represents another hidden subsidy to production in rich counties. The risk inherent in undertaking industrial development in a late developer in the context of relatively underdeveloped capital markets is generally too high to induce investment in the absence of subsidization and protection.

The challenge of late development is further enhanced when we take into account that the learned capabilities in an economy are slow-gestating, and mostly firm- and industry-specific (Chandler and Hikino 1997). While targeting industries may have some rent-seeking costs, it has certainly been at the heart of such success stories as South Korea and Taiwan (Amsden 1989, 2001; Chang 1994). The role of state-created rents in socializing risk for firms is a key component of their potentially positive role. The *dynamic* efficiency of learning rents is that their maintenance is contingent on performance criteria. This suggests that the efficiency costs of rent-seeking are not only the influencing costs but also the extent to which the rights that accompany rents promote learning and innovation.

That rent-seeking in general, and corruption in particular, entails potential costs and potential benefits implies that the appropriate evaluation of state intervention is to assess actual "inputs" and "outputs" of rent-seeking as a *process* (Khan 2000b). Simply put, rent-seeking is a rights-appropriating process, and rent-seeking is the cost of "producing" property rights that generate rents.[18] Following Mushtaq Khan (2000a, 74–89), the inputs into rent-seeking are the

---

18. As Eggertsson (1990, 92–101) points out, general equilibrium analysis ignores the costs to the state and to private individuals of *establishing* property rights: "It is important to realize that the costs of the resources that are diverted to the 'production' of property rights are often substantial. The worth of perfectly defined and enforced rights is represented by the net present value of the rents, and individuals would be willing to spend up to that amount to obtain those rights. If the rents can be obtained for less than this amount, net rents will be positive and society's output will be greater" (92).

rent-seeking costs, which include the resources used in influencing activities such as lobbying, and other political activity, such as campaign financing and bribery. There is no reason to assume that rent-seeking contests will generate a unique outcome. Rent outcomes include the economic rights created, maintained, destroyed, or transferred to create specific rents (e.g., licenses allocated, monopolies and subsidies granted, land rights created). Indeed, Douglass North (1990) has modeled the generation of socially beneficial change of institutions as a rent-seeking process. Rent-seeking models focus only on the constraining aspect of centralized authority without investigating the constitutive or enabling aspect of that same institutional structure.

What matters is the *net effect* of the rent-seeking process. The net effect is equivalent to the net social benefit associated with the rent outcome minus the costs of inputs used in rent-seeking, that is, the rent-seeking cost (Khan 2000a). For example, if the state offers a subsidy to promote learning in an infant industry, this is likely to induce rent-seeking to capture the rent associated with the subsidy.

Given the rent-seeking cost, the *outcome* of the rent-seeking process can differ depending on the conditions under which rents are allocated. Consider two scenarios. In Scenario A, rents are allocated to an inefficient crony, and subsidies are perpetuated regardless of performance. In Scenario B, rents are allocated to political insiders, but are open to competitive bribery, and are subject to performance criteria. In this case, it is not only likely that the more efficient crony capitalist will obtain the rent, but because the maintenance of subsidies is tied to some performance criterion (such as productivity growth or export targets), there are incentives for the rent recipient to maximize the growth potential of the firm.

It is likely that the net effect of the rent-seeking process will be more growth-enhancing in Scenario B even if the rent-seeking costs are *lower* in Scenario A. Indeed, the level of rent-seeking in Scenario A is likely to be much lower because the potential contestants are likely to perceive they have little chance to capture the rent. If there are known insiders in the rent market, rent-seeking costs are lower than if the rent market is more competitive.

The evidence suggests that countries attempting industrial policy differ far more in the *outcomes* of the rent-seeking than in the rent-seeking *costs*, that is, the levels of influencing and corruption (Khan 2000a). This implies that similar levels of influencing and corruption may be compatible with very different growth trajectories across countries. More generally, there is no reason to assume, as rentier theorists do, that the mere existence of mineral rents generates greater rent-seeking than in non-mineral-dominant economies. The amount and cost of rent-seeking depends on the political conditions that induce competition over rents in the first place and the relative power of the groups seeking those rents.

In order to begin to map the differential impacts centralized state rent deployment may have, it is necessary to incorporate politics. The mainstream theories of rent-seeking neglect the role politics plays in the use of state-created rents. Because political struggle and settlements are historically specific, deterministic models cannot accurately explain rent allocation in actual political systems. The extent to which mineral economies generate both higher rent-seeking costs and less developmental rent-seeking outcomes is ultimately an empirical issue.

4.4 Does Oil Abundance Cause Higher Corruption and Lower Economic Growth?

This section examines the comparative and historical data on the relationship between mineral abundance, corruption, and growth across a range of developing countries. As mentioned, the rentier state model posits that higher levels of mineral rents will increase rent-seeking and corruption, and that increases in rent-seeking and corruption will, in turn, generate lower growth. Paulo Mauro (1995) suggests that there is a negative correlation between corruption and investment and growth. I will examine the extent of these claims even after allowing for the selection bias inherent in the rentier state model.

Consider table 4.2, which examines the relationship between mineral abundance, corruption (one common proxy for measuring the extent of illegal rent-seeking), and growth in selected developing economies. The corruption index used in the study was developed by Transparency International on the basis of *subjective* answers by executives doing business in the country in question. It is important to note that it measures corruption faced by large (mostly multinational) corporations and does not measure political corruption and small-scale corruption.[19]

In the case of the selected late developers, there is little to indicate that mineral abundance systematically causes either higher corruption or slower growth.

19. The Transparency International index is a "poll of polls" that survey respondents on their subjective assessment of the level of corruption in a given country. While the rankings are based on subjective responses, there are important reasons for investigating the patterns such surveys reveal. First, the cross-national ratings tend to be highly correlated with each other and highly correlated across time. Moreover, the evaluations of different types of respondents (multinational executives, domestic entrepreneurs, country experts, and risk analysts) are highly correlated. Second, the perception that corruption is a problem may have serious consequences, since perceptions

TABLE 4.2 Mineral resource abundance, corruption, and growth in selected developing economies, 1965–1998

	Mineral and fuel exports[a] (1980)	Corruption index[b] (1980–85)	Corruption index[b] (1996)	Average annual percentage change in GDP (1965–90)	Average annual percentage change in GDP (1990–98)
Mineral-resource-rich countries					
Venezuela	98.0	3.2	2.5	2.6	1.9
Nigeria	97.0	1.0	0.7	4.2	2.7
Botswana	90.0	n.a.	6.1	12.4	4.8
South Africa	67.0	5.7	5.7	2.6	1.5
Egypt	66.0	1.1	2.8	6.5	3.9
Chile	64.0	8.3	6.8	2.8	7.2
Ecuador	60.0	4.5	3.1	6.0	1.8
Mexico	58.0	1.9	3.3	4.3	1.8
Indonesia	47.0	0.2	2.7	6.6	7.5
Malaysia	35.0	6.3	5.3	6.6	8.7
Average	69.4	3.6	3.9	5.7	4.2
Mineral-resource-poor countries					
Thailand	14.0	1.5	3.3	7.4	7.5
Brazil	11.0	4.7	3.0	6.4	3.1
India	7.0	3.7	2.6	4.5	5.9
Pakistan	7.0	1.5	1.0	5.6	4.4
Argentina	5.0	4.9	3.4	1.7	4.5
Philippines	3.0	1.0	2.7	3.8	3.3
Colombia	3.0	4.5	2.5	4.8	4.5
Uganda	2.0	0.7	2.7	1.5	7.2
Taiwan	1.0	6.0	5.0	7.9	7.0
South Korea	1.0	5.7	5.0	7.5	7.2
Bangladesh	0.0	0.8	2.3	2.7	4.5
Average	5.0	3.2	3.0	4.9	5.4

SOURCE: World Bank, *World Development Report*, 1991, 1997b; World Bank, *World Development Indicators*, 2000; Subjective Corruption indices from Transparency International.

[a] Mineral and fuel exports are given as a percentage of total exports.

[b] A corruption index of 10 indicates minimum corruption, an index of 0 indicates maximum corruption.

Moreover, similar rates of corruption seem to be compatible with widely divergent growth rates. There are several observations that substantiate these claims. First, the mineral-abundant economies have lower average corruption rates than the non-mineral-dominant economies. Second, within the mineral-dominant economies, corruption rates vary widely. Botswana, Chile, South Africa, and Malaysia are perceived as less corrupt than Venezuela and Nigeria. Third, within the non-mineral-dominant economies, corruption rates also vary widely. Thailand and the Philippines, for example, are much more corrupt than Taiwan. Fourth, corruption rates *change* within both groups of economies over time. For instance, corruption in mineral-resource-rich Venezuela worsens over time, but improves within non-mineral-resource-dominant Thailand over time. Finally, the average growth record between the mineral-dominant group and non-mineral-dominant group does not support the rentier state theory. In the period 1965–90, the mineral-dominant countries' average growth rate *exceeded* the non-mineral-dominant economies, though the reverse was true in the period 1990–97. Nor can the rentier state model explain why growth rates vary substantially *within* both groups of economies. Venezuela grew more slowly than Malaysia; similarly, the Philippines grew more slowly than Thailand and South Korea.

The available data also suggest the lack of any pattern in the relationship between levels of corruption and economic growth.[19] Some relatively "less corrupt" polities such as South Africa grew relatively slowly, while some relatively corrupt economies such as Thailand, Indonesia, and Colombia sustained among the most rapid growth rates over long periods. Some economies such as Chile, Thailand, India, and Argentina experienced increases in growth rates between the two periods despite the fact that corruption *increased* in each of these economies. These contingencies suggest that outcomes of the rent-seeking process vary much more than the rent-seeking costs stressed by rent-seeking theory.

While the previous sample is itself subject to selection bias, a comparison of all the developing countries for which there is corruption data also seems to indicate that there is an indeterminate relationship between levels of mineral abundance on the one hand, and corruption and growth, on the other. Consider table 4.3. In the period 1965–90, the median annual growth of the non-mineral-abundant developing economies (5.4 percent) did outpace that

---

shape expectations and thus can affect investment rates. See Treisman 2000 for a critical discussion of the methodology underlying the construction of the Transparency International indices.

19. Svensson (2005) also finds that there is *no* robust empirical support for the proposition that higher levels of corruption are either associated with or cause lower levels of economic growth.

of the mineral-abundant economies (4.3 percent). However, in the same period, the median corruption rate of the non-mineral-dominant economies was *slightly higher* than that of the mineral-dominant economies. In the 1990s, the mineral-dominant economies grew *slightly faster* and were slightly *less corrupt* than the non-mineral-dominant economies. None of this evidence provides much support for rentier state theory. As important, Khan (2002a) points out that it is not sufficient to claim that higher growth may be due to compensating factors such as high investment rates or efficient capital use in the high-growers since "the institutional argument is that good governance has an economic effect precisely by enhancing investment and its efficiency" (13).

Tables 4.2 and 4.3 suggest that there is little to indicate that mineral abundance systematically causes either higher corruption or slower growth. However, proponents of the rentier state theory, such as Karl (1997), argue that oil exporters

TABLE 4.3 Growth and corruption in mineral-abundant and non-mineral-abundant developing countries, 1965–2000

1965–90	1. Mineral-abundant[a] developing countries (2) (13 observations)	2. Non-mineral-abundant[a] developing countries (2) (19 observations)
Median GDP growth rate, 1965–90	4.3	5.6
(Range)	(2.5–12.4)	(1.5–9.5)
Median corruption index,[b] 1980–85 (1)	3.9	3.6
(Range)	(0.2–6.5)	(0.7–8.8)
1990–2000	1. Mineral-abundant developing countries (13 observations)	2. Non-mineral-abundant developing countries (19 observations)
Median GDP growth rate, 1990–2000	4.0	3.7
(Range)	(1.6–7.0)	(−0.6–10.3)
Median corruption index, 1996	3.3	3.2
(Range)	(0.7–6.8)	(1.0–5.0)

SOURCE: World Bank, *World Development Indicators*; Subjective Corruption indices from Transparency International.

[a] A "mineral-abundant" economy is one whose mineral/fuel exports were equal to or greater than 35 percent of total exports in 1980; a "non-mineral abundant" economy is one whose mineral/fuel exports were less than 35 percent of total exports in 1980.

[b] A corruption index of 10 indicates minimum corruption; an index of 0 indicates maximum corruption.

are a *special case* of mineral-abundant exporters because oil generates a level of mineral rent much higher than in other mineral-abundant economies. Because rents are likely to be higher in "petro-states," the level of rent-seeking will also be proportionally higher. As a result, rentier state proponents expect oil-exporting developing economies to perform particularly poorly compared to non-oil economies. Table 4.4 compares the growth rates of selected oil-exporting economies in the period 1965–98 to the growth rates of four developing regions (Latin America and the Caribbean, East Asia and the Pacific, sub-Saharan Africa, and South Asia), all developing countries, and the world economy.[21]

Once again, the evidence does not support the rentier state hypothesis. In the period 1965–80, the average annual growth in GDP of oil-exporting economies was 6.7 percent, which was faster than the average annual growth of all developing regions except East Asia and the Pacific, faster than the growth of all lower- and middle-income economies, and considerably faster than growth in the world economy. The period 1980–90 sees a partial reversal of this trend. In this period, the average annual growth in GDP of oil-exporting economies slows considerably to 1.8 percent, which was considerably lower than the average annual growth of East Asia and the Pacific (8.0 percent), and South Asia (5.7 percent), and only one-half the growth rate of all lower- and middle-income economies (3.5 percent), and considerably lower than growth in the world economy (3.2 percent). However, the growth rate of oil-exporters slows to the *same rate* as in all of Latin America and the Caribbean (1.6 percent) and sub-Saharan Africa (1.8 percent). Oil exporters are not particularly poor performers; they have plenty of company. Finally, in the period 1990–98, the average annual growth of oil exporters imploded to 0.6 percent, which was considerably below the growth rates of all developing regions, the lower- and middle-income economies, and the world economy. However, this result turns on the particularly poor growth performance of the four oil-exporting transition economies. In this period, the average annual growth rate of the four transition oil economies was *negative* 7.3 percent.[22] The average annual growth rate of oil exporters (excluding the transition economies), in the period 1990–98,

21. There is no available evidence on subjective indices of corruption from Transparency International for most of the oil exporters. In any case, tables 4.2 and 4.3 already suggest that there is no determinant relationship between mineral abundance and corruption rates or between corruption rates and economic growth.

22. Growth rates, however, are much *less* negative, even in these economies, if the period of growth is extended to 1990–2001. The average annual growth rate of the four transition economies is *negative* 1.8 percent over this period (average annual growth rates in the Kyrgyz Republic, Azerbaijan, Russia, and Kazakhstan were −2.9, −0.3, −3.7, and −2.8 respectively). Growth rates are based on data from World Bank, *World development indicators*.

TABLE 4.4 Economic growth of selected oil-exporting developing countries in comparative and historical perspective, 1965–1998

	Fuel exports in 1980 (% of total exports)	Average annual change in GDP (%)		
		1965–80	1980–90	1990–98
Latin America				
Venezuela	98	3.7	1.1	2.0
Trinidad & Tobago	93	5.0	−0.8	2.4
Ecuador	63	8.8	2.0	2.9
Mexico	58	6.5	1.8	4.2
Southeast Asia				
Indonesia	47	7.0	6.1	5.8
Middle East				
Kuwait	100	1.6	1.3	n.a.
Saudi Arabia	99	10.6	0.0	1.6
Oman	96	13.0	8.4	5.5
Iran	93	8.7	1.7	4.0
Iraq	>40	7.3	−8.8	n.a.
Africa				
Libya	100	4.2	2.5	1.4
Algeria	98	6.7	2.7	1.2
Nigeria	97	0.7	4.2	2.7
Gabon	88	9.5	0.9	4.0
Angola	78	n.a.	3.7	−0.4
Transition Economies				
Kyrgyz Republic	93[a]	n.a.	n.a.	−4.7
Azerbaijan	66[b]	n.a.	n.a.	−10.5
Russia	43[b]	n.a.	n.a.	−7.0
Kazakhstan	33[b]	n.a.	n.a.	−6.9
Average				
(excluding transition economies)	80.2	6.7	1.8	2.9
Overall average	80.2	6.7	1.8	0.6
Regions				
East Asia and the Pacific		7.3	8.0	8.1
Latin America		6.1	1.6	3.7
Sub-Saharan Africa		4.2	1.8	2.1
South Asia		3.7	5.7	5.7
Lower- and middle-income countries		5.8	3.5	3.3
World		4.1	3.2	2.4

SOURCE: World Bank, *World Development Indicators*.

[a] In 1990.

[b] In 1996.

increases to 2.9 percent, which while considerably slower than growth in East Asia and the Pacific (8.1 percent) and South Asia (5.7 percent), is similar to growth of all lower- and middle-income countries (3.3 percent) and Latin America and the Caribbean (3.7 percent), and indeed *faster* than growth in the world economy (2.4 percent). In sum, there is little evidence that oil abundance is necessarily a "curse." Moreover, the downturns in economic growth in oil exporters in the period 1980–98 follows closely the growth slowdowns in both Latin America and sub-Saharan Africa.

Finally, basic historical evidence on Venezuelan corruption scandals does not support the mainstream view that large-scale corruption is associated with economic stagnation. Andrei Shleifer and Robert Vishny's (1993) model of corruption, similar to the predatory model of the state (North 1981), predicts that a rent-seeking, or revenue-maximizing, state would organize bribe collection to favor large-scale, capital-intensive projects (even when financially and economically unviable) since the transaction costs involved in corrupt exchanges are minimized when the number of agents and projects are minimized. In twentieth-century Venezuela, the centralized nature of the state and the concentrated form of ownership among competing family-run conglomerates and the concentrated state-owned enterprise sector has meant that most corruption has taken place under centralized rent-deployment processes. The main corruption scandals in Venezuela have always involved *large-scale* projects and have coincided with eras of rapid growth (1930–50) and with eras of stagnation (1980–2003).[23] Moreover, as discussed in the previous chapter, there is no evidence that the degree of industrial concentration or capital intensity of industry, which may or may not be the result of corrupt state decision-making, explains differences across middle-income latecomers.

4.5 Limits of the Cognitive Failure Approach to Policy and Institutions in Venezuela

Was the failure to sustain growth due to "getting policy and institutions wrong," as the "good governance" paradigm (see, for example, World Bank 1997a) emphasizes? The mainstream view on rent-seeking and corruption is indeed

---

23. See note 12 of this chapter for references on the period 1900–1958. See Karl 1982, 444–512, for a discussion of corruption scandals of large-scale projects during the oil boom years (1974–79), and Naím and Francés 1995 for large-scale financial scandals during the beginning of the liberalization era (1989–93).

consistent with much of the recent work on institutions, which has identified the problem of state capacity simply in terms of cognitive failure (North 1990; Olson [1996] 2000). North (1990), for example, argues that "individuals make choices based on subjectively derived models that diverge among individuals and the information the actors receive is so incomplete that in most cases these divergent subjective models show no tendency to converge " (17). Similarly, Mancur Olson and Satu Kähkönen (2000) argue that "there is no great likelihood that most poor countries will soon come to understand what changes they need to make in their institutions and policies, much less be able to undertake collective action needed to make the appropriate changes" (11). The maintenance or creation of perverse incentives would appear to be, in this perspective, a "technical" mistake.

When considering the issue of cognitive failure, the static (or one-shot) nature of the rent-seeking model is inappropriate for understanding *long-run* processes of institutional formation, capital accumulation, and growth. Dynamic maximization would imply that agents, over time, could evaluate suboptimal policies and institutions. The process of assessing the effectiveness of institutions can take place whether the institutions in question evolved or were designed (Sen 1999, 254–57).[24]

The appeal to incorrect mental models as an explanation of long-run stagnation is nonetheless inadequate given the plethora of advice and examples of superior alternatives available to decision-makers in stagnant economies. The substantial historical evidence of domestic commissions identifying growth-restricting policies and institutions in poor economies suggests that actors in poor economies understand the obstacles to growth they face without the assistance of foreign expertise. Moreover, in a democratic polity such as Venezuela, it would also be relevant to ask why powerful interest groups, such as business chambers and labor unions, do not impose some *domestic conditionality* on state officials who sanction growth-restricting or growth-retarding institutions. Surely the World Bank was not the first organization to recognize ineffective institutions in Venezuela. In any case, a convincing answer to this question of

---

24. Of course, there will always be differences of opinion concerning the criteria for assessing what constitutes "effectiveness." For some, the overriding criterion may be the extent to which institutions and policies help distribute resources in an egalitarian manner. For others, the principal criterion could be the extent to which institutions and policies promote economic growth, technological development, and so on. The question of what constitutes "an effective institution" is the very essence of politics and political philosophy. In the examples considered here, we are interested in the extent to which institutions and policies contribute to the extent of corruption and to economic growth.

cognitive failure would depend on the degree to which there were knowledge failures among decision-makers in the first place.

There are indeed several important indications that the potential and real benefits *and* costs of centralized intervention were widely recognized in Venezuela. The "enclave" nature of the oil industry had convinced state planners of the need to diversify the production and export structure of the economy. As early as 1936, the term "sembrar el petróleo" (sow the oil) became an increasingly dominant theme in economic thinking among policymakers and academics.[24] At this stage, government policy was not driven by a coherent vision on the part of the private sector, nor was it driven, as in much of Latin America at the time, by concerns over the balance of payments or external debt. It was driven, rather, by a growing concern about the risks involved with relying on a single export.

With the transition toward democracy in 1958, a consensus emerged that state planning was needed in order to further develop an industrial base, which was viewed as a viable road to sustained economic growth. Beginning in 1958, a conscious strategy of import substitution was launched by the newly installed democratic regime (see Chapter 7 for details). The state set up a central planning ministry and provided protection, subsidies, and financing to private industry through the expansion of state-owned development banks, many of which were begun in 1940s under the increasingly inclusive and accommodating military regime of General Isaías Medina Angarita (1941–45).

The techniques and rationale of *centralized,* medium-term planning were well known in Venezuela by the time of the democratic transition in 1958. In particular, postwar French indicative planning had a significant influence on planning authorities in Venezuela (Hausmann 1989). In 1974, the planning minister Gumersindo Rodríguez explained: "We shall travel along the French road of concerted planning, a road which reunites the capitalist classes, organized labor, and the public sector in order to establish a series of concerted actions which commits these sectors to utilize their resources in agreement with the plans established by the authorities in the new institutions of concertation" (quoted in Karl 1982, 304).

---

25. The term "sembrar el petróleo" was coined by the Venezuelan writer Arturo Uslar Pietri on June 14, 1936, in the daily newspaper *Ahora*. In the editorial, he warned that Venezuela was in danger of becoming an unproductive and parasitic country, totally dependent on the oil industry, whose affluence would prove to be temporary (cited in Baptista and Mommer 1987, 15). Adriani ([1931] 1990) provides an early account of the need to diversify the production and export structure. Baptista and Mommer (1987) provide an intellectual economic history on the management of oil largesse for wider development aims in Venezuela.

As early as 1960, there was considerable debate and knowledge on improving oversight and control mechanisms of state-owned enterprises (Karl 1997, 289 n. 6). In 1974, during the first Pérez administration, the debates about public enterprise management reform demonstrated sophisticated knowledge of state planning techniques in this area. Alan Brewer-Carías, who had been an administrative adviser in the Caldera administration (1968–72), proposed the strengthening of oversight functions of democratic institutions, especially the ministries and the Congress, citing cases of France, the United Kingdom, and the United States as models to follow (Brewer-Carías 1975). Pedro Tinoco, the director of the Commission for the Integral Reform of Public Administration (CRIAP), formed in 1974, proposed reforms that ran counter to Brewer-Carías's ideas. Tinoco argued that the influence of political parties reduced the ability of democratic institutions to oversee the management of public enterprises and proposed a more centralized presidentialist model of managing state-owned enterprises. He called for the formation of holding companies organized by sector with one large entity to oversee all sectoral corporations, a plan that drew on extensive knowledge of the Italian and Spanish models (Karl 1997, 143–44).

At the same time, awareness of the evolving structural weaknesses in the Venezuelan manufacturing sector and the need to change the policies and institutions of industrial strategy and planning was widespread. In the first decades of consolidated democratic rule, several policy documents identifying weaknesses in the manufacturing sector were widely discussed in policy circles and had been brought to the attention of state planners (see, for example, Furtado 1957, Merhav [1971] 1990, and CORDIPLAN 1973). The main points highlighted in these documents were as follows. First, production was declining in nearly all manufacturing sectors (Merhav [1971] 1990). Industrial growth of 11.4 percent between 1950 and 1960 was among the highest in Latin America, although such growth rates were, in part, due to the fact that the starting point of rapid growth in Venezuela began from a very low base in 1950 (see table 7.1). Nevertheless, such growth slowed to 6.5 percent in the period 1965–70 (105), and there was at least awareness that attention needed to be paid to the factors underlying the slowing of manufacturing growth. Second, there was, in comparison with many developing countries, a relatively low level of public and private *long-run* credit financing to the industrial sector, which was viewed as hindering the development of long-gestating and risky ventures in import substitution and natural-resource-based industrial exports (World Bank 1970, 1973; Merhav [1971] 1990). Third, exclusive reliance on a small internal market (exacerbated by unequal income distribution) was seen as limiting the possibilities of capital-intensive industrial expansion at efficient scale economies, which resulted in chronically low levels of capacity utilization (Merhav [1971] 1990). As early as 1965, national development planners started to insist on the need to promote manufactured exports (Hausmann 1989, 8). Evidence of the limits of exclusive reliance on the inward-looking growth was seen in the comparatively low levels of manufacturing productivity growth (1.4 percent per annum) in the period 1964–69 (World Bank 1970, 12). Finally, shortages of skilled workers, insufficient research and development, and insufficient managerial capacity within the industrial structure were seen as inhibiting the possibility of industrial upgrading toward capital goods (Furtado 1957).

It is also revealing to consider that the dangers of corruption and other challenges of centralized discretionary authority over resource allocation were also widely debated and well known in policy circles (Betancourt 1978; Brewer-Carías 1975; COPRE 1989; Pérez Perdomo 1995, Little 1996). The first president of the democratic era in Venezuela, Rómulo Betancourt, was well aware of the need to monitor and discipline the recipients of "infant industry" protective rents. Reflecting upon his policies in the short-lived democratic administration in the period 1945–48, he warned: "It was not deemed necessary or consistent with national interest to distribute credits to parasitical industries which could be kept alive only behind a protective tariff wall. Industries which pretend to maintain irrational costs or which produce commodities of such poor quality that can only subsist through protective tariff barriers do not merit either interest or stimulation of a responsible government. . . . This had to be done under definite guidelines" (Betancourt 1978, 200, 324). This reasoning anticipates later arguments identifying state discipline of rent recipients and selectivity of rent creation as key features of successful industrial policy in particular (Amsden 1989) and developmental states more generally.

As early as 1945, there was a keen awareness of the dangers that corruption in the state bureaucracy could have on the legitimacy and predictability of government. In the early 1960s, the Tribunal de Responsabilidad Administrativa was set up to investigate the rampant cronyism that occurred in the authoritarian administration of Pérez Jiménez (1950–57), particularly concerning large-scale public construction contracts. In the 1978 elections, then former president Betancourt, still an active leader of Acción Democrática (AD, the dominant political party from 1958 to 1998), called for a battle against corruption prevalent in the Venezuelan legal system: "It is necessary to put an end to the calamity that the Judiciary is an archipelago in which each party has its own plot. . . . An end must be put to the fact that a good number of tribunals are grocery stores [*pulperías*] where verdicts are bought and sold" (quoted in Coronil 1998, 342). In the mid-1980s, a high-profile commission on reform of the public sector was set up to identify and diagnose classic problems of lack of coordination between ministries and the decentralized public sector, transparency, and meritocracy in Venezuelan public administration (COPRE 1989).

The problem of administrative corruption cannot be blamed on inadequate legislation. In addition to the general provision of the constitution of 1961, the 1982 Public Property Protection Law (Ley Orgánico de Salvaguardia del Patrimonio Público, or LOSPP) identifies the following grounds for prosecution: the procuring of illegal advantage in the administration of public affairs; any official act performed or not performed in return for cash whether received or not; private arrangements between public servants and private contractors concerning contracts, services, or goods which prejudice public monies; misuse of public funds; the issuance of false documentation; influence trafficking; the illegal use of permits; alteration of documents; and negligence or failure to support the proper actions of public bodies. The exhaustive nature of the provisions of the LOSPP suggests that its drafters were hardly in the dark concerning the extent and variety of corrupt activities in Venezuela (Little 1996, 67).

In sum, knowledge and assessment of the problems facing the Venezuelan industrialization drive and potential problems of corruption in state institutions were not absent. This suggests that the mere identification of inadequate policies is not a particularly relevant factor in explaining deteriorating economic performance over time. Moreover, the cognitive failure approach cannot answer the more relevant question of why many attempted changes were not implemented or sustained.

4.6 Conclusion

The proposition that oil abundance induces extraordinary corruption, rent-seeking, and centralized interventionism and that these processes are necessarily productivity- and growth-restricting is not supported by comparative or historical evidence. Similar levels of state centralization and corruption coincided with cycles of growth *and* stagnation in Venezuela. Explaining governance and state capacity in Venezuela needs to be consistent with this basic evidence. The rentier state model suggests that oil abundance *causes* growth-restricting patterns of state centralization, public ownership, and rent-seeking. However, rentier state theory fails to provide a more robust theory of incentives that explains a longer historiography in Venezuela and the association of centralization, public ownership, and rent-seeking with more successful industrial transformation elsewhere. Moreover, rentier state models and neoclassical models of corruption cannot explain why corruption worsened throughout Latin America in both mineral-rich and mineral-poor economies undertaking neoliberal reforms in the 1990s. Liberalization was supposed to reduce the extent to which the state can create rents and, by implication, was supposed to reduce levels of rent-seeking and corruption. Without denying that the quadrupling of oil prices in 1974 created enormous dangers for Venezuela, comparative evidence does point to the fact that few Latin American economies, apart from Chile, have efficiently managed massive resource inflows (Palma 1998). Moreover, rent-seeking models fail to explain why suboptimal institutions persist despite the fact that decision-makers knew their deficiencies.

5

ECONOMIC LIBERALIZATION, POLITICAL INSTABILITY, AND STATE CAPACITY IN VENEZUELA

This chapter examines the economic and political impact of economic liberalization programs in Venezuela from 1989 to 1998. In particular, it examines the extent to which liberalization reforms introduced in 1989 reduced corruption and enhanced growth and investment as rent-seeking and corruption models would suggest. The costs associated with centralized state authority and state intervention in the economy, as developed in models of rent-seeking and corruption, form the theoretical pillar behind the idea that reducing the state's role in resource allocation will enhance the prospects of economic growth. According to rent-seeking logic, economic liberalization reduces the discretionary authority of public officials to deploy valuable benefits and rents, which, in turn, should eliminate incentives for rent-seeking and corruption.[1]

Venezuela, a long-standing democracy, experienced a virtual political implosion in the period 1989–98. The rapid downward spiral saw an increasing crisis in governability, manifested by the collapse of the two main political parties, an increase in political polarization, more frequent coup attempts, alarming increases in voter absenteeism, increasingly frequent mass (and often violent) street demonstrations, dramatic increases in crime, and growing labor unrest (McCoy 1999; Stambouli 2002; Roberts 2003). Moreover, Venezuela has experienced, in the period 1988–2003, among the steepest declines in per capita income and among the largest increases in income inequality in Latin America.

---

1. The World Bank has recently argued that there is nothing more important than the fight against corruption (World Bank 2000).

Venezuela was an unlikely candidate for political and economic implosion. First, in the period 1958–88, Venezuela maintained through political pacts, and corporatist bargaining, one of the most stable democratic systems in Latin America (Levine 1973; Karl 1986; Rey 1991). Thus, crisis and breakdown in Venezuela occurred in a polity that had accumulated substantially strong mechanisms to regulate and contain conflict. Second, Venezuela was, for six decades before 1980, the second-fastest–growing economy in Latin America and the economy with the lowest inflation rate, the latter a sign of a polity that contains and regulates conflict. Given its favorable initial conditions, the Venezuelan case may prove to be an instructive case as to the stresses liberalization can unleash, not only in transition economies (Przeworski 1991), but in a late-developing, capitalist, and long-standing democratic polity.

The influence of rent-seeking models in development and governance policies is substantial. The basic idea is that liberalization would spur economic recovery and lead to more accountable, transparent, and effective governance. Even after the World Bank backed away from promoting radical structural adjustment programs, it reaffirmed these basic propositions about the role of liberalization in its so-called capability approach, which posited that "poor governance" is the result of an overextended state relative to its institutional capacity at a given moment in time (World Bank 1997a, 61–75). While the World Bank acknowledges that state-created rents and rent-management systems, such as in industrial policy, may have worked in a handful of countries, especially in Northeast Asia (World Bank 1993), the advice for poorly performing Latin American economies would be "not to try this at home," since the administrative capacity to effectively administer subsidies, export "contests," and investment coordination is generally lacking. Otherwise, the proponents of liberalization would argue, there wouldn't be so many infant industries which failed to grow up.

Moisés Naím (1993, 147–50), for instance, argued that targeted intervention of the South Korean or Taiwanese variety was unsuitable for Venezuela.[2] This is because the government lacked the bureaucratic capacity to deploy subsidies and rents in an effective manner. Naím believed that if Venezuela were to attempt to implement a selective industrial policy, "corruption [would] soar much higher and much faster than exports" (150).

This analysis of governance crucially assumes that *inherited* capacity constrains selective state intervention. And this constraint is what should orient

2. Naím became minister of industry in 1989.

the shape of administrative, institutional, and policy reform. The capability approach implies that if the state can focus its scarce administrative skill in the "fundamental" areas and leave production to the private sector, the overall management and governance of the economy will become more effective. The policy advice for poorly performing economies is to reduce the state's role in resource allocation decisions, especially in the areas of industrial policy and in direct production through public enterprises. Trade liberalization and privatization are advocated as means of insulating the state from growth-restricting political lobbying. In academic and policy circles within Venezuela, the main diagnoses of how to improve state capacity and governance in Venezuela were driven by the insights of rent-seeking theory and its related variant, the rentier state model (Naím 1993; Karl 1997).

In this chapter, I suggest that economic liberalization and political decentralization has not strengthened the state as the capability approach predicts. What is missing in the capability approach is a political analysis of how the capacity to reform is constructed and, in particular, the role that political strategies of conflict resolution and competition play in constructing legitimate alternatives to failed state-led development projects. The "good governance paradigm" promoted within the international development community downplays the task of reconstructing or building political organizations. This is because of the influence of the rent-seeking and corruption literature in informing policy on state capacity building and the negative view of politics that flows from that analysis. In what follows, it is suggested that political analysis allows us to develop a more adequate account of the risks that reforms generate and to discuss the political sustainability of reforms.

The rest of the chapter is organized as follows: Section 5.1 describes the liberalization reforms and examines the domestic political support based for such reforms. Section 5.2 examines the economic and distributional impact of the reforms and discusses the causes and consequences of the failure of the state to regulate the banking system. The evidence suggests that economic liberalization not only failed to revive private investment and economic growth, but also contributed to a worsening of the distribution of income, which contributed to growing polarization of politics. Section 5.3 examines the political processes through which neoliberal reforms contributed to political instability and the increased use of the corruption scandal as a weapon of political and economic competition. Particular attention is paid to the role of policy switches and party-neglecting strategies among the reformers. Section 5.4 briefly suggests why neoliberalism generated greater political instability and decline in state capacity in

Venezuela compared to some other Latin American reforming economies. The conclusion explores the theoretical and policy implications of the Venezuelan experience.

5.1 The Economic Liberalization Package of 1989

In 1989, Carlos Andres Pérez launched one of the most ambitious liberalization reforms in Latin America. The liberalization plan known as the "Great Turnaround" ("El Gran Viraje") included the unification and massive devaluation of the exchange rate, trade liberalization, privatization and financial deregulation, including freeing of interest rates, elimination of nearly all restrictions on foreign investment, and the introduction of tax reforms, including the introduction of value-added taxes (for details see Hausmann 1995, Corrales and Cisneros 1999, and M. Rodríguez 2002).[3]

Since the transition and consolidation of democracy in 1958, the Venezuelan polity had historically been legitimated by state-led developmentalism and economic nationalism (Coronil 1998), with centralized rent deployment patterns controlled by the executive and brokered by two hegemonic and highly centralized and clientelist political parties.[4] The program was intended to be an orthodox reform package along the lines of the Washington Consensus (J. Williamson 1990). Its chief architect, Planning Minister Miguel Rodríguez, envisioned that Venezuela, in *economic* terms, would follow the post-1982 Chilean model of neoliberal reform.[5] While big-push industrialization policies were discredited, there seemed to be little doubt among the Venezuelan economic reform team that an equally ambitious big-push liberalization program could be carried out. This lack of caution was bold, given that Venezuela had suffered in the previous decade from an increasingly ineffective state apparatus. Albert Hirschman (1981b) observed that a sense of *fracasomania* (or failure complex) prevailed among Latin American leaders. *Fracasomania* reflects the idea that

    3. Political reforms included the elimination of political party slates, the installation of direct elections for governors and legislators at the state and municipal levels, and the devolution of power in health and education sectors (De la Cruz 1997; Grindle 2000, 37–93).
    4. Ironically, in the first Pérez administration (1974–78), the government attempted one of the largest *state-led* big-push, natural-resource-based industrialization programs in Latin America, a program that was known as "La Gran Venezuela" (Karl 1997, 143–60). See Chapter 7 for details.
    5. Interview with Miguel Rodríguez, June, 2003. Rodríguez failed to point out two importance differences between 1989 Venezuela and post-1982 Chile. First, the latter was an authoritarian regime that had severely weakened the power of political parties and labor unions. Moreover, there was broad support for neoliberal reforms among the opposition parties in Chile, whereas there was little support in Venezuela for such reforms.

everything that has come before has been an utter failure, and thus, radical widespread reform is only viable path to revive economic development.

The impetus to these policies was a balance of payments crisis in 1988. The previous three years of the Lusinchi administration (1986–88) were characterized by populist macromanagement. The maintenance of several price controls and a multiple exchange rate regime were sources of corruption and distortions. In these years, the fiscal deficit averaged 7.6 percent of GDP. By 1988, inflation had reached 30 percent in an economy with historically very low inflation. The deficit on the current account had reached 9.9 percent of GDP and net international reserves were negative US$6.2 billion (M. Rodríguez 2002, 39). As with many structural adjustment programs, macroeconomic stabilization was combined with economic liberalization even though there is little evidence that it is either prudent or necessary to implement both simultaneously (Rodrik 1996). Restoring macroeconomic balances in the exchange rates or in fiscal accounts does not imply that trade liberalization will improve export performance or productivity growth.

Apart from the balance of payments crisis in 1989, there were economic and political factors that contributed to a sea change in Pérez's policy stance. First, the long-run growth performance of the Venezuelan economy and its state-led industrialization strategy was becoming increasingly poor. In the period 1973–88, GDP per capita declined 15 percent. Second, declines in oil exports per capita further restricted the resources for patronage and induced many actors, such as labor unions and business chambers, to clamor for changes as well as reduce their loyalty to the party system, which was central to the distribution of oil rents (Penfold-Becerra 2001). In the period 1950–80, real exports per capita averaged $1,550 (in 2000 U.S. dollars), but declined to $1,150 in the period 1980–90. Within this decline in oil exports per capita, oil prices collapsed by 50 percent in 1986, which translated into oil exports per capita plummeting to an annual average of less than $600 per capita in the period 1986–88. Sudden declines in commodity prices seemed to have added to the sense of *fracasomania* as it underlined the vulnerabilities of export dependence on oil.

Politically, there was increasing domestic pressure from the mid-1980s on from influential business groups, who were advocating economic liberalization as a way to increase economic efficiency and to reduce the dominance of the state and political parties over resource allocation.[6] One manifestation of this

---

6. While neoclassical political economy, and in particular, rent-seeking theory, links domestic pressure to the creation of state-led interventions, there is much less analysis of the role that interest groups play in inducing the state to withdraw from governing the economy (Schamis 1999).

pressure was the Grupo Roraima (an association of influential business leaders and academics), which produced a series of influential documents, particularly the *Proposición al país* (Proposal to the Country) in 1984, critiquing state-led development policies and calling for radical change (Grupo Roraima 1984).[7] Second, Pérez's policy conversion to neoliberalism may have been influenced by the large-scale campaign contributions and close relationships with economic groups who expected to benefit from the new rent opportunities that privatization and liberalization would bring.[8] Third, Pérez was greatly influenced by the emergence of an elite group of academics (mostly economists and business administration experts trained in U.S. universities) who had no party affiliation and generally were champions of radical, neoliberal reform (Navarro 1995, 127). Many of the key positions in Pérez's cabinet were given to these academics, most of whom worked in the country's leading business school, the Instituto de Estudios Superiores de Administración (Institute of Higher Administration Studies, IESA). This group of ministers was eventually called the "IESA Boys" in Venezuela, analogous to the "Chicago Boys" in Pinochet's Chile (Coppedge 2000, 128). Finally, while international agencies like the World Bank, the IMF, and Inter-American Development Bank greatly supported neoliberal reforms throughout the region, there is little evidence that these organizations were putting pressure on the Venezuelan state to contemplate an ambitious liberalization program. Thus, the reform team had a high degree of ownership over the program.

The most rapid liberalization reforms took place in trade policy. It became well known that the protection system did not have clear performance criteria guidelines. The perpetual protection of infant industries that failed to grow up became the norm (World Bank 1990, 19). Trade liberalization for industry was one of the most rapid and profound in Latin America at the time. The maximum tariff was reduced from 135 percent, one of the highest in the region, to 20 percent by 1992. By 1993, average tariffs declined further to 10 percent, one of the lowest among Latin America's major liberalizers (Corrales and Cisneros 1999, 2103). Average tariffs were reduced from 37 percent overall (and 61 percent for finished goods) in 1988 to 16 percent overall (and 26 percent for finished goods) in the period 1991–93 (ibid.). Non-tariff barrier coverage declined from

7. For a discussion of Grupo Roraima, see De la Cruz 1988, 71–80.
8. Interview with Gumersindo Rodríguez, June 2003. See Coppedge 2000 on the role of campaign contributions in influencing policy decisions. The two most influential figures were probably Pedro Tinoco, who owned the Banco Latino and became president of the Banco Central in 1989, and Diego Cisneros, head of the Cisneros Group, which was one of the most successful Venezuelan business groups (both domestically and internationally).

an average of 44 percent in the period 1985–87 to less than 5 percent, one of the lowest rates among major Latin American trade liberalizers, in 1991–92 (Edwards 1995, 200).[9]

The decision to implement the liberalization program was clearly based on the idea that the regulatory structure of the state had become the principal contributor to stagnation in Venezuela (Naím 1993; see also section 2.4). The abundance of resources available and relatively high investment rates in Venezuela over long periods would strongly suggest that the problem of poor long-run growth has been, at least in part, a problem of inefficient resource management and hence incentive structures at all levels of the Venezuelan economy. The institutional collapse of the state apparatus was argued to be a sufficient reason for the state to abdicate power and let decentralized private agents assume control of coordinating economic activity (CORDIPLAN 1990). This logic is broadly similar to the capability approach of the "good governance" paradigm (World Bank 1997a). What was *not* communicated to the public, however, was why a rich oil country with three years of rapid growth in the period 1986–88 needed such harsh structural adjustment.

5.2 Neoliberal Reforms and Economic Performance

The Venezuelan experience with economic liberalization from 1989 until the emergence of Hugo Chávez as president in 1998 calls into question many of the predictions of the good governance paradigm, the capability approach, and the models of rent-seeking and corruption upon which both are based. The period 1989–98 was marked not only by the continued stagnation of output and productivity growth, but a growing perception that corruption was increasing (see table 4.2).[10]

While some analysts have hailed the full implementation of trade liberalization as a remarkable political achievement (Corrales and Cisneros 1999, 2099), per capita GDP *declined* 2.7 percent over this period.[11] In the non-oil economy,

---

9. According to Corrales and Cisneros (1999, 2102), the number of import licenses granted by the state declined from 2,204 (representing nearly 48 percent of manufacturing production) to fewer than 140 in 1992 (affecting approximately 2 percent of manufacturing production).

10. Based on Transparency International subjective corruption indices (an index of 10 indicates minimum corruption, an index of 0 indicates maximum corruption), Venezuela's corruption index worsened from 3.3 in the period 1980–85 to 2.5 in the period 1992–95.

11. It should be noted, however, that growth rates did average nearly 10 percent in 1990 and 1991.

economic liberalization did not reverse the long-run decline in growth rates. Non-oil annual growth did increase from 1.1 percent in the period 1980–90 to 2.3 percent in the liberalization period of 1990–98; however, this rate was still well below the growth rate in the previous three decades and collapsed again to −3.5 percent in the period 1998–2003.

The decline in manufacturing growth has been particularly disappointing. In 1989, the first year of the "Great Turnaround" plan, there was a dramatic decline in manufacturing output of 14.5 percent. Manufacturing growth, which had been relatively rapid (although increasingly unproductive) in the period 1950–80, declined from an annual average of 4.3 percent in the period 1980–90 to 1.5 percent in the reform era of 1990–98, and collapsed to *minus* 5.1 percent in the period 1998–2003 (see table 2.1). In the period 1988–98, the decline in manufacturing growth was widespread across sectors as indicated in table 5.1. Only seven out of twenty-five sectors registered positive growth rates over the period, and most of the sectors that did register growth were in low-technology, limited growth technology sectors (i.e., food products, nonferrous metals), natural-resource processing, or turn-key assembly-line sectors with little technology transfer (e.g., autos).

Economic liberalization also induced changes in manufacturing product specialization that were unlikely to enhance long-run productivity growth and technological development. Table 5.2 illustrates the extent to which the relative shares of manufacturing sectors with different technological characteristics changed in the period 1970–98. Following the classification in Katz 2000, Group I are technology-intensive sectors where productivity growth is likely to be greatest. Group II is the motor-vehicle sector, where growth is generally dependent on the extent to which foreign direct investment is active. Groups III and IV are the capital-intensive sectors of food processing and natural-resource production. Group V are generally low-technology, labor-intensive sectors such as footwear.

Several important trends emerged. First, the technology-intensive sectors (I) increased from a low base *before* economic liberalization. Thereafter, their share in total manufacturing value-added decreased slightly from 12.4 percent in 1990 to 11.0 in 1998. Thus, all of the increase in technology-intensive sectors occurred in the era of protectionism. The share of group I sectors in Venezuela was similar to that of other medium-sized Latin American countries such as Chile and Colombia by 1996 (1592). Second, liberalization induced a further specialization in the capital-intensive sectors of foodstuffs and natural-resource processing. The share in manufacturing value-added of groups III and IV increased from

TABLE 5.1 Growth rates in Venezuelan manufacturing, 1988–1998 (average annual growth in gross output, %)

Sector	Average annual growth in gross output (%)
All manufacturing	−1.8
glass products	6.4
pottery, ceramics	5.7
industrial chemicals	3.2
food products	2.8
transport equipment	2.4
nonferrous products	1.8
plastic products	1.5
footwear	0.3
textiles	−0.4
beverages	−1.2
printing & publishing	−2.5
petroleum derivatives	−3.2
non-metallic minerals	−3.7
nonelectrical machinery	−4.1
iron and steel	−4.6
electrical machinery	−5.4
wood furniture	−6.0
pulp, paper	−6.1
rubber products	−6.5
other chemical products	−6.6
fabricated metal products	−7.6
leather products	−7.7
apparel	−8.4
other manufacturing	−9.0
tobacco	−9.3

SOURCE: OCEI, *Encuestra industrial*, various years; Banco Central de Venezuela, *Serie estadística*, various years.

55.4 in 1990 to 60.3 percent. The growing predominance of these sectors limited employment creation, increased the investment requirements of production, and was only modestly dynamic in terms of technological development.[12] Once again, Venezuela's focus on foodstuffs and natural-resource production is similar to the Latin American pattern in the 1990s (1592). Third, a somewhat surprising outcome of liberalization was a *decline* in the share of the labor-intensive sectors. The group V share in manufacturing value-added declined from 30.4

12. It should be noted that further industrialization of the mineral and fuel sectors could *potentially* provide a dynamic and positive growth pole in the economy (see Wright and Czelusta 2007).

TABLE 5.2 Structure of manufacturing value-added in Venezuela, 1970, 1990, 1996, and 1998

Sectors	1970	1990	1996	1998
Group I	5.3	12.4	11.2	11.0
Group II	5.1	1.7	2.3	3.0
Groups III+IV	52.8	55.4	61.3	60.3
Group V	33.2	30.4	25.3	25.8

SOURCE: OCEI, *Encuestra industrial*, various years.

NOTE: Following Katz 2000, 1591–92, the groups used here are as follows: Group I: Metalworking and engineering-intensive industries, except motor vehicles (ISIC 381, 382, 383, and 385); Group II: Motor vehicles (ISIC 384); Group III: Food, beverages, and tobacco (ISIC 311, 312, and 313); Group IV: Natural-resource processing industries (ISIC 341, 351, 354, 355, 356, 371, and 372); and Group V: Labor-intensive industries (ISIC 321, 322, 323, 324, 331, 332, 342, 352, 361, 362, 361, and 390).

percent in 1990 to 25.8 percent in 1998. One of the main ideas in the Washington Consensus was that devaluation, by reducing wage costs, would induce more labor-intensive production. This did not happen in Venezuela; nor did it happen more generally in Latin America in the 1990s (Katz 2000). In sum, the liberalization period did not generally promote shifts in production that were either productivity-enhancing or favorable for employment creation.

While economic liberalization policies may have provided increased *discipline* on producers compared with previous era of protectionism, it has not provided the *incentives* for firms to engage in restructuring. There were several reasons for this. First, rapid trade liberalization left many firms with little time to compete with lower-priced imports. Second, financial deregulation led to a drastic increase in interest rates. Bank loan rates, which were often fixed at negative real rates, averaged 12 percent in the 1980s (De Krivoy 2002, 308). In 1989, with financial deregulation, the loan rate jumped dramatically to 34 percent, and averaged 45 percent in the period 1990–98 (308). The average ratio of bank credits to GDP in the period 1989–93 declined to 31.3 percent, nearly one-half the ratio of 52.6 percent in 1988 (55–56). Third, credit to manufacturing firms (particularly to small- and medium-sized companies) declined dramatically. This was an important factor that contributed to the decline in political support for economic liberalization from parts of the business community. In the context of trade liberalization and economic stagnation, many of the family conglomerate groups engaged in manufacturing were dismantled (Francés 2001), and many smaller operations were forced into bankruptcy.

There is little evidence that competitive pressures were providing growth-enhancing and productivity-enhancing producer incentives. The best evidence for this is that there was a sharp decline in total investment and particularly private sector investment (see table 2.3). Both total and non-oil investment rates, particularly *private* investment rates, in the period 1990–98, were significantly *lower* than in *any* period since 1950. The non-oil public investment rates in the 1990s were also lower than at any period since 1950, which also suggests that the effectiveness of the state in mobilizing resources did not increase as a result of economic reforms. In particular, reducing the role of the state in the economy did not lead to a more secure environment in which to invest.

While investment stagnated, there was a massive shift in the factor distribution of income in favor of profits and away from wages in the liberalization period, as indicated in table 5.3. The share of corporate profits, rents, and dividends oscillated between 51 percent and 54 percent in the period 1950–88. However, in the liberalization period of 1989–98, capital owners appropriated an annual average of 64 percent of national income. Despite the fact that capitalist surplus appropriation and rents were increasing, private sector investment rates *declined* in the 1990s. In fact, much of the increased surplus appropriation went abroad as accumulated capital flight, which reached $14 billion in the period 1994–2000. Capital flight was nearly the same as the accumulated surplus in the current account of the balance-of-payments ($15 billion) in the same period (M. Rodríguez 2002, 67).[13]

While liberalization was associated with a dramatic decline in labor's factorial share, it is not possible to conclude definitively that liberalization caused this decline. There are important feedback effects to consider (F. Rodríguez

TABLE 5.3 Net factor distribution of national income in Venezuela, 1950–2002

	Mean annual share in national income (%)	
	Of wages and salaries	Of corporate profits, dividends, rents, and interest payments
1950–60	47	53
1960–70	46	54
1970–80	49	51
1980–88	46	54
1989–98	36	64

SOURCE: Banco Central de Venezuela, *Serie estadística*, various years.

13. Even during the first five years of the decidedly anti-neoliberal Chávez administration, the share of labor income in the GDP dropped to 35 percent in the period 1998–2002 (F. Rodríguez 2003, table 1). Substantial capital flight continued in this period as well (see table 9.4).

2000, 35–36). On the one hand, labor unions had weakened prior to the liberalization period. In 1975, the rate of unionization was 33 percent, but fell to 26.4 percent by 1988 owing to economic stagnation and decline. Moreover, the ties between the main labor-based party, AD, and the main labor federation, Confederación de Trabajadores de Venezuela (CTV), weakened throughout the 1980s (Roberts 2003, 60–61). Labor's weakened political power contributed to the decline in its bargaining power over wage shares. On the other hand, liberalization policies clearly exacerbated this trend, as lower wage shares along with weak demand for labor (owing to stagnant investment) further weakened labor's power. In the period 1988–95, the rate of unionization fell by nearly 50 percent, declining from 26.4 percent of the workforce in 1988 to 13.5 percent in 1995 (61).

Accompanying the decline in labor union membership was an increase in informal employment in the liberalization period. In the period 1980–90, the rate of informal employment of the nonagricultural labor force averaged 39.5 percent. However, the level of informal employment increased to an average of 44.5 percent in the period 1991–95 with a tendency toward continued increases as informality rates reached an average of 48.5 percent in 1994–95.[14]

Decentralization policies also may have contributed to the widening gulf between rich and poor. In Caracas, three of the wealthiest areas—Chacao, Baruta, and El Hatillo—broke away as separate municipalities. This process denied shantytown areas of significant tax revenues (Ellner 2008, 93), which likely fueled further resentment among low-income groups. This pattern was repeated throughout the country over the period 1989–98 where over "thirty municipalities were created at the expense of inner city and marginalized areas" (93).

The worsening of income distribution contributed to the growing polarization of politics. Such divisiveness was manifested in increasing factionalism within and between the political parties and declining support among the poor for economic reforms (Roberts 2003).[15] The severity of the growing

14. Based on data from corresponding editions of OCEI, *Encuesta de hogares* and *Encuesta de empleo*. Growing fragmentation and informalization of the labor and production process negatively affected the social bases of support for political parties, and hence contributed to the deinstitutionalization of conflict mediation capacities in the Venezuelan polity. It also meant that populist/outsider strategies became more effective. It is perhaps no accident that the two subsequent political leaders, Rafael Caldera and Hugo Chávez, relied on *anti-politico*/outsider discourses and less on the corporatist modes of intermediation that had characterized Venezuela's "pacted" democracy in the past (see sections 8.2.4 and 8.2.5).

15. Naím (1993) explains political instability in terms of the failure in the communication strategy of the Pérez administration to inform the poor and middle class of the benefits of reform. This line of reasoning underestimates the distributional impact the liberalization model itself.

polarization was manifested in the widespread support among the poor for two military coup attempts in 1992, the first of which was the military rebellion of Hugo Chávez, whose popularity was based on his attacks on the injustices of neoliberalism.[16] Growing inequality was also the focal point of Rafael Caldera's famous speech in Congress in 1992, in which he condemned the actions of the coup plotters, but emphasized that the discontent of the military officers was a fair reflection of the injustices of the neoliberal program.[17]

Finally, the inability of the state to effectively regulate the banking system was a powerful indication that a smaller state is not necessarily capable of managing "fundamental" regulatory functions effectively. Venezuela experienced a major collapse of the banking sector. The lack of supervisory and regulatory mechanisms and blatant theft of government bailout funds by bankers (estimated at nearly US$7 billion) in the form of capital flight led to large-scale bank closings and government takeover of many of the economy's largest commercial banks in 1994 and 1995. The bailout cost the government the equivalent of 18 percent of GDP, the fifth most severe banking crisis in the world during the period 1975–95 (World Bank 1997a, 68).

The main cause of this crisis was the weakness of the state relative to financial groups and its inability to impose effective banking supervision, regulation, and enforcement of fraudulent practices in the financial liberalization period (De Krivoy 2002). Some of the leverage that large financial groups had were the result of large campaign contributions many had made to the Pérez presidential campaign in 1988 (Coppedge 2000).[18] Moreover, financial groups were able to resist the opening of the banking system to foreign competition and takeover until 1995.[19] In the period 1989–92, the annual budget of the banking

---

Polling evidence from the period 1989–91 suggests that the poor strata were much less likely to support the reforms than upper-income groups (Roberts 2003, 63).

16. See Canache 2002 for evidence on the lower-income groups' support for the military coup as well as their continued support for Chávez in the period 1992–2000.

17. This speech revived the political stature of Caldera, who was president from 1968 to 1973. Caldera went on to win the presidency in 1994.

18. The appointment of Pedro Tinoco (one of Pérez's close friends) as president of the Banco Central was indicative of the power of the financial community and the degree of cronyism between the executive and some big financial groups. Tinoco owned Banco Latino, which went from the fifth-largest to second-largest bank in Venezuela in the period 1989–93. Banco Latino became the largest holder of government treasury bonds, one of the more lucrative enterprises in the banking sector. The lack of banking regulation permitted Latino and other banks to engage in self-loans and transfer large amounts of deposits offshore. By 1993, 70 percent of all bank's assets were self-loans (De Krivoy 2002, 165).

19. In the period 1995–98, a wave of foreign investment followed in the banking sector led by two leading Spanish banks, Banco Santander Central Hispano (BSCH) and Banco Bilbao Vizcaya

regulation board amounted to a paltry $8,000 per private financial institution (De Krivoy 2002, 30). This amount was equivalent to the annual salary of a middle-level manager (30). Weak state capacity in this case was not simply "inherited," as the capability approach would have it, but was the result of the political power of financial groups able to resist changes in the regulatory system and to resist implementation of already existing laws. In sum, there is little evidence that the leveling down of the state increased its capacity to intervene in the economy and polity more effectively than in the preliberalization era.

Proponents of economic liberalization might argue that reform policies were not followed consistently or that the reforms were not carried far enough. They would point to the fact that growth was very rapid in 1990 and 1991 and was derailed because the reforms failed to be deepened, particularly in the area of banking reform and labor regulations. There are at least two problems with this argument. First, there is little cross-country evidence that accelerations in the growth trajectory of countries were *preceded* by wide-ranging economic liberalization (Rodrik 2004a). Second, *comparative* regional evidence suggests that Latin American reforms were the most profound, yet the region's growth was among the lowest among less developed regions (ibid.). This brings into question the extent to which further liberalization is necessarily growth-enhancing. Finally, such arguments do not consider the extent to which the introduction of liberalization reforms is politically destabilizing.

5.3 Neoliberal Reforms, Corruption, and Political Instability

The liberalization period also produced uncertainty and political tension as well as a perception that corruption had worsened. Sudden economic deregulation led to a frenzy of what Naím (1993, 95–100) refers to as "oligopolistic wars" among business groups vying for control over raw materials, financing, and distribution channels.[20] The rapid dismantling of trade protection and a decline in state-business cooperation further reduced the already weak collective

---

Argentaria (BBVA). The increase in market share of foreign-owned banks went from less than 2 percent in 1994 to over 40 percent by 1999 (World Bank 2002, 89).

20. See also Ortiz 2004, 79–84, for a discussion of the "dirty wars" among big business groups, particularly in the banking and media sectors, during the early 1990s. Ortiz notes that these "struggles were high-profile conflicts, and the ensuing 'dirty wars' involved high-ranking government officials and political party leaders. This intensified the perception that the political regime was corrupt and paved the way for the two coup attempts in 1992" (81).

action capacity between the big business groups, and made strategies of conflict more likely than strategies of cooperation between competing family conglomerates.[21] In the context of weak judicial and regulatory mechanisms, these wars turned into nasty battles undertaken in the media as business groups aggressively invested in newspapers, magazines, and radio and television stations. According to Moisés Naím and Antonio Francés (1995), there was a point where no major media enterprise was independent from a major private conglomerate group. The limited social capital of business groups clearly intensified a "war of positions" within the private sector that added greatly to the atmosphere of political and social instability that marked the liberalization era of the 1990s. Neoliberalism, if anything, created the setting for increases in mafia-like activity to appropriate the large rents that suddenly emerged with deregulation.

The manner in which economic liberalization was introduced also destabilized the polity. Pérez, elected in a landslide, had been president from 1974 to 1979, and many voters associated him with a period of prosperity and state largesse. Neither Pérez nor his party (Acción Democrática) stressed during the campaign that rapid and profound reforms were planned, though there were policy documents that indicated that some market reforms would be initiated (Stambouli 2002, 175–76). By hiding his policy intentions, Pérez was one among many "first-generation" reformers in Latin America, such as Salinas (in Mexico), Fujimori (in Peru), and Menem (in Argentina), who introduced economic reforms in the early 1990s, despite running campaigns that belied their intentions. This misrepresentation of policy intention during the campaign (essentially deceiving the electorate) was to prove damaging to the long-standing consultation process in the Venezuelan polity.[22]

Hiding reform intentions exacerbated the "shock" to the public when economic liberalization therapy was actually introduced. A few weeks after the announcement of reforms, Venezuela experienced its bloodiest urban riots since urban guerrilla warfare in the 1960s. The riots, known as the *Caracazo*, occurred

---

21. The particularistic nature of the "politics of privilege" between the state and business groups meant that business and industrial chamber associations ceased to have effective collective institutions beginning in the early 1970s (see Chapter 8).

22. The second major policy switch in the period under study occurred in 1996 when Rafael Caldera, who won the presidency in 1993 running on an anticorruption, anti-neoliberal campaign, abandoned two years of price and capital controls and endorsed the Agenda Venezuela, a structural adjustment package with IMF support. The policy switch did not only not reverse economic stagnation, it led to a further disillusionment with the party system, and a deterioration of state institutions and public services (López Maya 2003, 83).

in late February 1989. A doubling of gasoline prices, which were passed on by private bus companies, induced the outbursts. The government had actually announced that bus fares were allowed to rise by 30 percent, but did not monitor the increases bus companies were charging. Moreover, bus drivers ignored discounts to students (Corrales 2002, 51). The riots that ensued were contained by a relatively undisciplined military response that left more than 350 dead in two days. Although never documented, there are many informal accounts that point to left-wing organizations that mobilized groups to incite riots when gas prices were increased.[23]

The way in which liberalization reforms were decided was also divisive. Pérez decided to completely abandon consultations with large rival factions within his party, Acción Democrática, and introduced reforms by relying on insulated technocratic decision-making. From 1958, Venezuela's democracy had been consolidated around a series of political pacts that relied on consensus building among the main political parties, labor unions and business associations (see section 8.2.2). Two-thirds of cabinet ministers in 1989 were from outside the governing political party, a move that created resentment and opposition in the legislative assemblies, including within Acción Democrática (Stambouli 2002, 179–80).[24] Moreover, Pérez and his ministers' discourse was confrontational and insulting to anti-Pérez factions within Acción Democrática. Miguel Rodríguez, the planning minister, labeled critics of reform critics "dinosaurs," "maladapted," "cowards," and "unschooled." Pérez did little to dissociate himself from such remarks (Corrales 2002, 122). Such a discourse contributed to the "activation of boundaries" between the self-proclaimed "modern" reformers and the "backward" old guard of the political parties. The creation and activation of boundaries contributes to the escalation of political conflict and violence (Tilly 2003, 96). The break with pact-making and consultation exacerbated the emerging factionalism between and within political parties and was largely responsible for the adversarial executive-party relations in the first three years of reform, and the massive increase in corruption scandals and accusations in the period.[25] As a result, political instability and investment risks increased.

23. Interviews with military officers and leaders of the Bolivarian Circles, June, 2003.
24. Kornblith (1995, 81) also notes that Pérez broke with a long-standing tradition in the democratic era of leaders building consensus over policy within and between political parties.
25. The rebellion of Pérez's own party, along with the 1989 riots, contributed to the isolation of the executive. Such isolation signaled a legitimacy crisis for the government, which, in turn, encouraged further attacks against the state (Corrales 2002, 167). The most notable examples were the two coup attempts in 1992 and the support among all political parties for the impeachment of Pérez (who was forced to resign in May 1993).

One way to gauge the increase in conflict is to examine inflation levels, which reflect increases in the intensity of distributive struggles (Rowthorn 1971) and the increasing inability of the state to manage such conflicts. While Venezuela's inflation rates were relatively low by Latin American standards throughout the period 1960–80, increases in the inflation rate in the 1980s, and particularly in the period 1990–98, were significant in terms of the country's own record of low inflation. Annual inflation, which averaged less than 6.6 percent in the 1970s, rose to 23.0 percent in the 1980s, and more than doubled to 50.1 percent in the period 1990–98.[26] Clearly, economic liberalization did not generate a constellation of political constituents capable of imposing stable macroeconomic management within the state.

The increase in executive-party tensions along with two coup attempts heightened the resistance to reforms. Indeed, the Pérez government was forced to *abandon* many reforms. The programs abandoned included banking reform, the creation of a Macroeconomic Stabilization Fund, and the liberalization of gas prices (Corrales 2002, 61). Abandoning some reforms contributed to further political destabilization. For example, the failure of the state to implement a meaningful and progressive tax reform signaled that the wealthier groups were able to effectively resist shouldering the burdens of reform. When tax reform was finally implemented in 1995, the main increases were in value-added taxes, which increased from 0.5 percent of GDP in 1990–94 to 3.9 percent of GDP in 1995–98 (Tanzi 2000, 22). Income and property taxes remained negligible and were among the lowest on the continent (Obregón and F. Rodríguez 2001, 25). The regressiveness of value-added taxes and the generally low tax burden on the rich further exacerbated the growing problem of income inequality and the polarization of politics. Regardless of the desirability of these reforms, the capability approach does not explain where the power and legitimacy of the state to implement reforms will come from to maintain the reform process.

The perception that corruption increased in the context of liberalization also refutes the predictions of the capability approach. The reasons for this increase are complex, though several factors were important. First, as mentioned, the insulated manner in which policy reforms were introduced ran contrary to the consultative processes that had characterized the political pacts upon which Venezuelan democracy had been built since 1958. Such insulation exacerbated factionalism within the governing party and between the government and opposition parties. This increase in factionalism increased the degree of

---

26. See table 9.3 for data on the evolution of Venezuelan inflation rates in historical and comparative perspective.

"whistle-blowing," as those left out of decision-making used the corruption scandal as a weapon of political contestation. Moreover, high levels of campaign financing by some of the prominent business groups for the Pérez presidency created animosities among rival contenders within Acción Democrática, and fueled allegations that Pérez supporters would benefit from reforms.[27] Second, the media increased and magnified the coverage of scandals, including the growing anticorruption discourse among politicians and rival economic groups, which increased public perception that corruption was increasing (Pérez Perdomo 1995). The fact that few scandals ever resulted in arrests or penalties further fueled public outrage. Third, the failure of the state to effectively regulate the banking system allowed banks owners and managers to engage in the illegal diversion of funds to offshore accounts and to illegally fund their related business interests in nonbanking ventures (De Krivoy 2002). Finally, the decline real wages in combination with growing inequality quite likely reduced the tolerance the majority of people had for corruption, and thus corruption scandals became politically more explosive and destabilizing.

Finally, the liberalization period coincided with the disintegration of the political party system's legitimacy and growing crisis of governability, which was manifested in growing social unrest, political violence, and a dramatic increase in voter abstentionism.[28] Apart from growing abstentionism, there are several other clear indicators of the decline in the legitimacy of the two dominant political parties since the consolidation of democracy in 1958. From 1958 to 1988 one of the two dominant parties (Acción Democrática or the Comité de Organización Política Electoral Independiente) won the presidency; since 1993, neither has. There were also two abortive military coups in 1992 and the impeachment of President Carlos Andres Pérez in 1994 on corruption charges.

In 1998, the landslide presidential victory of a popular, radical political outsider and former coup-plotter Hugo Chávez, along with the virtual disappearance of the once dominant political parties, manifested a political rupture in

---

27. As mentioned, Rafael Caldera emerged as the leading political opponent of neoliberal reforms in the early 1990s. His political revival was centered on an anticorruption platform.

28. Since the consolidation of democratic elections in 1958, voter abstention rates averaged 7.6 percent in the five presidential elections in the period 1958–83. From 1983, these rates increased dramatically: 18.1 percent in 1988, 39.8 percent in 1993, and 32.3 percent in 1998. At the regional and local levels, the average level of abstention in state and local elections since the institution of decentralized elections in 1979 also increased dramatically. The aggregate figures for state/local abstention rates are as follows: 1979 (27.1 percent), 1984 (40.7 percent), 1989 (55.0 percent), 1992 (52.8 percent), 1995 (53.9 percent). The data on all abstention rates from 1958 to 1995 are taken from Grindle 2000, 83. The figures for 1998 are taken from the Venezuelan National Electoral Commission (www. Elecciones.eud.com/absten.ntm).

a dramatic period of evolving crisis. Venezuelan democracy has, like many of its Andean neighbors experiencing economic crises, moved toward a stronger presidential system with a declining role for political parties, along with the rise of political outsiders (see sections 8.2.4 and 8.2.5).

5.4 The Venezuelan Liberalization Experience in Comparative Perspective

In comparative perspective, the breakdown and crisis of the Venezuelan state to govern the economy was among the most severe in Latin America in the 1990s. The capability approach, given its technical focus on governance, does not adequately address the extent to which a polity's "initial conditions" affect the political viability of radical reform proposals. While it is beyond the scope of this chapter to explore detailed comparisons, there are several factors that may have contributed to the more destabilizing effects of neoliberalism in Venezuela.

First, neoliberalism was associated with a worsening of income distribution. In the period 1970–90, Venezuela had among the *least* unequal distributions of income in Latin America: only Uruguay, Costa Rica, and Peru were less unequal (IADB 2000, 6). However, in the period 1990–97, the growth in income inequality in Venezuela was the fastest in the region (Székely and Hilgert 1999). In comparative terms, the *growth* in inequality was perhaps more destabilizing politically in Venezuela than in other reformers where the initiation of reforms began with among the highest levels of income inequality such as in Brazil, Chile, or Mexico. In the latter countries, there was little scope for income distribution to worsen further. This suggests that rapid *increases* in income inequality matter more for instability than initial *levels* of inequality (Brazil and Chile have had much higher levels of income inequality yet have proven much more stable politically).

Second, hiding reform intentions and then attempting policy switches can be particularly destabilizing in a long-standing democracy where economic programs were generally predictable and known. Given the comparatively inclusive and consultative tradition in Venezuelan politics, such a "policy switch" (Stokes 1999) may have proved more destabilizing than in many other Latin American reformers where either democratic politics was less salient (such as Chile, Mexico, and Argentina) or where previously more chaotic macroeconomic management (often resulting in hyperinflation), and policy switches were a common feature of recent pre-reform history (such as Argentina, Peru, Bolivia, and Brazil).

Third, the absence of *immediate threats* to the economy and polity also made a radical policy switch less justifiable and thus less legitimate to many interest groups. Venezuela was experiencing a crisis of long-run economic stagnation, but its effects were gradual. There are several factors to consider here. First, the economy was not experiencing hyperinflation, which damages the incomes of the poor and middle classes, and which requires immediate and draconian measures, as were implemented in Argentina and Peru. Second, there was no threat of a guerrilla insurgency as in Peru or Colombia. Such threats can give the executive greater legitimacy and leverage to act without legislative consent. Third, many people still believed Venezuela was a rich oil economy and did not need reforms (Naím 2001, 20).

Finally, the failure of the Pérez government to forge an effective alliance with all tiers of the military, especially in the context of its party-neglecting strategy, proved disastrous. Many of the middle-ranking military officers had been disgruntled with the control political parties had over promotions, a problem Pérez ignored (Aguero 1995). In comparative perspective, Fujimori in Peru, while also antagonistic to parties, was more successful in maintaining both political stability and the reform process, precisely because his government made an effective alliance with the military, and even increased its role in the government. Given the shifts in power balances and assets that liberalization can bring, it is perhaps no accident that the relatively successful reconstruction of public authority and state capacity in Chile and Mexico, two countries undertaking neoliberal reforms, was achieved in the context of *nondemocratic continuity* (Cavarozzi 1994, 138).

This discussion of how Venezuela's initial conditions interacted with the manner in which liberalization reforms were introduced improves on recent political analyses explaining the destabilization of the Venezuelan political system in the post-1998 period. Steven Ellner (2008, 89–108), for instance, argues that the depth of neoliberal reforms, the worsening income distribution that coincided with these reforms, and the lack of national debate over economic liberalization were responsible for the destabilization of the party system and the rise of radical left-wing leaders and a growing polarization of politics. While much of this analysis is plausible, it fails to consider how and why the introduction of neoliberal reforms generated *greater* political destabilization, political polarization, and antiparty politics (see section 8.2.5) and *less* impressive economic performance in Venezuela in comparison with many other Latin American economies that introduced similar or even more far-reaching economic reforms.

5.5 Conclusion

The logic of reducing the state's role in resource allocation is to reduce the discretionary control by government official over rent creation. In the mainstream rent-seeking logic, economic liberalization and greater democratization are supposed to make state action more accountable, transparent, and predictable. There are several problems with this view. First, there is no evidence that state abdication, by reducing some type of state-created rents, reduces rent-seeking and corruption. The regulatory structure of the state is still open to influencing under any type of economic regime. Second, democratic reforms, and in particular, decentralization, have not made the regulatory environment more secure. If anything, the institutional environment has become more uncertain with the decline in the legitimacy of political parties and government ministries in this period.

The Venezuelan experience also calls into question many other implicit assumptions with respect to the capability approach, and the rent-seeking theory upon which it is based. First, "building up" state capacities is marginalized in favor of "leveling down" the state's role in the economy. Second, it is suggested that the net benefits of living with market failures outweigh the net benefits of actual state interventions in most developing countries. Neither of the first two assumptions is valid for the Venezuelan experience. Downsizing the state has, far from improving state capacity, led to a collapse in the Venezuelan state regulatory capacity. There is little sensitivity in the capability approach to the historical factors behind the size of the state, and how a state's size may serve important functions of, for example, maintaining political stability and social cohesion. Third, there is little distinction made between low- and middle-income countries. This is an important lacuna since, for poor economies, resource constraints may limit effective state capacity whereas the relevant question for middle-income countries is *why* capacity is missing. Fourth, the capability approach endorses the idea that there are *prerequisite* capacity requirements to economic development. This approach is ahistorical. As Alexander Gerschenkron (1962) argued long ago, the very process of embarking on late industrialization occurred despite the lack of fully developed prerequisite capacities in late-developing economies. As such, late development involves the development of capacity and endowment *in the very course* of industrialization. In any case, there is little evidence that long-run growth paths were initiated in any late-developing country by liberalizing trade and reducing corruption (Rodrik 2004a).

Finally, the capacity of the state to effectively regulate market activity (i.e., competition policy, financial regulation) is judged to be unproblematic in a more liberalized market economy. One of the reasons for this may be that there is little discussion of the large-scale rents that neoliberal reforms create. The reality is that financial deregulation, large-scale privatizations, and private monopolies create large rents and thus rent-seeking/corruption opportunities. Moreover, the fact that much of these rents are *new* implies that the state will need *to immediately* develop regulatory capacities; otherwise, monopoly power can be exercised relatively easily. The inability of the Venezuelan state to effectively implement financial reforms and regulations is a case in point. "Getting institutions right" is not simply a technical exercise but requires an understanding of politics, and in particular, where and how the power and legitimacy of the state to enforce rules and defend property rights arises.

While declines in state capacity and economic decline have been a main factor behind the collapse of the old pacted democracy, it is important to incorporate the role specific political strategies played in the system's demise. This downfall was not an inevitable result of economic decline as some have argued (Penfold-Becerra 2001; Hausmann 2003). The inevitability of decline also prevails more generally in new institutional theories of path dependency (North 1990) that view degenerate development paths and stagnation as historically inevitable given the set of institutions a country inherits. Path dependency allows no room for politics or agency and would thus seem ill suited as a framework for understanding the contemporary trajectory of the Venezuelan polity.

The timing of the breakdown requires explanation and the contingent political strategies to implement reform, and their political and economic impacts need, at least, to be included in the multifaceted process that constitutes the collapse of the party system and the capacity of the state to govern the economy. The brief comparison with other Latin American economies suggests that policy switches and political party–neglecting strategies are potentially more destabilizing in long-standing democracies, where consultation is an inherited feature of the polity. The capability approach, given its technical focus, fails to provide an adequate framework for incorporating the political viability of reform strategies in concrete historical situations.

PART 3

AN ALTERNATIVE POLITICAL ECONOMY OF VENEZUELAN
GROWTH AND DECLINE

6

TOWARD A NEW POLITICAL ECONOMY OF
LATE INDUSTRIALIZATION

Explaining the "paradox of plenty" has dominated the economic and political economic historiography of Venezuela. The main thrust of the literature, and particularly the rentier state model, is the claim that oil abundance induces centralized public authority and excessive state interventionism and discretion, which in turn cause growth-restricting and productivity-restricting corruption and rent-seeking. None of the analyses on Venezuela, however, examine whether the state strategies and structures are *systematically different* from those of other late developers. The cross-country evidence and impressionistic accounts suggest that neither the structure of state intervention nor the scale of corruption in Venezuela have differed sufficiently from those of more successful late developers to explain the dramatic differences in long-run growth. In contrast to the mainstream "technical" view of state capacity (World Bank 1993, 1997), the "politics of privilege," and in particular, cronyism, appears to accompany industrial policy in both more *and* less successful conjunctures of centralized state intervention.

The comparative and historical evidence of Venezuela's economic growth performance raises serious concerns about the adequacy of the reigning explanations. First, oil abundance has been associated with both output and productivity growth and stagnation over long periods. Therefore, the existence of natural resource rents has not determined "incentives" in any systematic fashion. Second, a persistently centralized and clientelist state in Venezuela (whether run by *authoritarian* caudillos or *democratic* leaders) has coincided with very different levels of economic performance. Political regime type does not appear to adequately account for the variation and change in economic performance. Third, the period of neoliberalism (1989–98) failed to revive investment and growth in Venezuela (or in many other countries in Latin America), which suggests a need to improve on the rent-seeking paradigm of state failure. Explaining trends in economic performance, governance, and state capacity in Venezuela needs to be consistent with this basic comparative and historical evidence.

In contrast to the rentier state argument and neoliberal theories of rent-seeking, there are series of institutional arguments that point to the net *benefits* centralized rent deployment and management can provide. Developmental state theorists have pointed to the benefits that industrial policy can have in inducing learning and technology acquisition (Amsden 1989, 2001; Chang 1994; Aoki et al. 1997). This literature has provided an important contribution to the understanding of the salience of targeted subsidization, or rent creation, in driving the late industrialization process. As well, recent models of corruption (Shleifer and Vishny 1993) and the statist literature on the efficacy of state coordination (Amsden 1989, 2001; Evans 1995; Rodrik 1995; Aoki et al. 1997) have stressed the growth-*enhancing* incentives strong, centralized state intervention brings.[1]

In developmental state theories, the motivations and power of the state to undertake goals are assumed to be dominant within the polity. The strategies are chosen to achieve economic growth and development goals under circumstances dictated by existing managerial, labor, technological, and financial capacities. With industrial policy given priority, the capabilities of the state and firms becomes endogenous to industrial policy, rather than a constraint, not least because of the learning-by-doing effects involved in policymaking and implementation. The value of these analyses is that they provide a useful counterpoint to the opponents of industrial policy, who argue that countries that lack capability should not try industrial policies (the so-called capability approach; see, for example, World Bank 1993). The dynamic view of industrial policy views the creation of capacity as the central purpose of well-defined strategies and policies; and the formulation and implementation of policies can only be acquired through a learning-by-doing process guided by the exigencies

---

1. None of these authors, however, makes this claim for oil economies. Rentier state theorists would argue that the existence of rents is inherently greater in oil economies. This is because inflows accruing to the state are "free" (or "unearned"), and thus rents in control of the state are more readily available than in a mineral-resource–poor country such as South Korea where rents can only be created by policy interventions. The "free" nature of inflows may also create a "soft budget constraint" mentality on the part of state and private sector investors. While these are valid insights, I have pointed to the limitations of the rentier state argument in Chapter 4.

of industrial policy (Rodrik 2004b; Hausmann and Rodrik 2006). As Dani Rodrik argues:

> The conventional approach to industrial policy consists of enumerating technological and other externalities and then targeting policy interventions on these market failures. The discussion then revolves around administrative and fiscal feasibility of these policy interventions, their informational requirements, their political economy consequences, and so on. . . . The task of industrial policy is as much about eliciting information from the private sector on significant externalities and their remedies as it is about implementing appropriate policies. . . . Correspondingly, the analysis of industrial policy needs to focus not on the policy *outcomes*—which are inherently unknowable ex ante—but on getting the policy *process* right. We need to worry about how we design a setting in which private and public actors come together to solve problems in the productive sphere, each side learning about the opportunities and constraints faced by the other, and not about whether the right tool for industrial policy is, say, directed credit or R&D subsidies or whether it is the steel industry that ought to be promoted or the software industry. Hence the right way of thinking of industrial policy is as a discovery process— one where firms and the government learn about the underlying costs and opportunities and engage in strategic coordination. The traditional arguments against industrial policy lose much of their force when we view industrial policy in these terms. (2004b, 2–3; emphasis in the original)

In effect, such an approach underlines the symbiotic and reinforcing relationship between policymaking and capabilities.

While the potential benefits of state intervention may be accepted, the developmental state literature generally explains how a state intervenes to overcome the challenges of lateness, but pays little attention to the questions of where the power to implement policies and where the policy goals come from in the first place. As a result, the developmental state analysis does not explain why similar types of industrial policy fail in many contexts.

Comparative and historical evidence on state-led development strategies suggests that the same institution, such as centralized industrial policy, can produce different levels of rent-seeking, corruption, competitive pressure, growth, and so on. For instance, similar levels of rent-seeking *inputs,* whether through lobbying or corruption, can generate substantially different rent-seeking *outputs* in terms of the types of rights created, maintained, or changed (see section 4.3). This implies that the *outcomes* associated with the institutions of centralized rent deployment in Venezuela (or anywhere else) over time depend on other conditions. An examination of the *variation* in the effectiveness of Venezuela's centralized rent deployment and management, particularly its industrial policy, is thus interesting in terms of evaluating both the neoliberal and developmental state positions.

Another reason to seek a more defensible explanation of Venezuelan performance is that there has been widespread knowledge among Venezuelan decision-makers in the course of the twentieth century of deficiencies in policies and institutions at different junctures. This is documented in Chapter 4 (section 4.5), where it was shown that knowledge and assessment of the problems facing the Venezuelan industrialization drive and potential problems of corruption in state institutions were not absent. In fact, there is little evidence that the Venezuelan state decision-makers chose irrationally to develop unviable industries. The cognitive failure approach is inadequate in explaining why many attempted changes fail to be implemented or sustained.

The reigning interpretations of Venezuelan economic slowdown focus on isolating one policy and/or institutional failure of state-led centralized rent deployment. Such "single issue" approaches are inadequate. The main problem with them is that they do not consider that governance is an interdependent process of institutional design where such designs are simultaneously specifications of incentives and political power over property rights, policy agenda, and policy implementation. The mainstream notion of externality (the consequences of interdependence of action) should make it clear that institutions cannot be studied as independent variables but as interactions of processes (Papandreou 1994).

The most general interaction that institutions specify is the one between economic and political processes. At the most general level, the formal and informal property rights structures provide incentives and *simultaneously* specify asset distributions. For instance, establishing property rights around a fishery provides the *incentive* to manage the fish stock efficiently. However, in a world of scarce resources (that is, where not every individual or village has access to a fishery), establishing fishing rights *simultaneously* imparts a distributional or positional advantage that favors those able to become the owners of fisheries (on positional goods, see Hirsch 1976 and Pagano 1999).

It is not enough simply to *establish* property rights. Such rights need to be *enforced*, otherwise the problem of free entry into the fishery would result. The establishment and enforcement of property rights (in a world of scarce assets) is a formidable political challenge, since only a small part of the population can become privileged asset owners. In order for such an unequal (but potentially efficient) distribution to work, a property rights structure needs to be legitimated through the political process. When property rights structures are contested, the costs of enforcement can become prohibitive or even lead to violent political conflict.

It is thus not possible to separate issues of incentives and efficiency from issues of distribution and equity. As a result, conflict and conflict settlements, and thus the political process, are an intrinsic part of the process of building effective institutions. Therefore, incentive structures (whether destructive or productive) necessarily constitute historically specific settlements of conflict or compromise over the distribution of rights and resources. The study of institutions is a problem in political economy, and it is therefore necessary to endogenize processes of political contestation and strategy, conflict resolution, and legitimacy into the wider processes of capacity formation and change.

Because governance is an interdependent process, identifying one aspect of the resource allocation problem (such as resource abundance, corruption, exchange rate management, degree of centralization, or public ownership) in isolation is likely to be part of the reason why the prevailing theories explored in Chapters 3 and 4 are not compatible with the long-run evidence. If governance were *not* an integrative or interactive process, then it might be possible to *add* the insights of different explanations. This, however, is theoretically incompatible with the most general notion of externality. This suggests that a more promising strategy to reinterpret the Venezuelan growth experience would be to attempt to build a more *integrative* framework for explaining the variations and changes in governance and economic performance.

Given the *contingent* nature of political institutions and contests, there is no reason to expect that the appropriate institutional structure and politics will emerge to accommodate a country's stage of development and changing technological challenges. If different strategies require different levels of selectivity and concentration of economic and political power to be initiated and consolidated, then it becomes necessary to identify what political changes are required for effective implementation of such new development strategies. The mapping of political strategies and types of development trajectory followed opens up the possibility of explaining growth accelerations as the result of the *compatibility* between development strategy and political settlements and, analogously, growth slowdowns as the result of a growing *incompatibility*, or *mismatch* between political settlements and development strategies. The latter possibility is similar to the Marxian problem of the relations of production acting as "fetters" on the development of productive forces in a given conjuncture.[2]

The purpose of this chapter is to develop an alternative framework that will allow for an interpretation of Venezuelan economic history that is more compatible with comparative and historical evidence on economic output and productivity growth in the period 1920–2005 than reigning economic and political economy arguments. I suggest an alternative approach that conceives of institutional performance as a historically specific interactive process between political and economic development strategies. The chapter addresses the need to focus on bringing production (particularly the challenges of late development under different types of development strategies) and politics (in particular the shifting grounds of legitimate rule) into analyses of policy, institutions, and governance. None of the reigning explanations take into account the role that politics plays as a determinant of state capacity or the extent to which stage of development and different development strategies affect the nature of political conflicts. Such an integrative framework will also allow for a more adequate account of why well-intentioned policymakers in Venezuela fail to change inefficient regulatory and institutional arrangements even when they have knowledge about the problem.

Section 6.1 examines the nature of property rights and economic rents, and suggests that understanding institutions requires an incorporation of political analysis beyond what mainstream analysis considers. Section 6.2 further develops why political analysis is also central in explaining the outcomes of state-created rents. Section 6.3, the core analytical section, presents a theoretical framework that demonstrates why the "easy" stage of import substitution is likely to present different (and simpler) sets of political and economic challenges than the more advanced stage of import substitution, including big-push heavy industrialization strategies. Section 6.4 presents the main hypotheses of the argument that emanate from the discussion in the previous section. If the framework and evidence presented in this chapter are reasonable, then the argument here will lead to an examination of whether the political capacity of the Venezuelan state changed in ways needed to undertake more difficult state interventions.

---

2. For a discussion of Marx's theory of history, see Elster 1985, 235–317.

6.1 The Political Economy of Property Rights and State-Created Rents

At the most general level, the need to examine the interaction of economics and politics is most clearly evident when considering the nature of property rights and institutions. Institutions are the set of rules that structure the interaction of agents subject to these rules (North 1990; Knight 1992), among the most important being property rights. A property right specifies the power to consume, generate income from, and alienate a given asset (Barzel 1989, 2). The way in which property rights are defined and enforced fundamentally affects economic performance. There are two reasons for this. First, by assigning ownership rights to valuable assets and by designating who bears the benefits and costs of resource-use, property rights structure the incentives for economic behavior.[3] Second, by allocating decision-making authority, property rights arrangements determine who the actors are, and, in turn, the distribution of positions and relations of power and resource control in the economy.[4]

Property rights are *political* since the *state* is the set of institutions responsible for specifying such rights. The neoclassical and new institutional approaches to rights formation, however, do not adequately engage with issues of politics, power, or legitimacy. This shortcoming is tied up with the mainstream's reliance on methodological individualism: institutions are seen mainly as specifying incentives facing individuals and consequently are analyzed through optimization exercises based on notions of individual rationality. As such, social interaction is modeled as a process of bargaining between individuals that cannot capture the determinant role of *organizational* power. Because of its reliance on methodological individualism, new institutional economics (NIE) views institutions, including property rights, as specifying only "incentive" structures. NIE ignores (or at best downplays) the way institutions embody power relations.

The emphasis on individual optimization as the mechanism through which rights are created is central to general neoclassical and new institutional models. In both frameworks, resource allocation (and the governance structures

---

3. For an early sophisticated discussion of the benefits of secure property rights, see Bentham [1789] 1982.

4. Rousseau [1755] (1984) and Marx [1867] (1976) provide radical critiques of the institution of private property rights. In contrast to Bentham and the modern economic treatment of property rights (e.g., Coase 1960), which emphasize the role property rights play in establishing incentives for efficient resource allocation, Marx and Rousseau stressed the distributional advantages or *privileges* that owners of property enjoy. Moreover, they emphasized that the origin of private property rights was based on theft and exaction and thus such rights were morally indefensible.

underlying it) is modeled under the underlying assumption that it is useful to *separate* problems of economics (such as incentive formation and optimization) from problems of politics (such as the contingent and shifting grounds of what constitutes a legitimate distribution of assets). John Stuart Mill (1848) famously argued that productive processes can and should be insulated from attempts to accomplish distributional objectives (by manipulating production). According to Mill, government spending, rather than manipulations of the production process, can accomplish distributional objectives. More recently, Jack Hirshleifer (1994) has posited that the two main and *separate* strands of the resource allocation problem in economics are, on the one hand, the production of goods and, on the other hand, the conflicts over who appropriates wealth.

If institutions were simply a set of incentive structures, then it would be possible for an appropriate institutional *design* to improve the growth prospects of a failing economy. Indeed, the idea that market economies depend crucially on effective state institutions and state capability to create growth-enhancing "rules of the game," or "incentives," underlies the current interest in "good governance" (Stiglitz 1994; World Bank 1993, 1997a).[5] In this technical view, "getting institutions right" is tantamount to "getting incentives right."

The idea that rights specify only incentives is, however, incomplete. This is because property rights also, by definition, specify *a historically specific distribution of control and authority over assets* (Dahlman 1980, 213–14). Without an analysis of the nature of the polity, it is impossible to know, a priori, the legitimacy or enforcement costs of a given structure of property rights (Field 1981). Very high enforcement costs imply political resistance to a particular institution and rights, and will likely result in worsening the security of property rights associated with that particular institution. This, in turn, will likely lower the growth rate of the economy.[6]

Enforcement is central to effective governance because property rights are only effective in providing incentives to manage resources efficiently when asset owners can count on the *power to exclude* others from appropriating or attenuating their rights (Umbeck 1981). In developmental states, such power to enforce property rights is normally provided by the state. In this perspective,

---

5. One of the main insights of new institutional economics, which is the main theoretical foundation for discussions of "good governance" in agencies like the World Bank, is the crucial role of institutions in ensuring the level and efficiency of investments. Prominent works in the field include Coase 1960 and North 1990.

6. However, see Haber et al. 2003 for an analysis of how the selective (as opposed to universal) enforcement of property rights of business groups in Mexico in the period 1876–1929 was consistent with rapid economic growth even under situations of substantial political instability.

"a consistent application of the axioms of rational self-interested action makes imperative an analysis of power" (Papandreou 1994, 216). The dual nature of rights and institutions is consistent with the definition of governance proposed by Oliver Williamson (2000): "Governance is the means by which *order* is accomplished in a relation where potential *conflict* threatens to undo or upset opportunities to realize *mutual* gains" (95).[7]

If one is concerned with the capacity of the state to enforce or change property rights and implement development strategies, then the new institutional approach to this problem has important limitations. The standard "market" or Coasian approach to bargaining and exchange is to model an exchange between individuals over rights, given a prior structure of entitlements and rules.[8] However, if the problem concerns developmental state capacity, this usually involves a *change* in the entitlement/rule structure that the state sanctions. The process of changing ineffective development strategies involves a bargaining process *for* institutional change where there is *no prior structure* defining the bargaining process (Papandreou 1994, 222). Outcomes in the "market" *for* institutional change are sets of conflicts that are resolved by the relative power and coercive capacity of the competing factions with historically specific sets of demands and organizational power. This suggests that state capacity cannot be divorced from political mobilization and organization or from the balance of forces within society. It also suggests that whatever collective benefits may arise from increased state capacity do not emerge entirely by design but are often unintended by-products of conflict.

Since property rights become legitimate (and therefore enforceable at a reasonable cost) through historically specific political processes, there is no definitive efficiency implication underlying "an institution." It is also the reason behind the classic development problem of transferability of institutions across contexts. Without an analysis of politics, the outcomes of centralized rent deployment cannot be known, a priori, in Venezuela or elsewhere. In this perspective, a theory of governance requires a theory of power and legitimacy that new institutional economics and dominant approaches informing development

---

7. John R. Commons, whose work greatly influenced Williamson, recognized that "the ultimate unit of activity must ... contain in itself the three principles of conflict, mutuality, and order.... This unit is a transaction" (Commons 1932, 4).

8. Even in this case, the emphasis on individual bargaining is misleading. As Adelman (2000) notes; "Property matters not so much that it should not be reduced to the language of individual efforts to maximize returns but rather should be viewed as concerns around which actors mobilize *collectively* to create rules that favor them or disfavor others" (47, emphasis added). In this perspective, rights flow from relations and not vice-versa.

policy on governance do not currently have (Hirshleifer 1994; Khan 1995).[9] The distinction made by Mill and Hirshleifer between production and distribution is thus untenable in a world of competition over resources (that is, a world where there are opportunity costs to resource allocation).

The political nature of rights and rent creation takes on particular importance in poor economies where a transition to capitalism is taking place.[10] The central role of the state in the process of capital accumulation means that the *politics of privilege* are important features of primitive accumulation and late development (Hutchcroft 2000). The challenges of catching up with more advanced industrial economies have meant that latecomer states have had to "mobilize bias" in strategically selective ways. In particular, the state in latecomers has historically aided and accelerated the creation of a capitalist class, particularly in industry, through purposeful construction.[11]

This process of primitive accumulation means that the state is inherently involved in profoundly divisive decisions.[12] Competitiveness in many industries depends on the state being able to develop firms of sufficient size to achieve economies of scale in order to compete with more advanced competitors. The empirical record points to concentration of ownership and the predominance of a privileged few conglomerates controlling relatively large firms as the leading business form in successful late developers (Chandler et al. 1997). In the context of scarce resources, selectivity is inherently conflictual, divisive, and therefore contestable politically. This is because there are many more groups that may legitimately demand an opportunity to receive state subsidization than a late-developing state can afford to patronize without sacrificing efficiency or fiscal instability. This is why the capitalist transition in late-developing states is neither inevitable nor universally popular.

The central role of the state in processes of primitive accumulation and late development also implies that the *legitimacy* of all state policies is potentially

---

9. While there are significant problems of measuring "power," similar difficulties in measurement for terms like "transaction costs" and "asymmetric information" have not hindered their introduction and usefulness in economics (see Papandreou 1994, 209–16, for a discussion concerning the concept of "power" in the economic analysis of institutions). The similarly serious problems of measuring "class power" or "interest group pressure" has not prevented them from being usefully being incorporated into the study of transitions from feudalism to capitalism (Brenner 1976) or the comparative historical processes of democratization (Moore 1966).

10. This paragraph draws on Khan 2002b.

11. Such mechanisms include protective tariffs, credit subsidies, tax concessions, land grants, exchange and interest rate controls, and the manipulation of relative prices between agriculture and industry (see Amsden 2001 on discussion of the various mechanisms of subsidization in latecomers).

12. For a discussion of Marx's idea of primitive accumulation, see Cramer 2006, 199–219.

highly contestable. This is because there are not only large distributional consequences of state patronage and subsidization patterns in poor economies but also because establishing the legitimacy of an inequality of asset ownership takes *time*. The newness of the asset-creation process is a distinguishing feature of the early stage of capitalism, or primitive accumulation. Moreover, the underdeveloped nature of capital markets in late developers (Gerschenkron 1962; Amsden 2001) means that there is less pluralism in the mechanisms of financing accumulation. As a result, the state becomes a central focus of financing investment, which makes capital accumulation a more politicized process in late developers. If the right decisions are made, a class of producers will emerge in leading sectors that will greatly contribute to the general prosperity of a late developer. But regardless of their productivity, those who are the beneficiaries of state support generally become far wealthier than the average citizen.

It is worth noting that the distributional struggles for access to state-created resources takes place in several spheres. First, there is the struggle between privileged asset owners and laborers. Second, there is the struggle between different factions of capital (e.g., agricultural, industrial, merchant, and financial). Third, there is struggle between *diversified* business groups. This last type of struggle is particularly important because the predominant form of business organization in many late developers is the family-run conglomerate that dominates in industry, finance, agriculture, and finance simultaneously (Amsden 2001; Haggard et al. 1997). In Venezuela, the dominance of diversified family-owned conglomerate sectors in the economy (and therefore in the struggle to acquire state-created rents) is well documented (Rangel 1972; Duno 1975; Naím 1988; Naím and Francés 1995).

The blurring of the distinction between the public and private sector and the accompanying arbitrariness of governance in the context of underdevelopment was examined by Max Weber in his analysis of the early stage of capitalism, which he called "political capitalism" (1978, 166).[13] Weber alerted us to the possibility that arbitrariness and unpredictability may accompany the politics of privilege, thus blocking growth-enhancing exchange between agents, thus blocking economic development. Marx ([1867] 1976, 873–930) analyzed the politicized and coercive nature of primitive accumulation, which can generate intense distributional struggles. Antonio Gramsci (1971, 210–38) also examined the problems late-developing states may have in legitimating the rights of a small and nascent capitalist class. More recently, Douglass North and colleagues (2007)

---

13. Swedberg (1998, 45–53) discusses Weber's analysis of the different ways profits are appropriated under political capitalism.

analyze how economic development can proceed despite the very limited distribution of state patronage to privileged and powerful individuals and interest groups, a situation they see as typical of less developed economies.[14]

Nevertheless, the political economy of "primitive accumulation" "political capitalism," and "limited access orders," and the corruption and rent-seeking that accompany these processes has been *variable* with respect to developmental *outcomes*. As such, one important challenge in the political economy of late developers is to explain why certain economies maintain rapid growth *despite* the existence of very politicized, and corrupt state-led rent deployment and accumulation processes. This is very different from arguing, as rent-seeking and rentier state theorists do, that the prospects for rapid growth will be enhanced when state-created rents, rent-seeking, and corruption are kept to a minimum.

6.2 The Politics of Centralized Rent Deployment and Management in Late Developers

A developmental state may be defined as the set of institutions responsible for the creation, maintenance, and restructuring of property rights and rents that generates growth-enhancing activities among politically and economically privileged rent recipients. The extent to which centralized rent deployment induces productive investment will depend on at least three factors.[15] Let us consider each in turn.

The first factor is the *motives* of the state officials who control the deployment of rents. The key question is whether these officials deploy the rents and subsidies to promote economic goals or to establish or maintain networks of political clientelism, patronage, and support. The latter motive is much less likely to result in a productivity-enhancing outcome.

The problem of where leaders' motives come from is based on studying politics, which is largely about the struggles over what constitutes legitimate rule. The motives of leaders are contingent upon the political support base that embodies a leader's power. This support is generally a combination of the ideas and strategies that are used by would-be leaders to persuade party members and other relevant interest groups to render support, and may also consist of campaign financing by capitalist groups and business associations. Thus, ideological and material support forms important *bonds* for leaders.

14. These authors characterize less developed countries as "limited access orders."
15. See Hutchcroft 2000, 225–26, for a similar framework.

A leader's motives are also affected by the bargains made with rival political factions and parties to maintain stability. Because leaders are *agents of coalitions,* rational action is bonded to the political strategies and support base that embody the power of the leaders in a specific bargaining conjuncture. In this perspective, historically situated or what I call *bonded rationality* (as opposed to substantive, procedural, and bounded rationality) as a conception of action provides a more realistic view of rationality. Given the historically specific nature of bargains over institutions and rights underlying rent deployment, examining the bonded rationality of state leaders is necessary for systematically improving our understanding of their motives and thus, in turn, processes of institutional formation and change. Neither the neoliberal nor the developmental state literature makes politics central to an understanding of the variation and change in state capacity in general and rent deployment and management in particular.

The second factor is the extent to which the state has the capacity to *monitor* and impose *performance or productivity criteria* (such as export targets) on the rent recipients. The most severe costs of corruption and more general rent-seeking are not necessarily the costs of bribes but the perpetual protection of inefficient capitalists by the state. If the state can ensure that emerging capitalists are efficient, it does not matter significantly for economic dynamism if bureaucrats and politicians make some illegal money in the process. The real peril is when inefficient capitalists succeed in bribing and influencing the state to capture resources (Khan 2002a, 166). When this happens, the cost to the economy is much greater than the resources wasted in lobbying and corruption. The costs of subsidies permanently captured by infant industries that refuse or fail to grow up are likely to far outweigh the economic costs of bribery and corruption (166). A related capacity includes the ability of the state to be *selective* in allowing a firm entry into a sector, a factor that is decisive to performance criteria in cases where scale economies are relevant. The ability of business associations to impose performance criteria on their own members can also enhance the productivity of rent deployment. For this to operate successfully, business associations have to be effective in imposing collective decisions and in promoting encompassing goals. Overall, the ability of state decision-makers to be selective and enforce performance criteria depends on the distribution of power and authority between central state institutions and rent recipients in both the private sector and the public enterprise sector.

The third key factor is the extent to which the institutional environment is predictable or *secure.* A stable macroeconomic environment is essential to long-run planning and investment. With respect to the security of property rights of asset owners, the issue of security is more complicated to assess. The dominant view among neoclassical and new institutional economists (North 1990; Olson 1993; Acemoglu et al. 2001) and international development institutions (e.g., World Bank 1997a, 2002) is that the general security of property rights is a necessary condition for economic growth.[16] However, highlighting the extent to which the state enforces all property rights is misleading as an indicator of the regulatory effectiveness of the developmental state. What matters for growth, in practice, is not simply the security of property rights per se but the security of the property rights of those actors undertaking *growth-enhancing and productivity-enhancing* investments (Khan 2002a). This is because it matters *whose* property rights are secure and for *what purposes* assets are utilized. Very secure but fragmented control of the means of violence by local mafias is not likely to increase the likelihood of political stability or the maintenance of a predictable rule of law. If asset owners are secure in the pursuit of accumulating capital in low-value-added sectors or unproductive activities, then the likelihood of economic stagnation increases. The ongoing and secure protection of an infant industry that fails to grow up is equally likely to result in dynamic inefficiencies. As a result, developmental states in latecomers have been characterized by effective *ex-post flexibility* (Okuno-Fujiwara 1997) to both discipline rent recipients and adapt policies to changing economic conditions. Also relevant is the extent to which there is a degree of continuity in the direction of policy. Rapid changes in state prioritization of goals can lead to the proliferation of rights and rents, which may overextend the ability of the state to feasibly monitor the rent deployment process. Such proliferation often has its roots in political decisions and strategies, as I shall discuss in the Chapters 7 and 8.

6.3 Bringing Production Back In: Why Are Some Industrial Strategies More Difficult to Implement than Others?

Neither the developmental state nor rent-seeking or rentier state models take into account how stages of development or development strategies may affect the nature of political conflicts. If, however, the nature of the technology that

---

16. Haber et al. (2003), however, argue that what matters for growth is not necessarily the universal protection of property rights, but rather the selective protection by the state of the property rights of important political supporters.

accompanies a development strategy required different levels of selectivity, or concentration of economic and political power, to be initiated and consolidated, then the possibility arises that the historically specific nature of political settlements may *not* be compatible with the successful implementation of a given development strategy. This implies that variations and changes in state capacity may be a function of the *type* of economic challenges facing the state *and* the extent to which state-led strategies generate resistance and legitimacy problems from powerful actors excluded from the distributional benefits of those strategies. The idea that democracy, for example, may be incompatible with late industrialization has been developed by several authors, including Barrington Moore (1966), Karl de Schweinitz (1964), Guillermo O'Donnell (1973), and Dietrich Rueschemeyer and colleagues (1992). Hirschman (1971) and O'Donnell (1973) consider the extent to which different types of technologies or different stages of import substitution can generate different types of conflicts within the industrialization process itself, though their insights have been lost in current debates on the political economy of state intervention and governance.

Economic analyses of Venezuela have identified changes in stages of development and development strategies. Economic analyses focusing on the processes of rentier capitalism (Hausmann 1981; Baptista 1995) and the rentier state (Karl 1997) have pointed to the *change in development strategy* from a smaller-scale, import-substituting industrialization toward a larger-scale, capital-intensive industrial strategy that began tentatively in the late 1950s but became a central focus of policy after 1973 (see Chapter 7 for details). In this process, the state's role as a producer increased significantly. The value of these analyses is that there is an identification of a periodization of the ISI process as a process that is not homogenous in its technological or structural characteristics. However, these analyses have not explored what a policy-induced switch in the technological nature of development strategy might mean in terms of the political economy challenges facing the state.

In this section, I present an alternative framework for improving our understanding of the variation and change in state capacity and growth in middle-income economies (such as Venezuela) undergoing a transition from small-scale, technologically simple production strategies to large-scale, technologically more complex industrial strategies. The general thrust of reigning economic analysis is to identify one aspect of state capacity to manage resource allocation problems without taking into account the extent to which the challenges facing the state were changing. As Hirschman (1995) suggests, however, the nature and intensity of conflicts may differ depending on how *divisible* the benefits of a given institutional arrangement are.[17] Following in the spirit of this insight, the core framework developed here is that the political challenges facing a late-developing country implementing the easy stage of ISI differ from those facing a late-developing country implementing a more advanced stage of ISI, such as big-push, natural-resource-based heavy industrialization. This, I posit, is because the technological characteristics of the more advanced stage requires more selective and thus more indivisible subsidization and more coordination of investment than the earlier stages of ISI.[18]

Section 6.3.1 considers the political and economic challenges of early ISI strategies. Section 6.3.2 examines the political and economic challenges of more advanced ISI strategies, in particular, big-push, natural-resource-based industrialization strategies.

6.3.1 Economic and Political Challenges in the Early Stages of Import Substitution

The early stages of import substitution generally present far fewer challenges to late-developing economies and states than more ambitious big-push heavy industrialization strategies. The challenges are simpler both economically and politically. There are several reasons for this.

First, early ISI *technologies* differ in terms of the political and institutional problems decision-makers face and therefore in the effective responses required from the state. Let us consider the case of small-scale, low-value-added industrialization strategies that normally accompany the early stages of industrialization or what is referred to as the "easy" stage of import-substituting industrialization (see Hirschman 1971, Cárdenas et al. 2000, and Fitzgerald 2000 on the differences between different stages of the import-substitution process).[19] The easy, or initial, stage "is the substitution of mass non-durable consumer

---

17. Divisibility refers to the extent to which the conflict over a right is a question of "more or less" (such as in the capital-labor struggle) as opposed to "all-or nothing" (as in establishing the ownership of a factory in the first place).

18. Moreover, sudden and premature state-led changes in strategy to more complicated and scale-intensive technologies, induced, for example, by oil windfalls, can exacerbate these challenges. As Hirschman argues: "For the fiscal linkage to be an effective development mechanism, the ability to tax must be combined with the ability to invest productively" (1981a, 68–69).

19. Chenery et al. (1986) distinguish between early and late industrialization. The latter classification is concerned with the idea that there is a first "easy" phase of import substitution, during which mostly nondurable consumer goods are produced, and the later phases, in which import substitution "deepens," producing durable consumer, intermediate, and capital goods (see also Furtado 1970 and Balassa 1980 on the distinction between easy and advanced stages of ISI).

goods and basic inputs behind tariff protection, with continued supply of consumer durables, complex units and capital goods from abroad relying on an assured domestic market, ample credit and investment and familiar technology" (Fitzgerald 2000, 64).

In the easy phase of ISI, investment projects are normally substitutes, and the technology is small-scale. Coordination of investment decisions by the state is not central to rapid growth in early ISI strategies. In fact, centralization or coordination of agencies may hinder a growth-enhancing outcome. If projects are substitutes, then it may be preferable to have decentralized and competitive rivalry. This is because, in the context of uncertainty over market opportunities and technology, differentiated expectations and experimentations reduce the possibility of making a large-scale, coordinated mistake later (Khan 2000b, 135).

The successful growth of Thailand with accompanying patterns of competitive clientelism and fragmented corruption is a case in point (Doner and Ramsey 1997, 2000). The dynamism of Thai economic development has been explained by the compatibility of a small-scale technology strategy and fragmented and competitive patron-client networks (Khan 2000b):

> A key factor which ensured that this rent-seeking system was efficient was the nature of the technology which Thai capitalists were adopting in the 1960s and 1970s. In sectors such as textiles, logging, food processing, Thailand was adopting technologies which had limited economies of scale and which did not require long periods of learning. In particular, they did not require learning rents for successful adoption. Rents in this context were only useful in accelerating primitive accumulation, and, as soon as that happened, the competitive bidding down of rents was socially desirable. This would not have been the case if useful rents had had to be maintained and managed, for instance, if learning had been required or if sophisticated monitoring and regulatory mechanisms had to be developed. In the long run, as Thailand moves into higher-technology sectors, institutional reformers will have to face these issues. (134)[20]

As it turned out, the relative strength of big capitalists in the Thai polity enhanced the prospects that clientelist faction-fighting resulted in competition over efficiency in production as opposed to mere redistribution of resources as

---

20. See also Khan 2000b, 101–4.

occurs in political patron-led clientelism (Doner and Ramsey 1997, 2000; Khan and Jomo 2000, 15).[21]

In the context of a resource-rich economy, small-scale industrialization, particularly at the early stages, may also be less conflictual than large-scale, big-push industrialization strategies. This is due to several factors. First, smallness of scale implies that rights and rents are more *divisible*. The greater divisibility of subsidies and rights generally means that a greater portion of powerful political agents can be viably accommodated as rent recipients in the form of firm ownership and employment opportunities within industry. This accommodation would include traditional artisan industries. Greater divisibility of rights also means that *selectivity* is of less concern, which implies fewer groups need to be excluded from state-created and sanctioned rights and rents. Finally, the early stages of ISI do not immediately require a disciplining role of the state, given the substitution opportunities to meet a relatively *unsaturated* internal market. For instance, the imposition of export targets as a condition for renewed protection or subsidization is a less salient issue at the early stages of ISI.

Because early ISI industries tend to be small-scale and technologically simple, there are several other complex institutional responses that are *not* central to continued capital accumulation. First, the employment of public enterprises is often not required in manufacturing. This is because early ISI industries do not require large fixed costs and long gestation periods to become competitive. As we will see in the next section, effectively managing public enterprises is a formidable challenge. Second, the relative simplicity of the technology produces far fewer demands on the state and business organizations to spend on research and development (R&D). When the internal market becomes more saturated, such R&D expenditures will need to increase if domestic firms are to compete with world-class companies abroad. Finally, the degree of physical infrastructure is generally less in early ISI industries. This is, in part, because the energy requirements of production tend to be lower. Heavy industry requires a greater intensity of energy inputs and thus produces more fiscal demands on the state.

The less conflictual nature of early industrialization is, of course, premised on the formation of a society where the state has a reasonable degree of legitimacy and control, which clearly has been the case in Venezuela since a series of

---

21. The ability of the Thai state to impose a "hard budget constraint" on the development ministries was an important condition for the maintenance of macroeconomic stability and making the competitive clientelism value-enhancing (Doner and Ramsey 1997, 151–55). Evidence that clientelism was capitalist-led is provided by Sidel (1996), who shows that there were many more capitalists as members of parliament than in other Asian economies.

Andean caudillos consolidated power beginning with Cipriani Castro in 1899 (Rangel 1980).[22] As will be discussed in Chapters 7–9, the dynamic growth of the Venezuelan manufacturing sector between 1920 and 1958, under relatively centralized rent deployment, was due in large part to its small-scale nature and the low levels of industrialization from which growth was proceeding. From 1958 to 1968, the increasing factionalism of the state in industrial policies and implementation did negatively affect the effectiveness of industrial policies and investments but was nevertheless still compatible with rapid growth. This was because the technology remained small-scale (if concentrated) and mutually substitutable, and there was still substantial scope for substitution effect in consumer goods to be realized.

6.3.2 Economic and Political Challenges of More Advanced Import-Substitution and Big-Push Industrialization Strategies

Sustained growth in Venezuela, as in many latecomers, depends on the successful development of intermediate technology sectors, which are generally large-scale and capital-intensive and often require long gestation periods, complementary investments, and effective coordination between the state and business conglomerates (Amsden 1989, 2001; Chang 1994; Aoki et al. 1997; Chandler et al. 1997). In mineral-rich exporters, successful industrialization is, in large part, dependent on successful mineral-based big-push industrialization. Minerals represent both a comparative and competitive advantage for mineral-rich economies (Wright and Czelusta 2007; Blomström and Kokko 2007).

The more advanced stage of ISI, which includes big-push, natural-resource-based industrialization strategies, involves the development of consumer durables, intermediate goods, and capital goods, all of which generally exhibit increasing returns to scale, and which require more sophisticated technological capability. This stage also "would face a more demanding market formed by primary producers, existing consumer goods branches and elite consumers requiring heavy technological investment, extensive government support and quality competition with foreign suppliers" (Fitzgerald 2000, 64).[23] Finally, the

---

22. The early stages of import-substitution are by no means conflict free as the failure of many poor economies, both resource-rich and resource-poor, to achieve or sustain low-technology, labor-intensive industrial growth demonstrates. In this perspective, the net benefits of state consolidation and centralization in twentieth-century Venezuela were notable. See section 8.1 for details.

23. One advantage of "lateness" is that identifying industries with dynamic productivity and output growth prospects is not a major obstacle. Development banks in nearly all late developers tended to target credit to the following sectors: basic metals (iron and steel), chemicals (primarily

promotion and nurturing of large firms is crucial to competitiveness and learning in late developers, given the centrality of scale economies in intermediate technology sectors (Amsden 2001, 194–99).

The later or more advanced stage of ISI also involves more *risk* in acquiring technology and capturing new markets, including export markets. Thus state-led initiatives to socialize risk need to be undertaken by targeting firms and sectors to persuade new industrialists to invest in or upgrade to new technology and improve productivity. The Fourth and Fifth Plans of the Nation (1974–82) encapsulated the intention to transform the industrial structure of Venezuela through a big-push industrialization strategy (see Chapter 7). The emphasis the Venezuelan government placed on chemicals and basic metals as leading sectors were indeed similar to the industries targeted in all successful late developers (Amsden 2001, 138–39). As we shall see, the Achilles' heel of Venezuelan industrialization strategy was a growing inefficiency in the implementation of second-stage ISI and big-push, mineral-resource-based industrialization strategies.[24]

The greater learning costs and gestation periods of such second-stage ISI and big-push investments bring greater economic and political challenges and risks that distinguish this type of economic strategy from the small-scale and substitutable technology of easy ISI stage. Again, there remains considerable debate over whether Venezuela moved too quickly toward the development of heavy industrial sectors. While the move to the more complicated stage was not obviously necessary or desirable in 1974, the challenges of executing such a strategy certainly increased vis-à-vis the easier stage of ISI.

---

petrochemicals), machinery (electrical and nonelectrical), transport equipment (ships, automobiles, and automobile parts), and textiles (Amsden 2001, 138–39).

24. It is worth noting that the second stage of ISI in Latin America evolved in distinct ways from the East Asia NICs. After World War II, Latin American industrial policy generally focused much more on import-substituting industrialization (ISI), which was characterized by "inward-looking" development, whereas the East Asian economies emphasized the switch from ISI to export promotion more rapidly than their Latin American counterparts. Fishlow (1991) suggests that the greater export pessimism in Latin America owed to the greater effects the Great Depression had in Latin America in arousing skepticism with regard to the potential of international trade as a vent for manufacturing products. The second stage of ISI in many countries in postwar Latin America, in effect, took place in closed-economy contexts. The second stage of ISI was thus less risky in Latin America as capitalists in the region were never forced to confront the investment risks of their East Asian counterparts. The lower national savings rates and more onerous external debt servicing throughout most of Latin America in the 1980s and 1990s also prevented sustained big-push industrialization in comparison with the East Asian NICs. The lower level of risk-taking was consistent with the persistence of less productive Latin American oligopolists/monopolists in many sectors. In the Venezuelan case, the big push into natural-resource-based industries did, however, emphasize the importance of export growth, and oil resources allowed for investment rates that were similar to East Asian rates in the 1960s and 1970s.

Table 6.1 summarizes the main differences between the two types of development strategies under consideration. The institutional challenges of the big-push strategies and the more advanced stage of ISI center on the need for the state to maintain continuity of centralized investment coordination, effective monitoring of public enterprises, selectivity in subsidizing investment through rent creation, discipline of rent recipients, and collective action capacities of business associations. I will consider each of these factors.

First, the theoretical justification for centralized investment planning is well known. The big-push or "balanced-growth" models (Rosenstein-Rodan 1943; Nurkse 1953; Skitovsky 1954) stress the demand complementarity between different industries, which require prior investment coordination by the state.[25] In the context of underdeveloped capital markets and a small internal market demand, spontaneous coordination of decentralized market actors is unlikely, since an investor in a sector characterized by increasing returns to scale may not judge the investment profitable on the assumption that demand will not increase significantly elsewhere in the economy. For instance, the demand for steel may not be sufficient to induce investment in that sector without the simultaneous development of railways or an auto industry, both of which use steel as a major input into production. This will be the case even if the investor has access to sufficient numbers of skilled workers. The reason is that, because of scale economies, only a large-scale and simultaneous movement of resources is guaranteed to be profitable. As Rodrik (1995) notes: "From the perspective of an individual investor, it will not pay to invest in the modern sector unless others are doing so as well. The profitability of the modern sector depends on the simultaneous presence of the specialized inputs; but the profitability of producing these inputs in turn depends on the presence of demand from a pre-existing modern sector" (quoted in Maier and Rauch 2000, 196). The creation of demand sufficient to make large-scale investments in poor countries profitable is the crux of the coordination problem. And it is one of the main reasons why state-led investment coordination is central to the effective implementation of big-push industrialization strategies.

Second, the coordination problem in large-scale, late manufacturing development is indeed one of the main factors behind the greater use of public enterprises in the advanced stage of ISI. Public enterprises, for example, were

---

25. For a recent exposition of the "big push" argument, see Murphy et al. 1989. See Rodrik 1995 for an analysis of how specific state policies such as control over credit allocation, tax incentives, trade policy, and "administrative guidance" helped to solve coordination failures in South Korea and Taiwan in the 1960s and 1970s.

TABLE 6.1 Main differences between early import-substitution strategies and "big push" heavy industrialization/more advanced import-substitution strategies

	Early ISI	"Big push"/ advanced ISI
Technology	simpler	more complex
Learning costs	lower	higher
Scale economies	generally lower	generally higher
Investment requirements at firm level	lower	higher
Gestation period required to make investments financially and economically profitable	shorter	longer
Investment requirements in physical infrastructure	lower	higher
Skill requirements of labor force	generally lower	generally higher
Interdependence of firm investment across sectors	generally lower	generally higher
Industrial policy	general subsidization of infant industries	targeted subsidization of specific firms
Importance of selectivity in picking "winners"	less important	more important
Prevalence of public enterprises in manufacturing production	less prevalent	more prevalent
Disciplining and monitoring firm performance	less important	more important
Coordination of investment	not generally required (may even be counterproductive)	important
Cooperation within and between business associations and conglomerates	generally less important	generally more important
Public and private research and development investment requirements	lower	higher
Economic costs of subsidization errors	lower	higher
Political challenges in implementing policies and correcting policy failures	lower	higher

particularly important in the heavy industrial sectors during the critical take-off years of the 1960s in both Taiwan and South Korea (Rodrik 1995). Leroy Jones and Il Sakong (1980) examined the development of public enterprises in South Korea in the 1960s and 1970s and found that the Korean policymakers had developed a coherent set of preferences with respect to what public enterprises should be established. They summarize their results as follows: "The industries chosen for the public enterprise sector [were] characterized by high forward linkages, high capital intensity, large size, output-market concentration, and production of non-tradables or import-substitutes rather than exports" (quoted in Rodrik 1995, 90). These are precisely the characteristics associated with the coordination problem that emerges during the more advanced stage of ISI.

The greater presence of public enterprises does, however, increase the administrative and political challenges that the central government faces in coordinating and monitoring public sector investment. The greater the level of public sector investment implies that the failure of the central government to impose hard budget constraints or some type of performance contract on public sector managers will generate loss-making public enterprises. This, in turn, can generate fiscal drains and reduce the productivity of investments over time in the advanced stage of ISI or during big-push heavy industrialization.

The macroeconomic challenges of big-push, natural-resource-based industrial strategies are thus generally greater than those of earlier ISI strategies. This is because big-push industrialization drives generate a greater "investment hunger" than the early stages of ISI. The common use of public enterprises in large-scale, natural-resource-based manufacturing often contributes a major part of this investment hunger in late developers. The dangers of overexpansion in a big-push strategy, for example, are typically the buildup of unsustainable and destabilizing external debt positions that can occur if large-scale investments do not lead to improved productivity, greater competitiveness, and higher export earnings.[26]

26. As Amsden (2001) notes: "That conditions of 'lateness' are inherently conducive to overexpansion is suggested by the fact that when a debt crisis occurs, it almost always occurs in a latecomer country.... This is because ... diversification in the presence of already well-established global industries involves moving from labor-intensive to capital-intensive sectors characterized by economies of scale" (252). The scale of big-push investments were, indeed, instrumental in the debt crises in Latin America in 1982 and in East Asia in 1997, as both were preceded by a surge in investment (253). Of course, the duration of a crisis will depend on the capacity of the state to change their institutions in a way that revives investment, growth, competitiveness, and exports. In this respect, East Asian developmental states were more successful than their Latin America counterparts (252).

Third, the exigencies of big-push industrialization strategies also present significant political and economic challenges to state-business relations. The advanced stage of ISI, and particularly big-push investment programs, entail complementary investments that have significant spillover effects in pecuniary and production terms (e.g., steel and electricity). Given the larger size of such complementary investments in the advanced stage of ISI, the failure of the state and business associations to coordinate investments may result in larger costs than in the earlier stage of ISI, where the scale of investments are generally lower.

The costs of failure to coordinate investments between the state and business associations can lead to low levels of capacity utilization, which has large negative effects on investment returns. Even if investment coordination is already achieved, the failure of the state or business associations to effectively monitor investments in a single large-scale manufacturing sector can be costly. For instance, if the state fails to impose conditions on rents or subsidies or permits excessive entry in a particular sector (for example, there have been fifteen car assemblers in Venezuela since the mid-1970s), the outcome is more costly for industries exhibiting increasing returns to scale than it is for smaller manufacturing operations. Moreover, excessive entry may lead to the development of firms with too small an average plant size to compete with imports or in export markets, especially in industries characterized by increasing returns to scale. Where proprietary knowledge and product development and hence research and development (R&D) is important to competitive success, large numbers of small, inefficient producers constrains the capacity of latecomer industries to compete. This is because small firms do not generate the turnover to direct sufficient resources to R&D (Kim and Ma 1997, 122–23; Amsden 2001, 277–81).

The case of high-value-added, small-scale technology is an important subset of the more advanced stage of ISI, but is distinct from the industries subject to large fixed costs in plant development such as steel and shipbuilding. An example of such sectors would be computers, auto parts, or special medical equipment. This type of technology also generally requires state coordination, since capitalists may be producing a small part of a sophisticated final product. For example, computers are often built in many stages by many different small producers. In these industries learning costs are significant, but capitalists are usually able to absorb them without state financial support. However, the viability of such industries depends on the ability of the state to develop national technology sectors, which lowers the cost of research, and technology-acquisition costs, as has occurred in the computer industry in Taiwan (Wade 1990, 179; Fields 1997) or in the electronics industry in Malaysia (Rasiah 1996). Thus, the role of the state and firms in coordinating the process of technology acquisition is decisive to takeoff in these sectors, whereas once the industry is set up, competition can ensure that productivity growth is maintained.

The greater risks entailed in big-push or advanced-stage industrialization strategies tend also to necessitate a greater degree of cooperation and collective action *among* business associations and conglomerates. Effective collective action can help socialize the risks of developing export markets, acquiring technology, managing firm entry and exit, and negotiating with the state to provide services that are in the collective interests of a given sector (Wade 1990; Evans 1995; Maxfield and Schneider 1997; Haggard et al. 1997; Doner and Ramsey 1997; Fields 1997). Therefore, *fragmentation* of business associations can also lead to more costly coordination failures as the scale and scope of investments increase. This is because fragmentation can lead to *particularistic* demands and bargaining that hinders the competitiveness of sector as a whole when, for example, there is a level of bargaining with the state for firm entry that leads to the creation of too many firms at substandard scale.

Fourth, for large-scale, resource-based and other heavy industrialization strategies, selectivity in the subsidization process is essential for the effective design and implementation of industrial policy (Lindbeck 1981). This is especially the case where increasing returns to scale are relevant. As such, the failure of the state to be selective will negatively affect productivity performance to a much greater extent in the advanced as opposed to the easy stage of ISI. Because the "prize" of receiving a license or subsidy to develop a large plant is rather more valuable than a similar license or subsidy to develop a small plant, the political challenge of selectively in deploying such licenses and subsidies tends to be higher in the more advanced stage of ISI (compared with the initial stages of ISI, where plant sizes tend to be smaller). If state decision-makers deploy licenses and subsidies to more firms than the protected domestic market can accommodate, there is a danger that promoted firms will develop at either suboptimal size and/or at high levels of spare capacity. While immediately engaging in export activities can permit the promotion of more firms than the domestic market can accommodate, it is generally the case that firms in technologically demanding sectors require a substantial period of "learning-by-doing" by selling to a protected domestic market. In any case, the potential for large-scale subsidization to degenerate into "cronyist" collusion is also a real danger even when state officials have been selective.

Fifth, selectivity needs to be complemented with state capacity to discipline rent recipients. The scale of the subsidization implies that big-push strategies require state decision-makers to make rents conditional on performance criteria (Kim and Ma 1997). The failure of the state to condition subsidy support in performance criteria can generate costly "white elephants" in the large-scale industrial sector. A similar failure to discipline small-scale infant industries may produce "baby white elephants," but the costs of such failure are likely to be much less. Hence there has been significant emphasis on state discipline of producers as one key to the efficiency of industrial policy (Amsden 1989).

It is difficult to know beforehand whether the producer chosen will develop the capability to compete and reduce costs. Once chosen, however, the recipient of a subsidy will have significant incentives to lobby for the maintenance of those (substantial) rents regardless of performance. This is because the rights generating those rents are, by definition, activity and relation-specific, and thus irreplaceable at least in the medium run.[27] Since there are few alternatives to the state in underwriting infant industry protection, given the weakness of capital markets in late developers, state-firm relations are decisive to the maintenance of profits. When a relation or activity is decisive, the intensity of struggle and conflict over maintaining valuable assets and the rights associated with them is likely to increase (Hirshleifer 1994, 7).

The complementary nature of big-push strategies also means that decision-making and investment are more interdependent than they are in early ISI strategies. The greater the interdependence of investments, the greater the need there is for an institutional design that ensures reliability and discipline (Demsetz 1997, 30–34). This generally means that a legitimate and powerful centralized authority is required. A common example of the importance of the reliability and discipline that centralized authority provides is a military unit during war. In this perspective, it is not surprising that states that have faced internal and external threats have used such exigencies to increase their bargaining power relative to business groups to mobilize resources, to convince or force industrial producers to forsake short-run profits, and to make subsidies conditional on performance criteria. For instance, the willingness of business groups to

27. The importance of asset-specificity is that physical and human capital cannot be reallocated to another task or sector without a significant loss in the value of such assets (O. Williamson 1985). The generation of profits (and rents) in the economy is, in part, based on the development of "specific" or specialized assets where the value of the assets depends on a *particular* relationship (Klein et al. 1978; Milgrom and Roberts 1990, 62–63). See Hausmann and Rodrik 2006 for a discussion on the extent to which industrial investments depend on specific inputs.

cooperate with state initiatives occurs because perceived and real threat raises the price of failure (Vartianen 1999, 224).[28]

In light of the exigencies of large-scale subsidization and interdependence, it is not surprising that big-push development strategies have enjoyed success only in a very few late-developing countries such as Japan, Taiwan, South Korea, Singapore, Brazil, Israel, and Malaysia.[29] Moreover, these late developers have been characterized by relatively legitimate, secure *centralized* states in terms of both political and economic organization (Woo-Cumings 1999). Relatively centralized business associations and state–big business links tend to accompany more complex and risky industrial strategies (Maxfield and Schneider 1997). This implies that fragmented or highly factionalized political contestation may, on the one hand, lead to significant coordination failures macroeconomically and in industrial policy, but on the other, may be compatible with growth in the case of small-scale, labor-intensive manufacturing or in the "easy phase" of import-substitution, provided the central state can provide basic macro and institutional stability.

In sum, I have suggested that stage of development and the type of development strategy undertaken do matter because the political and economic challenges are likely to differ depending on technological and other market challenges facing an economy. Neither the neoliberal nor the developmental state theorists systematically consider the different political challenges that stage of ISI can make. Moreover, they do not generally examine the interaction of economic and politics in any more general manner.

---

28. When the very survival of a nation becomes a legitimate ground for state action and reaction, then nationalism (that is, an increased sense of what Anderson [1993] calls an "imagined community") or more precisely, defending the nation, can become a defining parameter of state legitimacy and power (Putzel 1995). In this respect, it may not be a coincidence that some of the most successful industrial policy states (Japan, South Korea, Taiwan, Singapore, Finland, and Austria) were facing significant external threats to the survival of their respective nation-states. In postwar France, another successful industrial policy state, national reconstruction was organized around meeting the "American challenge." Katzenstein (1985) argued that external threat was central to the construction of effective corporatism: "Since the 1930s the small European states have experienced in economic openness and international vulnerability at least a partial substitute for the 'moral equivalent of war.'... cooperative politics has been one consequence" (208). On the historical relationship between the threat of war and state-building (and in particular the development of tax collection capacity), see Schumpeter [1918] 1958 and Tilly 1990. For an examination of why the potential stimulus of war did *not* transform Latin American states in the nineteenth century, see Centeno 1997 and López Alves 2001. One exception was the role the War of the Pacific in 1879 played in consolidating the Chilean state (Palma 2000).

29. It is perhaps no coincidence that many of these polities were subject to substantial external threat.

6.4 Growth as Compatibility of Development Strategy and Political Settlement: Main Propositions

The failure of dominant paradigms on state intervention to consider the changing nature of political conflict that different stages of ISI and different development strategies generate implies that such models will fail to *map* the extent to which contingent political settlements are compatible with different economic challenges of catching up over time. The main propositions concerning the importance of the compatibility of development strategy and political settlements for explaining the variation and change in economic growth rates are summarized in table 6.2

First, consider the case of polities where state formation is weak and unconsolidated. Here, the state does not have legitimate monopoly over the means of violence and is unable to effectively defend property rights. These types of polities often exhibit extreme forms of patrimonial and predatory rule where the separation between the public and private realms is essentially nonexistent. Corruption tends to be endemic and unpredictable in such states. Apart from the lack of formal institutions, these polities are often characterized by high levels of regional and ethnic conflict and contestation and are particularly prone to large-scale political violence, including civil war. Such states are often among the poorest in revenue collection and per capita income. The prospects for growth under any type of development strategy are remote in weak and unconsolidated states. Examples would include Afghanistan and Somalia. This scenario is not relevant for examining the Venezuelan experience in the twentieth century, though the Venezuelan case approximated this case in much of the nineteenth century (Gilmore 1964; Brito Figueroa 1966).

TABLE 6.2 Interactions of polities and development strategies and their effects on economic growth in late developers

Type of polity	Type of development strategy	
	Early ISI	"Big push"/advanced ISI
Unconsolidated, weak state	Low or negative growth	Unlikely to be attempted
Consolidated state with fragmented political organizations	High growth likely	High growth possible
Consolidated state with fragmented political organizations	High growth possible	Low or negative growth likely

The second category of polity is very similar to what has been characterized as a developmental state. This type of state (also relatively scarce among less developed countries) is what I categorize as a consolidated state with centralized political organizations. Such states resemble what Atul Kohli (2004) calls cohesive-capitalist states.[30] In this case, the state has a legitimate control over the means of violence. Public institutions are centralized and well institutionalized, with the executive generally holding significant power and authority. While corruption and clientelism may be prevalent, the distribution of patronage is generally centralized. These polities are often characterized by well-disciplined political organizations, including political parties. In such polities, the power of the state to mobilize resources for development, coordinate investment, and effectively monitor industrial policy is generally high. Political competition varies in such states, some being more authoritarian than others. The best-known of such states are South Korea, Taiwan, Singapore, Malaysia, Israel, and China. However, there are other examples, including Chile, and as I will argue in Chapter 8, Venezuela over the period 1920–65.

With respect to the issue of compatibility between development strategies and political settlements and organization, four main propositions emerge. The first two concern the growth prospects of economies characterized by consolidated states with centralized political organizations:

P1: Such polities are likely to promote economic growth through the early stage of ISI;

P2: Such polities are likely to be able to meet the political and economic challenges of big-push/advanced ISI development strategies.

With respect to the first proposition, consolidated states with centralized political organizations are most likely to be able to oversee rapid growth since complex industrial strategies are not required for infant industries to grow,

---

30. Cohesive capitalist states are characterized by centralized and purposive authority structures that often penetrate deep into society; the policy dominance of economic growth to ensure national security; tight control over labor; and often authoritarian and repressive politics (Kohli 2004, 10). It is important to point out that capitalist classes in such polities are not likely to have identical interests. It is plausible to assume that different factions of capital (financial, merchant, agricultural, and industrial) will likely lobby for different state policies and will compete for state resources. The relevant point is that the competition for state resources takes place in the context of a centralized state apparatus where it is difficult for particular capitalist factions to capture sections of state policymaking. Any such capture must be sanctioned by the executive. As a result, competing capitalists' lobbying is less likely to balkanize the state and less likely to result in a proliferation of fiscally and economically unviable subsidization.

particularly from low levels of industrialization. Moreover, such polities are likely to be able to maintain macroeconomic stability, and reasonably protect property rights, both of which are needed if easy ISI industries are to flourish. With respect to the second proposition, much of the developmental state literature (Evans 1995; Amsden 1989; Wade 1990) suggests that such polities have the bureaucratic capacity and political power to coordinate investments and successfully monitor the subsidization process, both of which are necessary if the more difficult advanced ISI and big-push industrialization strategies are to succeed.

It should be noted that a polity with centralized political organizations may be a *necessary* condition for the implementation of big-push industrialization strategies, but it is certainly not a *sufficient* condition. One notable example is the experience of the Soviet Union, which certainly was characterized by centralized and authoritarian political organizations. While big-push heavy industrialization was successful in promoting rapid output growth for many decades, the informational constraints embedded in central planning, the suspension of the use of the market mechanism as a source of information to decentralized agents with limited knowledge, and the absence of any type of private property rights limited the extent to which the Soviet economy was able to modernize its manufacturing sector after the mid-1970s.[31]

In this perspective, polities with centralized political organizations are most likely to successfully implement big-push industrial strategies, and industrial policy more generally (as in France, Japan, South Korea, and Taiwan), when they coincide with the selective use of the market mechanism (including integration into export markets), the protection of the private property rights of successful firms, the capacity of business groups to absorb technology, and institutional mechanisms for public and private cooperation.[32] It should also be noted that, as economies become more advanced, specialized, and sophisticated in technological and financial terms, the need for state action guiding industrial policy lessens. This is because advanced economies contain large, multinational corporations, which are themselves technologically and financially capable of undertaking industrial policy in the sectors where they dominate production. One indication of this capacity is that the sales of the largest multinationals are greater than the GDP of many low-income countries.[33]

31. See Hayek 1935a, 1935b, and 1945, and von Mises 1936 on the limits of central planning.
32. For a discussion of these processes in Japan and France respectively, see Johnson 1982 and Hall 1986.
33. See Nolan et al. 2007 for a discussion of the extent to which there has been industrial consolidation and concentration of business power at the global level, and how firms with powerful,

The third category of polity (perhaps the most common among less developed countries) is what I categorize as a consolidated state with fragmented political organizations.[34] This categorization is similar to what Kohli (2004, 10–11) calls fragmented multiclass states.[35] In such polities, the state may not be under severe contestation most of the time, but the contestation among groups and classes for access to state resources and other property rights is sufficiently high to force a more disperse patronage structure than in developmental states. Contending groups and classes capture the state to a greater degree than in developmental states and such capture often leads to a fragmentation of state institutions. Well-known political processes such as clientelism, patrimonialism, and populism contribute to the development of state fragmentation as well as the fragmentation of business associations and labor unions. Moreover, the fluid and intense contestation for access to state resources often generates a fragmentation of political organizations, including political parties. I will argue in Chapter 8 that political contestations in Venezuela led to a progressive fragmentation of state institutions, business associations, and political parties over the period 1968–2005.

The third and fourth propositions concern the growth prospects of economies characterized by consolidated states with fragmented political organizations:

P3: Such polities may be capable of generating relatively rapid growth when attempting to implement early-stage ISI strategies.

---

globally recognized technologies or brands constitute the "systems integrators" at the apex of global supply chains.

34. It is important to remember that these categories of polities are not fixed, nor are they the product of purely technical organizational designs. State and other political structures are largely by-products of class and other group contestation and the policy responses of leaders to challenges and threats to the economic and political viability of the polity in question (Moore 1966). On the one hand, dysfunctional state policies and class and other group struggles can undermine a state's legitimacy and fragment or render less effective centralized, well-disciplined political organizations. Equally, prudent political leadership and the class and group struggles can transform a dysfunctional weak state with fragmented political organizations into a more effective state with more encompassing political organizations. It is only possible to understand how and whether these transformations in political settlements happen by examining the historical political economy of a polity. See Kohli 2004 for an excellent examination of the comparative historical political economy of state effectiveness and industrialization in South Korea, Brazil, India, and Nigeria.

35. Fragmented, multiclass states are characterized by more fragmented public authority; coalitions of state support that rely on a broader class alliance, including middle- and working-class groups; an inability to narrowly define policy goals around growth and industrial strategy because of the high level of competing demands from noncapitalist but mobilized groups; and the politicization of policy formulation and implementation because of either intra-elite conflicts or because state authority does not penetrate deep enough into the society to incorporate or control lower- and middle-class groups (Kohli 2004, 11).

This is because the political and economic challenges of this strategy are manageable (as discussed in section 6.3.1) as long as the state is capable of maintaining a relatively stable macroeconomic environment. Thailand, for example, has a fragmented polity, but achieved relatively rapid growth because its development strategy was based on light manufacturing that did not require significant levels of investment coordination. As such, its polity and development strategy were broadly compatible.[36]

The relatively steady long-run growth experience in Colombia is another case that fits into this category. First, the state has had a well-known capacity to maintain macroeconomic stability, including relatively low inflation and a competitive exchange rate and was the one large Latin American economy that did not have a debt crisis in the 1980s because of prudent borrowing levels.[37] As well, the Colombian economy maintained one of the fastest rates of GDP and manufacturing growth, and manufacturing productivity growth, in Latin America over the period 1970–96 (Katz 2000, 1592). Second, the polity is characterized by the fragmented and clientelist nature of political organizations, including political parties. In Colombia, the party system has been based on the country's tradition of decentralized regional elites, which has resulted in the internal fragmentation and clientelism of the party system and particularly the Liberal party (Archer and Shugart 1997). The most salient feature of the party system is that the proportional-representation seat-allocation procedure is applied to each district to factional lists (representing alliances of regional elites), not party lists. Each list stands alone for the purpose of allocating seats; the party is not the main focal point in the allocation of seats (ibid.). Third, the Colombian development strategy for most of the twentieth century was based on a relatively decentralized, small-scale production structure (Ocampo and Tovar 2000). The leading sector, coffee, was produced by nearly two million smallholders. The production, marketing, and distribution of coffee were coordinated by the National Coffee Federation, which developed into one of the developing world's most successful examples of collective action (Thorp 2000).[38] The industrial development strategy also focused on regionally

---

36. However, see Amsden 2005, 223–30, on the extent of industrial policies in Thailand.

37. The maintenance of sustainable macroeconomic policy over the long run is unusual in these types of polities, since clientelist and populist pressures often generate debt-led financing to accommodate competing demands.

38. Financed by a small surcharge on members' coffee sales, the Federation has been instrumental in organizing improvements in quality, pest-control, technical advice, credit, and international marketing and branding for nearly 400,000 smallholder producers. The Coffee Fund under its auspices guaranteed internal minimum prices for coffee, acting as a buyer of last resort, which

dispersed, small-scale, labor-intensive sectors with the highest relative shares in foodstuffs, tobacco, and textiles and apparel over the period 1925–74 (Ocampo and Tovar 2000, 251). This strategy was maintained through the mid-1990s.[39] As in Thailand, the imposition of fiscal discipline at the macro level ensured that open-ended subsidies to infant industries were not forthcoming; that is, macroeconomic discipline created a relatively "hard budget constraint" on infant industry subsidy recipients. This limited the extent to which failed infant industry initiatives were prolonged.[40] Because its long-run development strategy has been based on light manufacturing, there was little requirement for significant levels of investment coordination. As such, the fragmented and competitive nature of the clientelist party system was broadly *compatible* with its development strategy.

It is, however, important to note that the maintenance of sustainable macroeconomic policy over the long run is unusual in these types of polities since clientelist and populist pressures often generate debt-led financing to accommodate competing demands (Fishlow 1991). While there have been many growth accelerations in these types of polities at low income levels, few of these polities are capable of maintaining such growth because of their inability to maintain sustainable macroeconomic policies. The emergence of debt and inflation crises across many developing countries is one of the main manifestations of the fragility of these types of polities.

The final proposition is as follows:

P4: A consolidated state with fragmented political organizations is much less likely to successfully manage the more difficult economic and political challenges of big-push/advanced ISI development strategies.

Such strategies require the state to have the bureaucratic capacity and political power to coordinate investments and successfully monitor the subsidization

---

reduced the risks of production. The political influence of the Federation was also influential as a powerful lobby historically to demand prudent macroeconomic management, particularly the avoidance of exchange-rate overvaluation. This note draws extensively on Thorp 2000.

39. The relative shares of high-technology and heavy industrial sectors were lower in Colombia compared to most of the other large Latin American economies over the period 1970–96 (Katz 2000, table 6).

40. The faster growth in Thailand relative to Colombia is due to at least two factors: first, the Thai economy has had a much higher national savings rate, which financed higher investment; and second, political stability was generally higher as Thailand has not experienced a prolonged guerrilla insurgency that has weakened the security of property rights in Colombia, particularly in some non-urban areas of the country.

process; however, because of the need to accommodate voracious and numerous contending collective actors, and because such groups fragment state institutions, such states are unable to exclude contending rent-seeking groups from state patronage, discipline groups already receiving state patronage, or coordinate investments and subsidies, especially when that would require reducing political patronage to relatively powerful business conglomerates and labor unions.[41] This incompatibility between the development strategy and the fragmented nature of the polity results in poor economic performance.[42] The following chapters argue that this incompatibility is precisely what evolved in Venezuela over the period 1968–2005.

Poor economic performance in the more advanced stage of ISI and during big-push industrialization strategies will be manifested in low output growth and/or low labor and total factor productivity growth. The fact that Venezuela maintained high manufacturing growth rates in the 1970s means that fragmenting political organizations do not eliminate the prospect of capital accumulation in large-scale industries for a while. This was facilitated in Venezuela by high levels of public investment in manufacturing and by subsidies and protectionism to the public and private sector infant industries, both of which were financed by oil boom revenues. However, the sustainability of such growth depends on the extent to which there is enough productivity growth to keep these industries viable. The very poor productivity growth levels of most heavy industrial industries was apparent from the early 1970s and thus compromised the sustainability of manufacturing growth, which indeed collapsed in the period 1980–2003.[43]

---

41. In Latin America, one of the main ways subsidization *has* been withdrawn is through wholesale economic liberalization. While trade liberalization has imposed discipline on domestic firms, it has also eliminated the possibility of providing incentives to these firms in overcoming the risks inherent in late industrialization.

42. While Kohli (2004) provides an excellent examination of the comparative historical political economy of state effectiveness and industrialization in South Korea, Brazil, India, and Nigeria, he does not attempt to analyze the extent to which differences in development strategies within the industrialization process matter for the types of political settlements that might make these strategies compatible with politics. As such, he cannot explain why states that have relatively fragmented and clientelist political structures such as Thailand have achieved relatively rapid rates of industrial growth, or why countries attempting big-push heavy industrialization strategies such as Malaysia have been successful despite the absence of a cohesive capitalist class (see Chapter 10 for a discussion of Malaysia).

43. The period 2004–7 has also been compatible with rapid growth because of the renewed boom in oil revenues, though there is no evidence that this boom has been associated with an improved competitiveness of the manufacturing sector. In this later period there has been no diversification of the export structure of the economy.

Recent work on the types of economies that have experienced growth accelerations in the past fifty years suggest that the propositions made are reasonable. Ricardo Hausmann, Lant Pritchett, and Dani Rodrik (2004) examine growth accelerations in the period 1957–92. A "growth acceleration" is an increase in a country's per capita growth rate by at least two percentage points or more (with most of the observed increases exceeding this threshold by a wide margin). A particular episode is considered a "growth acceleration" when the growth increase is sustained for at least eight years and the post-acceleration growth rate is at least 3.5 percent per year. In addition, to rule out cases of pure recovery (that is, rebounds after prolonged crisis), it is required that the post-acceleration output exceed the pre-episode peak of income. Following these criteria, the authors identified a surprising number (eighty-three) of growth accelerations in the period 1957–92 and determined that they were concentrated in the 1960s and 1970s in sub-Saharan Africa and that their frequency has declined steadily ever since.

The most striking fact about these accelerations is that most of them take place in countries at low or lower-middle levels of per capita income. The high number of growth accelerations in sub-Saharan Africa (a low-income region where industrial strategies were mainly focused on small-scale, light manufacturing) also indicates that the existence of a strong, centralized state is *not* a requirement for growth during the early stages of ISI. There were *few* cases of growth accelerations in upper-middle-income countries: South Korea, Malaysia, and Thailand in the 1980s and 1990s; Chile, Uruguay, and Argentina in the 1980s and 1990s. Within Latin America, Chile is the only country that has sustained its annual per capita growth rate above 2 percent for a further ten years after the initial eight-year growth acceleration period, and the Chilean industrial strategy in the period 1980–2000 focused more on the development of medium-sized agribusinesses such as wine and food processing than it did in heavy industrial sectors.

Most of the other countries in the sample (thirty-seven) that sustained growth rates above 2 percent for a further ten years were either low-income countries (e.g., China, India, and Egypt) or high-income OECD countries (e.g., Finland, Ireland, Spain, and Israel). The only middle-income countries (apart from Chile, which did not undertake advanced ISI/big-push strategies) to sustain high levels of growth after initial growth accelerations were in East Asia (South Korea, Malaysia). This is consistent with the proposition that the earlier stages of industrialization present less economic and political challenges than strategies to upgrade into larger-scale and higher-technology sectors.

6.5 Conclusion

The reigning explanations for the economic slowdown in Venezuela—whether from a neoliberal or more statist perspective—have proved inadequate, primarily because they neglect the distributional and therefore political nature of policies, institutions, and property rights. They also fail to distinguish between the economic and political challenges of the early stage of ISI, compared with the economic and political challenges of more advanced ISI and big-push industrialization strategies. As we shall see, Venezuela adopted more advanced ISI strategies in the post-1985 period. Explanations that fail to speak to the political economy of this more challenging stage of late development fail to speak adequately to the Venezuelan case.

It is, in fact, not enough to assert that the establishment of property rights and rents is inherently political. To understand late development one must direct one's attention to the *specific* nature of political conflicts of rights and rent creation under different technological conditions and challenges. The reigning explanations do not consider the extent to which the political settlements and boundaries of legitimate rule in Venezuela may or may not have transformed in a way that made the development and political strategies feasible or compatible. I posit that growth cannot be adequately understood without the application of an analytical narrative that maps the extent to which development strategies and political settlement are compatible.

7

PERIODIZATION OF INDUSTRIALIZATION STAGES AND STRATEGIES IN VENEZUELA

In order to assess the propositions made concerning compatibility and incompatibility, it is necessary to first trace the evolution of development strategies and political settlements. The next step in this historical political economy involves identifying periods when compatibilities and incompatibilities may exist. Having established these periods of compatibility and incompatibility, we can then examine the extent to which the evidence on economic growth and productivity growth is consistent with the propositions made.

In this brief chapter I attempt to periodize the changing nature of the development strategies and the changing nature of scale economies and technological complexity in the Venezuelan manufacturing sector from 1920 to 2005. It is not my intention to focus on the details of economic policymaking since there is already a substantial literature on macroeconomic and industrial policymaking during this period (see, for example, Hausmann 1995, Astorga 2000, M. Rodríguez 2002, F. Rodríguez 2003, and Weisbrot and Sandoval 2007). Instead, my aim is to provide evidence that important changes in the technological characteristics of the development strategy were in fact operative from the mid-1960s. Before I undertake this task, it is important to highlight several important caveats. First, I will not aim to pinpoint the precise year of the switch in development strategy and stage of ISI. Rather, I will identify significant *shifts* in the evolution and patterns of both strategy and stage of ISI over the period 1960–1973. Second, I do not argue that the "small-scale, easier" stage of ISI does not coincide with the development of some large-scale, technologically complex industries, or that the "large-scale, big push" stage does not contain important elements of small-scale niche producers. Rather, the idea is to identify important differences in the *dominant* pattern of technological and scale characteristics and import-substitution possibilities that reign in a given historical conjuncture.

The question of whether a development stage or strategy is exogenous to other factors is not the primary focus of the analysis. As with corruption, it is reasonable to assume that a development strategy is not something that simply happens irrespective of the demands and influences of powerful interest groups. In fact, I document below some of the contingent political factors, internal and external, that have influenced the adoption of different strategies and the extent to which these, in turn, have either accelerated or delayed changes in the development strategy or the stage of ISI under consideration. For instance, the role that political parties, urbanization, and the growth of labor unions played in the adoption of more interventionist ISI after 1958 is well documented in the Venezuelan case (see Hausmann 1981). Rather, it is simply my intention to identify that in fact there has been a marked change in the pattern and nature of ISI in Venezuela over the course of the twentieth century.

7.1 Indications of Changes in Stages of Import-Substituting Industrialization, 1920–2005

It generally agreed that there was an important increase in the capital intensity and scale of industrial output and composition in Venezuela beginning in the early 1960s (Hausmann 1981; Baptista 1995; Astorga 2000). The development of the state steel mill, SIDOR, in 1962 and the state-led attempt to promote the metal-transforming and transport sectors, particularly the automobile and auto parts sectors from the early 1960s onwards were notable examples. The capital intensity and scale of many leading industrial sectors increased further with the Fifth Plan of the Nation. Thus, for our purposes, it is possible to develop an approximate periodization of the "small-scale, easy phase of ISI" from the period 1910–20 until the period 1960–73.[1] After 1973 development switched to a natural-resource-based big-push strategy, and ISI entered a more advanced phase. What follows is some additional empirical evidence to support this claim.

---

1. See Karlsson 1975 and Naím and Francés 1995 for evidence on the beginning of small-scale and artisan factories at the beginning of the twentieth century.

The first indication that the easy stage of ISI in Venezuela was dominant from the 1920s to the period 1960–73 can be seen by comparing the share of manufacturing in GDP with other medium-sized Latin American economies. In a late industrializing economy, a lower share of manufacturing in GDP would indicate that the import-substitution process is still in its "easier," small-scale phase. As indicated in table 7.1, the share of manufacturing in GDP in Venezuela was the lowest in the sample.[2] Since economic historians have classified the period 1940–60 as the easier stage of ISI in Argentina, Colombia, and Chile (Cárdenas et al. 2000, 10), and since the share of manufacturing in GDP in Venezuela was below all of these economies, it is reasonable to assume that Venezuela was pursuing early ISI strategies during the period 1940–70.

TABLE 7.1 Share of manufacturing in GDP: Venezuela in Latin American perspective (1970 prices) (%)

	Venezuela	Chile	Colombia	Argentina	Peru
1940	7.8	19.7	9.1	22.6	n.a.
1950	6.3	23.3	13.1	23.8	14.1
1960	11.3	25.5	16.2	26.7	16.9
1970	13.7	28.0	17.5	30.6	20.7

SOURCE: Thorp 1998, table 6.1.

It is also possible to identify transitions in the stage of development and development strategy by examining the structure of manufacturing. Table 7.2 includes information about the structure of output, employment, and capital assets in manufacturing, excluding oil refining—where significant development had taken place (see Karlsson 1975, 71–115, on the development of the oil-refining industry). In output terms, the traditional industries of food, beverages and tobacco, and textiles comprised over 63 percent of manufacturing output. By 1971, the share of traditional industries declines to approximately 40 percent, a share that is maintained through 1996. The decline in the share of traditional industries in capital assets is also notable. While the traditional industries comprised respectively 52.8 percent of all manufacturing assets in 1936 and 62 percent in 1953, their share declines to 37.3 percent and 21.6 percent in 1961 and 1996. The traditional industries also show a steady, though more gradual decline in the share of employment over the whole period.

2. The relatively low share of Venezuelan manufacturing provides some support for the idea that there was still more scope to develop "easy-stage" industries with the oil windfalls rather than commit so many resources to capital-intensive sectors in the period 1974–85.

TABLE 7.2 Venezuela: Shares of selected Sectors in manufacturing value-added (excluding oil refining), 1936–1998 (%)

(ISIC)	Food, beverages, tobacco (311–14)	Textiles (321)	Paper (341)	Chemicals (351)	Non-metallic minerals (361–62, 369)	Basic metals (371–72)	Metal-transforming industries (381–84)
			Output (= value-added)				
1936	52.3	10.6	0.1	4.5	3.8	0.3	2.4
1953	41.6	8.1	1.2	8.5	9.7	0.4	3.7
1961	41.0	15.6	3.3	8.3	6.5	0.9	11.1
1971	29.2	12.1	4.3	10.2	6.7	6.0	17.7
1981	27.2	8.1	3.9	10.8	6.8	7.0	19.6
1988	32.2	4.7	3.5	10.0	4.5	10.3	18.6
1998	36.0	5.2	2.3	9.8	5.2	10.4	18.0
			Labor (= number of employees)				
1936	59.3	11.9	0.2	2.5	4.9	0.2	1.9
1953	48.9	8.4	1.0	4.1	6.8	0.3	3.8
1961	27.2	23.0	2.8	5.0	6.6	1.2	14.8
1971	24.7	15.4	3.4	7.5	6.9	4.6	16.8
1981	18.1	11.2	2.9	7.6	6.6	7.2	18.4
1988	25.6	7.3	0.8	7.9	6.6	9.8	13.7
1998	22.8	4.9	0.7	9.1	7.4	6.7	12.1
			Capital (= fixed assets)				
1936	43.2	9.6	1.3	—	—	—	—
1953	53.0	9.0	1.4	5.9	12.7	0.5	2.3
1961	28.2	9.1	3.2	10.6	7.5	26.5	6.5
1971	23.7	9.4	4.8	7.9	9.5	23.3	10.8
1981	18.8	3.8	2.8	6.0	10.4	34.7	13.9
1988	16.6	3.9	2.2	8.0	8.1	45.7	13.6
1998	18.2	3.4	3.4	17.7	9.7	31.0	10.3

SOURCE: Astorga 2000, table 8.1; OCEI, *Encuestra industrial*, 1988, 1998; Banco Central de Venezuela, *Serie estadística*, various years.

NOTE: The figures used to obtain the percentages of capital and value-added in 1936 and 1953 are in current bolívares, thereafter in bolívares at 1968 prices.

While the traditional industries were declining in their share of output, capital, and employment, the shares of heavier, large-scale industrial sectors (chemicals, non-metallic minerals, basic metals, and metal-transforming industries) were growing significantly. It is telling that the relative increase of these four sectors is greatest in terms in their share of total manufacturing capital assets.[3]

3. Given the well-known problems in measuring capital (Fine 2003), the data on fixed assets (K) should be treated with caution. In the standard one-sector production function model, capital

Periodization of Industrialization Stages and Strategies 173

This was due to a combination of the increased capital intensity and scale of these industries and, as we shall see, their relatively disappointing output and export growth performance. The share of fixed assets of the basic metals sector (as a proportion of all manufacturing fixed assets) increased from 0.5 percent in 1953 to 26.5 percent in 1961, and then increased much more substantially to reach a peak share of 45.7 percent in 1988. The increase in the relative share of fixed capital assets of the basic metals sector was the most pronounced of any sector in the period 1961–88. The relative share of the metal-transforming sector (which includes the capital goods and transport sectors) increased significantly from 2.3 percent in 1953 to 6.5 percent in 1961 and reached a peak share of 13.9 percent in 1981. In aggregate terms, the share in total assets of the four heavy industrial sectors increased from 20.9 in 1953 to 51.1 in 1961, to 54.6 in 1971, to 65.0 in 1981, and finally reached a peak of 75.4 in 1988.

Given that the largest increase in the aggregate share of capital assets of the heavy industrial sectors occurred between 1953 and 1961, the period 1960–73 can plausibly be seen as marking a period of transition of development strategy. An important caveat is the dramatic increase in 1981 of the share in total manufacturing assets of basic metals, which was the result of a state-led big-push program in steel and aluminum in the 1970s. Thus, the late 1970s represent an important increase in the big-push industrial strategy in the context of a more generalized shift in capital intensity and scale that took place in the period 1960–73.

The largest increases in the output of heavy industry as a share of total output occur in the period 1961–71. Heavy industry's share of total output rose from 22.4 percent in 1953 to 26.8 in 1961, but then leapt significantly to 40.6 percent by 1971. By 1996, this share had reached 53.2 percent of total output. The largest increases in heavy industrial employment as a share of total employment also occur in the period 1961–71. In 1953, heavy industry's share of total employment stood at 15.0 percent and rose to 27.6 percent in 1961. Given the

---

cannot be measured as pointed out in the Cambridge Controversy (see Harcourt 1972). In reality, the measurement of capital would require knowing the extent to which changes in capital were caused by changes in its physical aspect (quantity of machines) and to what extent by its price aspect. In a one-sector model, there are no relative prices, so any changes in capital measurement due to changes in the relative prices of new machinery cannot be incorporated in the measurement of capital. In neoclassical growth theory, all changes in the quantity of capital owe to physical changes. In the data presented in the table 7.2, the perpetual inventory method was used to measure fixed assets. This is at least the most acceptable method, since there is no attempt to price capital over time. Increases in capital consist of the addition of new machinery to the fixed stock in 1936, with a standard depreciation factor of 5 percent incorporated. There is no need to price second-hand capital in this method.

importance of the state in promoting the heavy industrial strategy in the form of protectionism, credit allocation, and public enterprise production, the switch in the development strategies was not the result of spontaneous market forces but the result of a state-led industrial strategy.

A third indication of the changing nature of the development strategy in the period 1960–73 can be seen in the evolution of the output and employment of large firms as shares of total output and employment. Table 7.3 presents data on the share of different sizes of firms in total manufacturing output. As can be seen from table 7.3, the share of large firms (defined as firms with at least one hundred employees) in total output increases significantly from 54.0 percent in 1961 to 68.1 percent by 1971. The increase in the importance of large-scale firms in their share of output steadily increases thereafter and reaches a peak of 85.6 percent in 1996. As a corollary, the period 1961–71 represents the most dramatic drop in the share of small and medium-small firms' share in total output. In 1961, small and medium-small firms accounted for 37.7 percent of manufacturing production. By 1971, their combined share had dropped to 21.7 percent; by 1996, to 8.5 percent.

A fourth indication that the period 1960–73 marked the end of the small-scale, "easier" stage of ISI can be seen in the decline in demand for imported goods. In the manufacturing sector, the weight of imports in the internal final

TABLE 7.3 The relative importance of large, medium, and small firms in the Venezuelan manufacturing sector, 1961–2003 (excluding oil refining; percentage of value-added by firm size)

	Large	Medium-large	Medium-small	Small
1961	54.0	8.3	15.0	22.7
1966	54.3	10.3	17.3	18.1
1971	68.1	10.3	10.6	11.1
1974	70.2	11.4	10.2	8.1
1979	70.5	10.1	10.7	8.7
1983	71.4	9.5	9.8	9.2
1988	79.1	7.0	7.7	6.2
1989	79.1	7.8	8.0	5.1
1996	85.6	5.9	4.9	3.6
2003	85.0	6.1	5.5	3.5

SOURCE: OCEI, *Encuestra industrial*, various years; INE, *Principales indicatores de la industria manufacturera*, various years.

NOTE: Large firms have more than 100 employees; medium-large, 51–100 employees; medium-small, 21–50 employees; and small, 5–20 employees.

demand for manufactured goods, or *import coefficient,* declined from 46.5 percent in 1950 to 31.1 percent in 1960 and further to 25.5 percent in 1968. According to Sergio Bitar and Eduardo Troncoso (1983, 260), the import coefficient declined even more significantly in traditional industries, which were generally subject to smaller-scale economies and are technologically less complex.[4] By 1968, the import coefficient of the traditional consumer goods industries had declined to 6.8 percent (Bitar and Troncoso 1983, table A-4). By 1968, the import coefficient of intermediate and capital good had fallen much less: to 26.7 percent and 54.5 percent respectively (ibid.). However, the large-scale auto sector's import coefficient had fallen as low as 10.6 percent by 1968 as the result of the state protectionist policy (Hausmann 1981, table 8.18).

A fifth indicator of the change in development strategy away from the easy phase of ISI is a growing policy emphasis on export growth and diversification. As mentioned earlier, the planning ministry had recognized that many import-substituting industries needed to export to continue to grow as they were saturating the possibilities of revenue growth in the relatively small domestic market. In 1973, the Fondo de Exportaciones (Venezuelan state export credit fund, FINEXPO) was set up to provide export credits to help firms offset the risks of entering the external market. This was the first export financing facility developed by the state since independence. From 1974 to 1989, there were important, if erratic, increases in export credits.[5] Manufacturing firms, especially the chemical, aluminum, and steel sectors, received the bulk of the credits. Export credits rose from an annual average of 75 million bolívares in the period 1975–77, to an average of over 400 million bolívares in the period 1980–82, and further to an average of 1.6 billion bolívares in the period 1986–89. The increase in 1986–89 was largely to compensate for the anti-export bias embedded in the multiple exchange rate regime (RECADI), which was in place from 1985 to 1988.

An examination of the structure and type of exports after 1975 also provides an indication of the *increasing* importance in the Venezuelan economy of large-scale, technologically more complex manufactures for export from the early 1970s on.[6] In the period 1970–75, oil and oil products accounted for nearly 98 percent of total exports. However, in the period 1975–2003, the fastest-growing *non-oil* exports, both in volume and value, were aluminum, steel, chemicals,

---

4. Bitar and Troncoso (1983) define traditional industries as those producing food, beverages and tobacco, textiles, leather goods, and wood products.

5. See Banco Central de Venezuela 1992, vol. 3, tables IV-17 and IV-18, for data on export credits.

6. The data on exports for this paragraph and the subsequent one are taken from tables V-3 and V-4, Banco Central de Venezuela, *Informe anual* (various years).

and transport equipment. Consider table 7.4. Non-oil exports increased from a negligible share in the 1970s to 12.6 percent of total exports in the period 1985–88. The four main non-oil export sectors were all large-scale, heavy industrial sectors: aluminum, iron and steel, chemicals, and to a lesser extent, transport equipment. Over the period 1989–98, average annual non-oil exports increased significantly to US$4,189 million, with the same four sectors dominating non-oil export revenues.[7] While non-oil export revenue performance did not increase much further in the period 1999–2003, the share of these four sectors remained dominant, with the share of the transport sector in total non-oil exports increasing at the fastest rate.

7.2 Transformations in Venezuelan Industrial Policies After 1968

The general economic history of Venezuela confirms that there was a transformation in development strategy between 1960 and 1973. While the period 1920–58 was characterized by relatively liberal trade policy, state-created rents became much more important after 1960. The most important mechanism

TABLE 7.4 Non-oil export revenues and composition in Venezuela, 1985–2003 (average annual values in millions of current U.S. dollars)

	1985–88	1989–98	1999–2003
Total export revenues	$10,833	17,560	26,130
Non-oil exports			
Revenues	1,644	4,189	4,568
As a percentage of total exports	12.6	23.6	18.4
Main non-oil exports as percentages of total non-oil exports			
Nonferrous metals	33.3	19.4	16.6
Iron and steel	20.6	14.4	19.7
Chemical products	11.2	13.6	18.8
Transport equipment	4.6	8.9	10.5
Road vehicles	2.0	5.7	5.3
Auto parts	1.3	2.2	3.3

SOURCE: United Nations, *International Trade Statistics Yearbook*, various years.

7. The three dominant sectors (aluminum, iron and steel, and chemicals) were negligible in production terms in the early 1970s. However, by the mid-1990s, average annual aluminum production reached 550 thousand metric tons, average annual steel production reached 3.15 million metric tons, and average annual chemical production reached 295 thousand metric tons (Banco Central de Venezuela, *Informe anual* [various years]).

of promoting industrialization through state-created rents was the system of protectionism through tariffs and non-tariff barriers. As in many other countries, a differentiated tariff regime was set up to make some imports more expensive than others. Differentiated tariffs allow the government to raise the import prices of goods that can be produced at home. This discourages domestic consumers from importing these goods and, at the same time, encourages them to buy domestic products, whose higher prices or lower quality may have prevented them from competing with foreign products at the early stage of industrialization. From the mid-1960s to the late 1989, average tariffs in manufacturing industry were 60 percent for consumer goods, 30 percent for intermediate goods, and 27 percent for capital goods (World Bank 1990, 16–18). The average level of tariff protection and effective rates of protection as well as the tariff dispersion was broadly similar to those of many Latin American countries (Edwards 1995, 198–203).[8]

A second component of the protection policy was the use of various forms of non-tariff protection such as import licenses and foreign exchange rationing, the latter used most extensively in the era of multiple exchange rates (1983–88). From 1968 to 1989, the average share of manufactured goods subject to non-tariff barrier coverage, that is, import licenses and prohibitions, ranged between 40 and 50 percent (World Bank 1990, 16).[9] Comparative data for Latin America in the 1980s indicates that Venezuela's share of non-tariff barrier coverage were similar to the regional average (Edwards 1995, 200).

Apart from state-led promotion of private industry, the administration of Carlos Andres Pérez (1974–79) envisioned a national project called La Gran Venezuela. The cornerstone of this "vision" was the heralded Fifth Plan of the Nation (1976–80), which was to set Venezuela on a path of a more pronounced state-owned, enterprise-led, natural resource-based, big-push heavy industrialization policy. In addition to the nationalization of the oil and iron ore industries in 1976, numerous public enterprises were expanded in heavy industries (steel, iron ore, aluminum, bauxite, petrochemicals, oil refining, and hydroelectric power) to provide inputs for domestic industry in an attempt to vertically

---

8. It is important to note that much of Venezuelan trade policy was much more liberal prior to 1973 because it was restricted by the reciprocal trade agreement with the United States (see section 8.1). Protectionism increased substantially as a reaction to the 1983 balance-of-payments crisis, which resulted in the adoption of multiple exchange rates and import controls from 1983 to 1989 (Hausmann 1995).

9. Import licenses were used most heavily for processed food, consumer goods, and transport equipment. Import monopoly licenses were also instrumental in establishing state-owned enterprises in steel, aluminum, and petrochemicals (World Bank 1990, 15).

integrate the import-substitution process and to improve the technological capacity and diversification of the industrial and export structure (Karl 1982, 194–208).

The industrial state holding, Corporación Venezuelona de Guyana (CVG), was responsible for managing natural-resource-based industrialization; and the national oil company, Petróleos de Venezuela (PDVSA), was responsible for the expansion of oil exploration and the modernization of oil refining. In the period 1970–80, the non-oil investment rate increased to an annual average of 36.8 percent of non-oil GDP over the previous decade where non-oil investments averaged an already vibrant 26.1 percent of non-oil GDP. The intention of the Fifth Plan of the Nation was decidedly more developmental than populist. As Pérez remarked: "Venezuela will export automobiles, steel, and aluminum to countries like the United States. One day you Americans will be driving cars with bumpers made from our bauxite, our aluminum, our labor. But first we have to create the productive apparatus. We have to create wealth before it can be redistributed" (interview with Carlos Andrés Pérez, summer 1979, quoted in Karl 1982, 196).

The big-push, natural-resource-based industrialization project marked a purposeful effort to expand the role of the state in direct production. In 1970, state-owned enterprises accounted for 5 percent of manufacturing value-added (excluding oil refining). Their share rose to 8 percent in 1980 and reached 18 percent by 1986. When oil refining is included, these figures rise to 4 percent, 36 percent, and 42 percent. In 1987, public firms were concentrated in nonferrous metals (largely aluminum), where the state sector accounted for 91 percent of production; iron and steel, where the state sector accounted for 61 percent; and chemicals and petrochemicals products, where the state sector accounted for 45 percent (World Bank 1990, 30). Public manufacturing investment (excluding oil refining) accounted for 24 percent of total manufacturing investment in the period 1968–71. In the period 1972–80, the public enterprise share of total manufacturing investment rose to 41 percent.[10]

With the nationalization of oil in 1976 and the development of non-oil state-owned enterprises, the state became the main producer in the economy, the largest generator of foreign exchange, and the largest employer. The main objective of natural-resource-based industrialization was to diversify the export structure of the economy and to "deepen" the manufacturing sector by increasing the share of intermediate and capital goods production. Heavy industry

10. Calculations based on data from corresponding editions of Banco Central de Venezuela, *Serie estadística* (various years), and OCEI, *Encuestra industrial*.

was thought to improve the economy's chances of catching up with developed countries, since such heavy industries were thought by Venezuelan policymakers to have greater potential for high productivity growth than light industry.

Big-push industrialization refers to a synchronized expansion of industrial sectors, coordinated by the state. The advocacy of this policy is based on the idea that markets in underdeveloped economies are subject to coordination failures and thus require coordination of investment decisions (see section 5.3). Paul Rosenstein-Rodan (1943), who developed the idea of the "big push," argued that, in the context of underdeveloped capital and consumer markets, large-scale investments in the modern industrial sectors may not be profitable to individual investors. This is because "the profitability of such investments is contingent on the availability of specialized inputs; but the profitability of producing these inputs depends on the presence of demand from a pre-existing modern sector. . . . It is this interdependence of production and investment decisions that creates a coordination problem" (Rodrik 1995, 80–81). The classic example is the demand complementarity between a steel mill and power suppliers.[11]

The big-push, natural-resource-based industrialization in Venezuela was heavily influenced by structuralist theories, which are mainly associated with the Economic Commission of Latin America (ECLA), and in particular with the work of Raul Prebisch.[12] The ECLA-led structuralism made the case that "peripheral," labor-surplus economies would, given the free reign of market forces, tend to develop a production path specializing in commodity production and away from manufacturing, which was considered the engine of growth in the economy. Moreover, structuralists argued that manufacturing products were more income elastic than commodities, and thus provided greater export growth potential. Consequently, a set of measures was proposed with the intention of developing a purposeful or "forced" industrialization. In addition to protectionist policies, ECLA-inspired structuralist thinking advocated direct state production in those areas, such as natural-resource-based industrialization, where large amounts of slow maturing investment was needed.

A final important influence of ECLA-inspired thinking in the 1970s was the emphasis on Soviet-style planning and in particular, the development of very large-scale, capital-intensive, state-owned enterprises and public projects.

---

11. Of course, the existence of coordination failures does not necessarily mean that state intervention is the most efficient way of overcoming them (Matsuyama 1997).

12. See Prebisch 1950. For discussions of ECLA's structuralist ideas, see Palma 1987 and Fitzgerald 2000.

The idea was that capital-intensive industrialization would be sufficient to create the export growth and production linkages, which in turn, would suffice to create the growth and employment opportunities needed to raise average living standards. Other potential indirect benefits of natural-resource-based industrialization include regional development, technology transfer, and employment and production linkages (Auty 1990, 22).

The cornerstone of natural-resource-based industrialization was to use the comparative advantage of *cheap energy* (in the form of state-subsidized hydropowered electricity and gas) to develop the steel, aluminum, and petrochemical industries. It is estimated, for instance, that, throughout the 1980s, SIDOR, the Venezuelan state steel producer, benefited from the pricing of electricity that was only 24 percent of the marginal cost of generating hydroelectricity (World Bank 1990, 77). In the mid-1980s, the two main aluminum smelters in Venezuela (VENALUM and ALCASA) paid one-half the price for electricity that competing smelters in other developing countries did (Auty 1990, table 9.6). Roberto Rigobón (1993) estimates that gas subsidy for Venezuelan users averaged US$1.05 per liter in the period 1986–91. In 1991, the average price of gas in Venezuela was US$0.07 per liter compared with 32 cents per liter in the United States, 83 cents per liter in the United Kingdom, and 92 cents per liter in France.

This "capital-intensive growth first" strategy downgraded the possibilities of promoting growth by emphasizing human capital through either ensuring basic needs such as adequate housing for the whole population or by developing human capital by prioritizing health and education spending. While the comparative advantage in Venezuela lies in developing downstream industries that utilize cheap electricity and oil, the opportunity cost of committing such large sums of state resources to natural-resource-based industrialization in the form of massive plants can be questioned.[13] While there has been considerable criticism of the state-led, natural-resource-based industrialization project in Venezuela (see section 3.4), the fact remains that steel, aluminum, and petrochemicals, all public-enterprise-dominated sectors, were among the few sectors to maintain relatively high productivity over time (see tables 3.4a–c).

Following the logic of ECLA's analysis, investment in oil production and refining were increased dramatically. The large-scale nature and technological

---

13. To gain an idea of the changing nature of planning priorities of the Pérez administration, the Fifth Plan of the Nation earmarked 20 percent of public spending to education, housing, health, urbanization, and government services, whereas the Fourth Plan (in the previous Caldera administration) had targeted over 35 percent of public spending to those same areas (Karl 1982, 198–99).

complexity of modernizing the oil-refining sector accentuated the shift in development strategies. There were several sound reasons for the rapid expansion of oil and oil refining. First, by the mid-1960s, the Venezuelan state had adopted a staunch policy of not granting any further concession to foreign oil companies. This was part of a shift in oil strategy away from profit-sharing with oil multinationals and toward maximizing output growth by deemphasizing production incentives and focusing on the maximization of oil rent extraction by the state (Espinasa and Mommer 1992). This strategy led to a long-run decline in production and investment by the oil companies. Oil production had declined from a peak of 2.8 million barrels per day (b/d) to 2.4 million b/d by 1973. Moreover, the net investment rate in the industry was negative in the period 1958–73, which lowered confirmed reserves to less than ten years of current production by the end of the period. As a result of the Fifth Plan, net investment in oil went from a rate of −0.02 percent of GDP in the period 1970–75 to an average rate of 3.4 percent of GDP in the period 1978–83 (M. Rodríguez 1991, 253).[14]

While oil-refining production had steadily grown from the mid 1940s, the lack of significant investment in refining left the sector unable to cope with changes in the world and domestic market for refined products. Venezuela's refining capacity was mostly in heavy fuels, whereas the world refining market was mostly geared toward lighter fuels (Boué 1993, 111–27). The refining capacity modernizations that took place in the late 1970s were "extremely innovative at the time, in the sense that its centerpiece (the Amuay Refinery Project, costing $770 million) was a type of processing plant (the flexi Coker), which had never been built on such a scale before" (113–14).

The development of the iron and steel, nonferrous metals, and energy sectors further reflected a transition toward a big-push industrialization strategy in the post-1973 period. First, the steel sector also was considered a development priority. While the state had begun to develop a steel industry, particularly with the establishment of the state steel company, SIDOR, in 1962, productive efficiency had been largely disappointing (Karl 1997, 125). The Fifth Plan called for

---

14. See Brossard 1993 and Wright and Czelusta 2003, 16–17, for a discussion of the major investments made by PDVSA for developing technologies to exploit the concentration of heavy oil in the Orinoco Belt. Such investments led to increased oil discoveries and thus an increase in reported reserves through the 1980s. Moreover, PDVSA, in collaboration with BO Petroleum (a company with experience extracting heavy oil in Canada), developed a new fuel (orimulsion) for use by power utilities and heavy industry (Wright and Czelusta 2003, 16). Orimulsion, which has favorable market prospects (because it has a potential for gasification, can be used in a combined fuel cycle, and is environmentally friendly), will be one of the main growth areas of the Venezuelan oil industry in the twenty-first century (Brossard 1993, 170–77).

an increase from a capacity of 1.2 million tons to 9.8 million tons in two new plants, SIDOR Plan IV and ZULIA. The plan to expand steel represented an important continuity in state policy to exclude the private sector from the largest steel projects (Coronil 1998, 195–200). Second, there were was a project to develop the expansion of the massive Guri Dam to develop hydroelectric capacity. The program aimed at increasing capacity fourfold, an increase that would create the second-largest dam in South America after the Itau Dam in Brazil. Third, the discovery of large bauxite reserves in the late 1960s led to the creation of a national bauxite firm (BAUXIVEN) and two state aluminum firms (VENALUM and ALCASA), which by the mid-1980s became the leading sources of non-oil export earnings. Apart from these state-led investments, smaller state investment projects were developed for nickel, cement, and the assembly of small aircraft, all relatively large scale and technologically complex.[15]

The Sixth Plan of the Nation (1981–85) continued the emphasis on large-scale, capital-intensive projects (Hausmann 1989, 10). Large allocations of investment were made for oil expansion (US$16 billion) and the conclusion of the electricity and manufacturing projects started by the Fifth Plan (US$5.4 billion and US$4.2 billion respectively). Moreover, nearly US$3 billion was allocated for transport and communication, including nearly US$2 billion for the Caracas Metro.

The other important trend to note over this period was the dominance of the public sector as the main exporter. While the nationalization of oil and iron ore had placed the state as the main exporter in the economy in 1976, state-owned enterprises were also significant in the generation of aluminum, steel, and chemical exports (state-owned enterprises accounted for 95 percent of aluminum exports, 45 percent of steel exports, and 20 percent of chemical exports). By the period 1993–95, the public sector was responsible for 83 percent of total exports and 36 percent of non-oil exports. Thus, the gradual shift toward large-scale, high-value-added technology was also a shift toward increased state ownership of that technology. This trend largely continued through 2005.

It is important to note that the nature of industrial strategy changed substantially between 1989 and 2005. One important change occurred with the liberalization reforms initiated in 1989 (see Chapter 5). In the period 1989–98, the role of the state in the promotion of heavy, natural-resourced-based industries declined somewhat with privatizations in the telecommunication and steel

---

15. One of the most important state-led initiatives to develop private sector industry occurred in the automobile and auto parts sector (Enright et al. 1994, 202–20).

sectors, though hydroelectric generation, bauxite, aluminum, and petrochemical production remained in state hands. Aluminum, steel, and petrochemicals were still the main non-oil export sectors by 2005.

There were also brief attempts to open marginal oil fields to foreign investors as well as to establish joint ventures for the development of the substantial reserves of extra-heavy oil (or tar sands) in the Faja de Orinoco (Oil Belt of the Orinoco) in the late 1990s under the Caldera administration. In the first seven years of the Chávez administration (1999–2005), there has been a reversal of the joint venture formula as the state has bargained for a higher share of the profits in the Faja de Orinoco, resulting in the departure of several large oil multinationals, including Exxon-Mobil and Conoco-Phillips. The telecommunications sector was renationalized in 2007 as well. Finally, there has been a decidedly less purposeful heavy industrial strategy or indeed any other explicit economic diversification strategy in the period 1999–2005 (Weisbrot and Sandoval 2007). This is because the Chávez administration has focused more on using increasingly high oil revenues (particularly after 2004) to fund social programs and also because the government has maintained a highly overvalued exchange rate, which, in the short run, reduces the incentives for non-oil tradable exports (ibid.).[16]

Nevertheless, there is little to indicate that the dominance of heavy industrial assets in the manufacturing sector was less in 2005 than it was a decade before.[17] There are three pieces of evidence that support this view. First, the large-scale industries of steel, aluminum, chemicals, and transport equipment still accounted for 75 percent of non-oil exports in 2005. Second, the sectoral growth rates in the *private* manufacturing sector over the period 1997–2005 (as indicated in table 7.5) suggest that there have been no major shifts in the structure of production.

In the period 1997–2007, the most dynamic sectors were chemicals, non-metallic minerals, and basic metals, all of which are heavy industrial sectors. The metal-transforming sector (e.g., transport equipment and capital goods) lagged behind the general private manufacturing production index, but this is offset by the even greater decline in the textile sector, which is a light manufacturing sector. What is also clear is that the dominance of large firms was as high in 2003 as it was in 1996 (table 7.3). This suggests that the economy was still very much in the advanced stage of ISI in the period 1989–2005.

16. Also see section 8.2.5.
17. Recall that heavy industrial sectors dominated the share of industrial assets in 1998 (table 7.2).

TABLE 7.5 Venezuela: Production volumes in private manufacturing industry, 1997–2005 (base 1997 = 100)

	1997	2003	2005
General index	100.0	73.4	105.3
Food, beverages, tobacco	100.0	88.1	102.4
Textiles	100.0	39.3	69.7
Paper	100.0	76.2	92.5
Chemicals	100.0	75.8	117.6
Non-metallic minerals	100.0	81.0	108.3
Basic metals	100.0	147.2	180.4
Metal-transforming industries	100.0	41.1	78.6

SOURCE: INE, *Principales indicatores de la industria manufacturera*, various years.

7.3 Conclusion

In sum, the data broadly support the proposition that a switch from the easy stage of ISI to a more advanced stage of ISI occurred in Venezuela over the period 1960–73. This represented a significant shift in the technological nature of the economy's development strategy. Moreover, government policy further accentuated this switch by using the oil windfalls of the 1970s to undertake a big-push, mineral-based heavy industrialization strategy dominated by state-owned enterprises. Table 7.6 summarizes the periodization of development strategies and stages of ISI in Venezuela. This shift in development strategy required (as argued in the previous chapter) the state to become more centralized, selective, and disciplinary in the rent deployment and monitoring process.

Continued growth became more challenging because the industries capable of import substitution (namely intermediate and capital goods) were characterized by increasing scale economies and required more complex state coordination and monitoring of investments (section 6.3). Even if one believes that the economy should not have moved into more capital-intensive industries or

TABLE 7.6 Periodization of development strategies and stages of ISI in Venezuela, 1920–2005

Type of development strategy	Period
Early ISI	1920–60
Transition to more advanced stage of ISI	1960–73
"Big push" heavy industrialization / more advanced stage of ISI	1973–2005

the production of capital goods so rapidly, further industrialization certainly involved greater risks and required more selective targeting of capital for longer periods as new expertise had to be acquired and new markets captured. This would also be true for the traditional, small-scale industries that developed from the early stages of ISI. Conquering export markets in consumer goods is a risky and long-gestating project that requires marketing, investment in distribution channels, and increases in scale economies. The declining import coefficients in traditional industries meant that further growth in these industries would require export-led growth. In this sense, the transformation of easy ISI production into exporting is part of the transition to a more complex development strategy. At the same time, this chapter has also established that large-scale heavy industrial investments dominated the industrial strategy after 1968.

8

THE STRUCTURE OF AND CHANGES IN POLITICAL SETTLEMENTS IN VENEZUELA

From 1920 to 1968, Venezuela was, by and large, a consolidated state with centralized political organizations, both under authoritarian rule from 1920 to 1958, then from 1958 to 1968 in the period of democratic transition and consolidation. Then, it transformed from a consolidated state with centralized political organizations to a consolidated state with fragmented political organizations in which political contestation became progressively more clientelist, populist, and polarized, particularly after 1973.

The importance of establishing the periodization of political settlements is that it gives us a general idea of how and when the class and group nature of the alliance that formed the support base of the state changed. This periodization will achieve two further aims. First, it will enable us to explain why the ability of the state to effectively implement industrial strategies was generally higher before 1968 than it has been since. After 1968, contention among populist and clientelist coalitions led to the increasing fragmentation of the state, which, in turn, limited the ability of the state to make coherent policy, coordinate investments, be selective in subsidization, and discipline the recipients of rents and subsidies in infant industries. It is in the actual politics of a polity that the conditions under which rents are deployed and used become known. Second, it will enable us to map these settlements onto the periodization of the development strategies and stages of ISI identified in Chapter 7.

8.1 The Construction and Consolidation of a Centralized State with Centralized Political Organizations, 1900–1958

In historical perspective, centralized state authority has been the twentieth-century response to the failure of Simon Bolívar and his supporters to create a consolidated and legitimate central state in Venezuela. The battle for centralization of the Venezuelan polity was finally won when a regional group appropriated power. This group, known as the Grupo Táchira, a network of caudillos from the coffee-producing region in the West, began nearly a half-century of rule when General Cipriano Castro led a *coup d'état* in 1899.

The rise of Juan Vicente Gómez in 1908 marked the beginning of effective state consolidation and centralization. Venezuela under Gómez, whose reign lasted until his death in 1935, was effectively a military-caudillo state.[1] Before Gómez, the "national" army was merely a coalition of regional armies that owed primary allegiance to regional *caudillos*. Gómez managed to consolidate the army through Machiavellian divide-and-conquer tactics. His regime centralized public finances by 1910, a decade *before* the arrival of the oil multinationals. During his reign, the regime proscribed the two main parties since independence (Conservatives and Liberals), developed a national state bureaucracy, and promoted the integration of a national economy through a national road system (Lieuwen 1954).

The consolidation of the Venezuelan state in the early part of the twentieth century was instrumental to the prospects of capital accumulation and growth.[2]

---

1. In a military-caudillo state the armed forces control most of the state institutions and the leader controls state resources and patronage.
2. I use consolidation in the Weberian sense of monopolization of the means of violence in the central state. The extent to which the state was "legitimate" is a more complex issue and beyond the scope of this analysis. "Much of what happens in politics is actually a battle of establishing the grounds of legitimacy" (Putzel 1995, 242). As Putzel (1997) notes: "While political legitimacy is often defined in very strong, positive and even democratic terms, this limits the usefulness of the concept in analyzing different regime types and political change. It is more usefully defined as 'acceptance of the right to rule.' The more legitimate a political regime is in the eyes of a population the less a regime requires coercion or the threat of coercion. It could be said that to maintain authority, a regime requires 'active legitimacy' (strong acceptance) from those actors who command economic and political power, and at least 'passive acceptance' (weak acceptance) from the society at large. In a very basic way, legitimacy is determined by economic performance. Other factors, however, such as national integrity, social order, promoting particular ethnic groups, and inequality may assume more or less importance to regime legitimacy at a given moment in history" (201). In this perspective, the consolidation of the Venezuelan central state at this juncture enhanced the prospect of legitimate rule because it reduced warlordism and social disorder and enhanced national integrity, both of which increased the possibility of capital accumulation and economic growth.

Moreover, this consolidation was achieved through very centralized and cronyist processes of rent deployment. The period of centralized authoritarianism lasted largely uninterrupted from 1900 to 1958. The regimes of the successors to Gómez, General Eleazar López Contreras (1936–41) and General Isaías Medina Angarita (1941–45), distanced themselves somewhat from his repressive rule by permitting political parties. However, after a brief and unstable period of democratic politics from 1945 to 1948, General Marco Pérez Jiménez seized power and banned them once again.

The centralized Venezuelan state was built on corruption and coercion. The defining characteristics of the period 1920–35 was a pact among oil multinationals, the Gómez family, and a network of cronyist landowners. The discovery of oil just before 1920 brought the most powerful multinationals to Venezuela. The foreign companies preferred to deal with one central authority when buying land in the oil-rich Maracáibo basin in the west of the country. To avoid multiple legal and political battles with the landowners there, they provided the Gómez regime with the resources necessary to defeat its regional opponents.

Since nearly all business activity required the backing of the Gómez family, a web of relations between the oil companies, the state, and prominent family conglomerates formed the military-caudillo state where primitive accumulation was sustained by large-scale, but centralized rent deployment and corruption. Gómez maintained "the most liberal oil policy in all Latin America" (Sullivan 1992, 258). The 1920s saw a "dance of concessions" among oil companies. The liberal treatment of the oil companies was a consequence of the mercantile relationship between the military elite, the merchant groups, the landed oligarchy, and the oil companies. The concessions policy run by Gómez was characterized by transactions of unparalleled corruption (Karl 1982, 70–80). Gómez granted land leases in the oil-rich Maracaibo basin to his favorite cronies in return for a "fee," and they resold those leases to oil companies at exorbitant profits (Karlsson 1975, 73).[3]

Rural elites in Gómez's favor sold their lands to the oil companies and thus turned themselves into a mercantile bourgeoisie. The traditional agrarian oligarchy and the commercial bourgeoisie, which once shared common interests in export of coffee and cocoa, now became oriented toward activities in urban

---

3. Land ownership became dramatically more unequal during the Gómez period. Members of the regime and those with political contacts received more than 50,000 hectares of public land in 1917, over 105,000 hectares in 1920, and more than 50,000 hectares in 1921 (Brito Figueroa 1966, 379–86).

The Structure of and Changes in Political Settlements 189

commerce, real estate, and light manufacturing. The transformation of agricultural lands into urban real estate and construction became the central basis of wealth for the formation of major *grupos económicos* (diversified family conglomerates). This process did not set an emerging bourgeoisie against a traditional landowning class as occurred in many late developers such as Chile and Argentina. However, it is important to note that the Gómez period saw a further weakening of the domestic bourgeoisie in agriculture and manufacturing.[4] This weakness of the bourgeoisie vis-à-vis the state is a recurrent theme throughout the Venezuelan industrialization process and, in part, explains the weakness of private sector capacity in non-oil tradable production.

In contrast to what rent-seeking theorists argue, corruption proved beneficial to the development and consolidation of capitalism in Venezuela in this period. Clearly, the impact of corruption depends on contingent factors of political contests and threats, and on the technological challenges of production. A state that manages to monopolize violence and manage conflict may not be a sufficient condition for all types of development strategies; nevertheless, it is a necessary condition for the possibility of promoting the 'easier" stages of import-substituting industrialization, which were relevant for the period 1900–1958 (see Chapter 7). While developmental state theorists point to the benefits of centralized state authority and intervention, it is still necessary to explain (rather than simply assume) why centralized rent deployment facilitated capital accumulation.

The main reason that Gómez and subsequent authoritarian leaders could consolidate power so thoroughly was the lack of effective political and business organizations in the country.[5] There were several factors that helped shift power to Gómez and his supporters. First, the legacy of conflict in the nineteenth century had undermined traditional patron-client links in the countryside. Miriam Kornblith and Daniel Levine (1995) remark that "Venezuela stands out for how little the colonial period and the nineteenth century shaped the modern scene" (41). Every modern census showed a high proportion of Venezuelans living outside their birthplace (41). In order to avoid endemic warfare,

---

4. The weakness of the domestic bourgeoisie has its historical origins in the nineteenth-century period of caudillo warlordism. The erosion of the economic base of the landed oligarchy through war and civil conflict had turned the military into the major source of social and political power. "The landed aristocracy was decimated in the nineteenth century civil wars: property and position depended on political power, not the reverse" (Levine 1978, 65). In the Latin American context, Venezuela is unique for having had a *weak* landed oligarchy for most of the nineteenth century.

5. In comparative perspective, the lack of political power of populist political parties, middle-class groups, and labor unions are some of the conditions underlying the construction of developmental states in Korea and Taiwan (Fishlow 1991; Woo-Cumings 1997).

many migrated toward the Andean region, which emerged as the center of the coffee economy in the mid-nineteenth century. Second, Gómez created alliances with regional caudillos and used oil money to buy support among merchant capitalists and landowners. This severely weakened the support base of political parties, whose conflicts had dominated nineteenth-century politics. Ultimately, political parties were destroyed. Third, the rise of the oil industry, in combination with the failure to promote and protect agriculture, induced a mass exodus to the cities, which further undermined rural political organization.[6] Political repression also delayed the development of mass political movements until after Gómez's death in 1935. Fourth, labor unions were weak in the face of state repression (Godio 1980).

Finally, the Gómez period is market by the absence of organization or political action by the business sector. The rapid decline of agriculture in the Gómez regime led to a decline in the landlord class as a political actor (Karl 1986, 200). In the "dance of oil concessions," landlords sold their property to oil companies and used the proceeds to develop commercial and financial enterprises in the large urban centers. Pressure for protecting agriculture was thus absent. In the cities, the deployment of patronage and contracts induced particularistic bargaining among the family-owned conglomerates. This delayed the construction of business associations until the mid-1940s and reduced the role of organized business groups in formal political organization (Giacopini Zárraga 1993).

The construction of the authority and power of the central state nevertheless permitted the development of those state institutions needed to support the development of the early stage of ISI. From 1936 to 1958, the state played an ever larger role in guiding the process of capital accumulation. While the largest investments and profits for the main economic groups occurred in construction, real estate, finance, and importing (Acedo de Machado et al. 1981), the state made important institutional changes that affected macroeconomic management capacity, resource mobilization, and industrialization. President López Contreras (1936–41) initiated the first modest programs to encourage industrial development. In 1937, the first state industrial development bank (Banco Industrial de Venezuela) was created. The government also implemented policies to provide special credit arrangements (through the private banking system) to leading firms so they could achieve public production priorities.

6. The share of the labor force engaged in rural employment declined from 71.6 percent in 1920 to 33.5 percent by 1961 (Karl 1986, 200). The percentage of the population living in urban areas increased from 16 percent in 1920 to 36 percent in 1935 (Baptista 1997, table I-2, 29).

There were other important institutional creations that helped shape state formation and provided key instruments for intervention. One of the most important was the set of laws that helped the state capture a greater share of the oil revenues from multinationals: the Customs Law in 1936, the Income Tax Law in 1942, and a new Hydrocarbons Law in 1943, which increased the share of the state in oil profits to 50 percent.[7]

The Hydrocarbons Law of 1943 was also instrumental in the development of oil-refining manufacturing capacity (Karlsson 1975, 71–115). This was one of the few large-scale, technologically more complex investments undertaken in what was otherwise an era of dominated by relatively simple, small-scale industries. Despite being the world's second-largest oil exporter after 1928, oil-refining capacity before 1945 was far below that of any other major oil producer at the time. The Gómez regime actually encouraged most refining capacity stay offshore, on the islands of Aruba and Curaçao, in order to *prevent* the development of organized industrial interest groups in the Maracáibo basin in western Venezuela (Lieuwen 1954, 17). In particular, Gómez prohibited the establishment of oil refineries in the national territory to avoid large concentrations of workers (17). This policy both contributed to and reflected the weakness of the domestic interest groups at the time.

Nevertheless, subsequent state-led initiatives paved the way for dramatic oil-refining growth in Venezuela. In 1938, Venezuela refined 26 thousand barrels per day. From 1945 to 1960, refining capacity increased tenfold, from 100 thousands barrels a day in 1945 to nearly 1.4 million in 1960. By 1974, refining capacity reached 1.5 million barrels a day, which meant that nearly 35 percent of the oil extracted was refined domestically.

One of the other salient features of the period of authoritarian rule from 1900 to 1958 was the maintenance of a relatively liberal economic policy with respect to manufacturing. The principal manifestations of this were the maintenance of relatively low average tariffs and limited non-tariff protection of manufacturing before 1958. From 1920 to 1945, tariff protection was limited to basic consumer goods such as food and beverages, tobacco, and textiles (average tariff rates for the whole of manufacturing are not available in this period).[8]

---

7. The percentage of state revenues from petroleum increased to 78 percent by 1970 and paved the way for oil nationalization in 1976 (Ascher 1999, 211).

8. For particular industries with significant domestic production, nominal tariff protection was relatively high. For instance, average tariffs in the textile sector rose from 15 percent in 1929 to 30.5 percent in 1936. In comparison, textile tariffs in Colombia and Brazil averaged 9.5 and 19.0 percent respectively in 1929, and 6.8 percent and 28 percent respectively in 1936. However, the effective protection rates between Venezuela and the rest of Latin America differed. Many countries in Latin

From 1945 to 1958, significant tariff protection was extended only to steel and petrochemical production (Ortiz Ramirez 1992, 79).

Historically, import restrictions have played a more important role than tariffs in protecting domestic manufacturing sectors in Venezuela. There is clear evidence that there was a significant proliferation of import licenses within subsectors over time. Between 1939 and 1960, 35 tariff items were subject to import licensing (World Bank 1973, 24). By 1969, the number of tariff items subject to import licensing reached 599 (24). By 1989, the number of tariff items subject to import restrictions increased to 5,749 (World Bank 1990, table 2.5A)! In historical perspective, the domestic manufacturing was much less protected from 1920 to 1958 than it was from 1958 to 1989.

The policy stance of the state was not the result of a general preference for laissez-faire, since the state, as we have seen, gradually created institutions to facilitate a more interventionist role in financing industrial and other economic activities through the creation of development banks. Rather, the relatively limited protectionist trade policies were due to contingent historical and political factors. One important factor was the *absence* of family conglomerates pressuring the government for widespread industrial protectionism. This was, in turn, due to two factors. First, the weak and fragmented nature of the main business associations during the authoritarian period reduced the capacity of big business for collective action (Gil Yepes 1981). Second, there was only a nascent perception among economic and political elites that explicit import substitution was necessary. This was because imports were easily financed by oil revenues. The lack of any political urgency to industrialization as a national project meant that the family conglomerate groups searched for other routes to accumulate capital. In the context of a rapidly growing internal market as a result of the dynamic growth of the oil industry, they found ample investment opportunities in real estate, construction, and banking (Acedo de Machado et al. 1981). Construction and real estate were a more feasible project to attract the bulk of public spending in the authoritarian period *after* Gómez—from

---

America in the 1930s and 1940s devalued their currencies and imposed import restrictions, which increased the effective rate of protection (Karlsson 1975, 132–33). However, in Venezuela, the local currency, the bolívar, appreciated from Bs5.06/$ to 3.06/$ (as a result of the U.S. devaluation of the dollar). This reduced the effective rate of protection throughout the 1940s and 1950s. By 1960, average nominal tariffs on primary commodities, intermediate goods and durable consumer goods, and nondurable consumer goods were 52.2 percent, 22.3 percent, and 111.6 percent, respectively. By 1989, average tariffs on primary commodities, intermediate goods and durable consumer goods, and nondurable consumer goods were 36 percent, 42 percent, and 61 percent, respectively (World Bank 1990, table 2.5A).

1935 until the end of President Pérez Jiménez's reign in 1957.[9] It was thus no coincidence that public works spending was by far the largest item in the budget of the economic development ministry from 1940 to 1960.[10]

A second important factor in the maintenance of liberal trade policies was the external pressure from the U.S. government. In exchange for providing military and intelligence assistance, successive US administrations insisted that Venezuela maintain markets open to U.S. products, especially in manufacturing. In 1939, the United States and Venezuela signed the Reciprocal Trade Agreement, which was to last, with some modification, until 1971. The treaty effectively gave special tariff concessions to most manufacturing and agricultural products. U.S. officials wanted to capture the internal market of a dynamic economy at a time when protectionism was growing in the region. The very high penetration of the U.S. light industrial products underscored that the first stage of ISI was still in its infancy compared with many Latin American economies at the time.

The motivation of the Venezuelan leaders to sign the trade agreement was to guarantee the sales of Venezuelan oil in its main market by preventing the imposition of U.S. import controls on oil. As a result of this agreement, Venezuela gave duty concessions on eighty-eight items covering 35 percent of total imports from the United States (Astorga 2000, 211). The list of items included most basic consumer goods. In turn, 90 percent of Venezuelan exports, consisting of mainly oil, coffee, and cocoa, were included in the agreement. Throughout the 1940s, the Venezuelan state leaders perceived that protecting the interest of oil exports was more important than promoting a more aggressive state-led industrialization drive (211).

A contingent yet important growth-enhancing consequence of the Reciprocal Trade Agreement was that it forestalled the possibility of "blanket," or indiscriminate tariff protection. That the Reciprocal Trade Agreement (unintentionally) generated a bias toward selective protection is not emphasized in any of the economic histories of Venezuela. Instead, it is treated as manifestation of a dependency relationship with the United States (Petras et al. 1977). The Venezuelan state did negotiate an escape clause, which allowed it to restrict imports of items, such as textiles, food and beverages, chemicals, and rubber

---

9. The share of public spending in all government spending increased from an average of 18 percent in the mid-1930s to over 34 percent in the period 1945–57 (Acedo de Machado et al. 1981, table I-19).

10. From 1940 to 1960, public works spending averaged nearly 50 percent of all spending in the economic ministries (Hausmann 1981, 320).

products, where it deemed there to be significant local production or potential production (Arenas 1990, 28–29).[11]

Even with the escape clause, the Reciprocal Trade Agreement provided a sense of limits on what areas could be protected. This in turn had important effects for resource allocation. First, there was an inevitable selectivity to the protection system simply because so many sectors were barred from protection. Second, selectivity in protection created higher rents in particular sectors and thus could potentially attract resources more effectively than if there had been more blanket protection. The two leading industrial sectors, textiles and oil refining, benefited greatly from this selective protection in the period 1930–58 (Karlsson 1975).

In sum, this section has briefly traced the construction and consolidation of a centralized state with centralized political organizations in the period 1900–1958. The main centralizing forces in the polity over this period were the military and central state economic institutions such as the central bank, the ministry of finance, and state development banks. The consolidation of a centralized state made possible the dynamic development of the oil and non-oil economy, including the manufacturing sector, which emerged later than in many Latin American economies.

8.2 From Centralized to Fragmented Political Organizations: Populist Clientelism in Venezuela, 1958–2005

The nature of the political settlement changed dramatically after 1958. The most important change that occurred was the transition to and the consolidation of democracy. The purpose of this section is to suggest how the political nature of the democratization process was to affect the efficiency of the rent deployment process and macroeconomic management. While rent deployment remained centralized during the democratic period, the motivations of the leadership changed dramatically as the social support base of the state and the grounds for what constituted a legitimate distribution of rents changed. After 1968 the changing nature of the state came to affect negatively the ability of state leaders to manage the rent deployment and management process in line with the economic challenges that came with the switch to policies to develop second-stage ISI industries and big-push, natural-resource-based industries.

---

11. The increasing use of the escape clause through to 1971 meant that the Venezuelan protection system in the 1940s was more import quota–driven than tariff-driven.

In particular, the nature and logic of the Venezuelan polity as it evolved disrupted the ability of the state to maintain selectivity in industrial promotion and made disciplining rent recipients (particularly capitalists in the private manufacturing sector, public enterprise managers, and public sector labor unions) difficult. Moreover, the increasing intensity of electoral rivalry further damaged continuity in policy and implementation that had devastating consequences for economic efficiency. One of the principal processes accompanying electoral rivalry was a growing factionalism within and between the main two political parties. As result, relations between business and the state assumed a growing factionalism and cronyism that undermined collective efforts at industrial restructuring.

The history of the Venezuelan party system in the twentieth century and the broad state-society relations that developed in relation to the political parties can be divided into five distinct periods. The first (1928–48) saw the development of a radical populist mobilization of peasant leagues, labor unions, and middle-class groups challenging the dominant military, landed, and commercial elites. While authoritarian rule dominates most of this period, the origins of a more inclusive populist and clientelist political settlement can be found here as well. The second (1958–68) saw the system of radical populist mobilization evolve into a more conciliatory and consensus-based system known as the "pacted democracy," or what Juan Carlos Rey (1991) refers to as the system of elite conciliation. In the third period, 1973–93, the system of pacted democracy breaks down as right-wing and left-wing threats to democratic regime fades and is replaced by a system of two-party populist electoral rivalry where factionalism and particularistic clientelism become the dominant forms of political competition.

The fourth period (1993–98) witnessed the decline in legitimacy of the two leading political parties, AD and COPEI, and is thus characterized by the increase in multiple-party electoral rivalries. The emergence of multiparty coalitions supporting candidates independent of the two previously hegemonic parties characterized this period, which resulted in the election of an outsider to the political party system, Hugo Chávez, in 1998. The growing increase of a more radical populism is the dominant tendency in this period.

The most recent period (1999–2005) saw the emergence of antipolitics—a political strategy based on anti-establishment, anti–political party, and anti-corruption discourse. Under the Chávez administration, there has been a significant rise in the polarization of politics. The radical and populist nature of the regime creates important antagonisms between the government and its main political support base (the urban and rural poor) and big business, traditional labor unions, and middle-class groups. This period saw a further deterioration in the willingness and capacity of the main political and economic actors to negotiate and reach compromises over the formulation and implementation of coherent economic policies. The general characteristics of this period seem to have continued beyond 2005.

8.2.1 The Origins of Populist Clientelism, 1936–1948

The "system of populist mobilization and conciliation" (Rey 1991) that briefly brought to power a radical democratic junta in 1945 and formed the multiclass populist basis of democratic transition was a long-gestating historical process. The battle cries of the populist mobilization were democratization, economic nationalism, and social justice (Powell 1971, 28). Urban middle-class political party leaders (the so-called Generation of 1928) mobilized labor unions, middle-class urban constituents, and peasant leagues (Powell 1971; Levine 1973, 1978; Karl 1986; Ellner 1999).[12] This process generated significant populist political inclusion not uncommon in Latin America (Di Tella 1970; Kaufman 1977; Cardoso and Faletto 1979).

There were several economic and political processes behind the construction of populist mobilization in Venezuela in the period 1936–48. First, oil mining provided a focal point for working-class mobilization and radical labor unions by middle-class intelligentsia in the late 1930s. Second, the rapid growth in the oil industry, combined with a liberal trade policy with respect to agriculture and manufacturing, induced a rapid exodus from the countryside as coffee and cocoa production collapsed. The rapid decline of strong traditional rural ties, induced by the attraction of urban service sector, and (to a lesser extent) manufacturing jobs, created an opportunity to organize the peasantry. Third, the dominance of oil sector revenues accruing to the state weakened the economic and political power of large landowners. This gave middle-class party leaders the opportunity to form anti-imperialist/anti-oligarchic alliances with working-class groups and the peasantry. The decline in agriculture also meant that large plantation owners lost their political clout and, in fact, started to move substantial amounts of capital into investments in commerce, finance, and real estate. Fourth, oil revenues accruing to the state through taxes on oil multinationals gave the state a degree of autonomy from domestic economic elites and provided the possibility for middle-class party leaders who could

---

12. Coronil (1998, 93–109) discusses the ideas of the "Generation of 28."

control the state to pursue a reformist and more inclusive agenda. Fifth, rapid urbanization in the period 1920–50 (before any substantial manufacturing development was under way) generated pressures for social mobility through state employment. Venezuela was among the most urbanized countries in the world by 1960 (Lattes 2000, 129). Sixth, there was a rapid "tertiarization" of the employment structure, that is, a rise in the share of service sector workers. A group of middle-class service workers emerged; this group was composed primarily of propertied and salaried small artisans and shopkeepers, complemented by a rapidly expanding state bureaucracy, which expanded from 13,500 to 56,000 over the period 1930–45 (Karl 1986, 201). By 1950, over 55 percent of the labor force worked in services.

A classic work on Latin American populism, *Dependency and Development in Latin America* by Fernando Henrique Cardoso and Enzo Faletto, emphasizes that accelerated urbanization and premature tertiarization in employment (*preceding* the emergence of an autonomous and dynamic manufacturing sector) were important preconditions for the development of radical multiclass populist parties (Cardoso and Faletto 1979, 11, 127–40). In Venezuela, the small and largely unprotected manufacturing sector absorbed relatively few rural migrants. As a result, most of the migrants found employment in the service sector. Because of this, the formation of a large, urban, salaried middle class preceded that of a significant and politically powerful industrial working class (Karl 1986, 201). In comparative perspective, the socioeconomic changes occurring in Venezuela in the 1930s and 1940s approximated the patterns that were to underpin populist politics in many Latin American countries. The disruptions and dislocations of rapid urbanization, and the beginnings of industrialization, rather than its absence, tended to produce populism (Di Tella 1970).

Acción Democrática (AD) was the most successful political party leading the mobilization of populist clientelism. Rómulo Betancourt, one of AD's founders (along with other exiled student activists), defined the terms of political thought that prevailed among noncommunist members of the "Generation of 28." All classes, he argued, had a common enemy: imperialism and its local allies, and the feudal structures embodied in *gomecismo*.[13] These premises became the center of AD's self-definition as a multiclass nationalist coalition whose mission was to unite the "masses" against the landed oligarchy and the foreign oil companies.[14]

---

13. These ideas are similar to Andre Gunter Frank's dependency analysis (see, for example, Frank 1969).

14. The class background of the AD leadership in the 1940s was mainly petit bourgeois and working class (Martz 1966, 256). The majority came from families headed by a manual worker,

AD developed a very centralized, almost Leninist-type organizational structure. This was necessary in order to compete with the centralized military state, which was re-enforced by an oil-fueled, centralized patronage pattern created under the Gómez regime. As Moreno and F. Rodríguez (2005) argue:

> Such a state was a formidable opponent for the dictatorship's adversaries in their attempts to promote the adoption of democratic institutions. The innovation of the emerging political leaders was the creation of political parties with broad memberships that could defeat the patronage-based structure of the state by reproducing it. The success of Acción Democrática and COPEI, the two dominant political parties in post-1958 twentieth century, came from being able to substitute the patronage-based web constructed and strengthened by the governments of the Andean hegemony by an eerily similar system of loyalties and favors articulated through populist political discourse and practices. (13–14)

In light of its political adversaries, it was no accident that AD cultivated a radical and populist image.[15] AD became a member of the Socialist International in the early 1940s. It called itself a "revolutionary" party in its statutes; its unofficial name was El Partido del Pueblo ("The People's Party"); and in its early days, conservatives suspected that it was a communist front. In fact, landowners and business conglomerates believed in the 1940s that AD represented a radical future that would lead to the demise of private property, and therefore named the party "Adeco," which meant AD-communist (Karl 1986, 204).

The organizational strength of AD and other political parties forced the creation of a military-civilian junta in 1945 that introduced electoral politics to Venezuela. The ensuing three-year period, known as the *trienio,* marked the decisive introduction of mass politics into national life. Apart from AD, the dominant party in terms of electoral strength, other new parties were formed in this period. The most important parties to form were Committee for the Political Organization and Independent Election, COPEI (a Christian Democratic party whose influence was initially concentrated in the conservative Andean states), the Democratic Republican Union (Unión Republicana

---

farm laborer, or shopkeeper (Coppedge 1994, 12). This pattern was to continue through to the late 1980s (256).

15. AD called itself the "party of the *choludos*" (those who wear sandals), a word which conjured up an image as evocative as the Peronist term *descamisados* (the shirtless ones) (Ellner 1999, 128).

Democrática, or URD, which aligned the urban middle classes with a faction of the military loyal to General Medina), and the Communist Party of Venezuela (Partido Communista Venezolano, a party that focused on oil workers and light manufacturing workers).

The *trienio* brought long-lasting changes.[16] Suffrage was significantly expanded from 5 percent of the population to 36 percent in 1945 (Kornblith and Levine 1995, 42). Elections in 1946 brought AD to power with a landslide electoral victory.[17] In three years, AD raised the number of organized peasant members from 3.959 to 43,302 and increased the number of labor unions from 252 to 1,014 (Powell 1971, 79). AD legislated wages raises and subsidies for consumer goods. Real wages (in 1957 bolívares) increased from 7.15 bolívares in 1944 to 11.71 bolívares in 1948 (Hausmann 1981, 323). Labor union mobilization increased as witnessed in the increase in the number of petitions and strikes, which increased from 15 and 4 respectively in the period 1936–45 to 203 and 70 respectively in the period 1945–48 (see table 8.1). State spending in education, health, water, and communications was extended for the first time to poor, socially excluded groups and regions (Hausmann 1981, 313–56). Ruth Berins Collier and David Collier (1991, 196–201), in their important comparative study of twentieth-century Latin American politics, regard the *trienio* as the most radical experiment by a post–World War II reformist democratic government in Latin America.[18]

The greatest legacy of the *trienio*, however, was the bitter conflicts that ensued over the radical reforms attempted by AD. As Daniel Levine (1989) notes: "It is hard to overstate the depth of the changes *trienio* politics brought" (252). Radical labor reforms and an extensive program of agrarian reform were introduced by an AD elite who did not perceive the need, given its electoral dominance, to consult other organized groups concerning their plans. Economic and social elites began to fear that the radical politics introduced would entirely destroy long-standing privileges and the existing social order. Opposition gathered on the right, represented by the Catholic Church, COPEI, conservative elements in the military, U.S. oil companies and domestic business groups, and the U.S. embassy (Karl 1986, 205–6; Kornblith and Levine 1995, 42–43).

---

16. See Ellner 2008, 41–44, on policies of the López Contreras and Medina administrations that laid the foundations for the progressive policies of the *trienio*.

17. As Ellner (1999) notes: "Like other Latin American populist parties, AD's emergence in 1941 occurred at a time when large number of non-privileged groups were being initiated and mobilized into political activity.... This created a legitimacy crisis for traditional military elites, landlords and business elites" (127).

18. See Ellner 2008, 46.

The three-year experiment with democracy gave way to a decade of repressive authoritarian rule. Under the leadership of General Marcos Pérez Jiménez, AD was banned. Labor, agrarian, and educational reforms were revoked; labor unions were repressed; political opposition was crushed; and more concession-friendly policies toward oil companies were instituted (Kornblith and Levine 1995, 43–44). General Marco Pérez Jiménez was to remain in power until the beginning of 1958.[19] The experience of the *trienio* was to have a profound effect on the nature of political strategy in the post-1958 era of democratic transition and consolidation.

8.2.2 Pact-Making and Democratization: Political Party Centralization and Consensus-Building, 1958–1968

The focal point of democratic transition was the 1958 Pact of Punto Fijo, which provided the institutional mechanisms for consensus building and cooperation between the dominant political party, AD, and the leading opposition party, COPEI. The success at political party organization and the dominance that the two leading political parties maintained in economic, social, and cultural aspects of Venezuelan society in the democratic era have led some to characterize the Venezuelan polity as a *"partidocracia,"* that is, "partyarchy," or "party system" (Coppedge 1994; Kornblith and Levine 1995, 37–38). In the Latin American context, Venezuelan parties were considered among the most institutionalized in terms of co-opting access to the state and mediating the demands of interest groups in civil society (Mainwaring and Scully 1995, 1–34).

Levine (1973) suggests that the political penetration of civil society was so advanced in Venezuela because the parties were founded in an "organizational vacuum." Mass-based political parties began organizing in the late 1930s, following a series of repressive dictatorships that banned the creation of autonomous social organizations, especially popular organizations. Many organizations (including practically all trade unions and peasant leagues) were thus *created* by the new political parties. These organizations were subordinate to political parties because they were never intended to be autonomous.

In terms of our framework, this period is characterized by the transition and consolidation of a (democratic) polity with centralized political organizations capable of achieving consensus on many political and economic issues among competing interest groups. The transition to a consolidated democratic

---

19. For more information about the Jiménez era, see Coronil 1998, 177–200.

system in 1958 was accompanied by significant changes in the nature of the political settlements and economic policymaking in Venezuela. Political change in 1958 was driven by the lessons that political actors perceived from the failed attempt at democracy during the *trienio*.

The main lesson most of the leaders of AD, COPEI, and URD drew from the *trienio* was that polarization, acting without building consensus, and the alienation of powerful minorities would lead to the return of authoritarian rule. Within AD, leaders realized that abusing electoral strength led to its downfall in 1948. The overwhelming electoral majorities in the mid-1940s contributed to AD ignoring the need to form consensus in a politically mobilized society that years of radical populism had generated. Accompanying the willingness of AD to reach compromise, there was substantial public pressure among powerful interest groups to limit the power of AD in the likely event of new elections. Conservative interest groups (including oil companies, the U.S. government, factions of the military supporting Pérez Jiménez, the Catholic Church, and domestic business groups) forced AD into formal and informal agreements to ensure their economic and institutional survival (Karl 1986, 211). This meant that the alliances that underpinned populism would follow a less radical and more consensual path, particularly in the period 1958–68.

The preservation of democratic rule became the primary political objective of political party leaders. As well, the beginning of the transition to democracy was shaped by the idea that preserving democracy was dependent on building coalitions. Compromise and political pact-making among contending groups was deemed not only desirable, but also essential to effective rule.

The perception among party cadres, unions, and business groups that there was a need to secure the fragile alliance for democracy profoundly shaped the actions of political leaders. The recognition that the army, oil companies, and traditional dominant business groups were capable of unraveling democracy produced what Rey (1986) has called an "obsessive preoccupation" with appeasement and accommodation. This concern with fragility was met through formal and informally negotiated compromises.[20]

---

20. The fragility of the transition period was exacerbated by a severe economic crisis confronting Venezuela in the period 1959–69. The fall of Marcos Pérez Jiménez resulted in a growing uncertainty over future national oil policy. Moreover, large-scale public works projects during the Pérez Jiménez regime left large-scale state debts that induced a severe banking crisis. The manifestations of the crisis, as discussed by Baptista (2003, 61–63), were as follows. First, total fixed investment declined by 45 percent in the period 1957–61. Second, oil investment declined by 80 percent in the period 1959–67. Third, the price of oil exports declined from an average of $2.50/barrel in the period 1957–59 to $2.00/barrel in the period 1961–69, reaching a low of $1.81/barrel in 1969.

Political science and historical studies on Venezuela have indeed emphasized that the democratic regime was maintained by a series of political pacts and clientelist links that were an important part of building a multiclass and (largely) urban populist alliance (Karl 1986; Levine 1973; Rey 1991). In the period 1958–68, and indeed for most of the period 1958–93, the Venezuelan political party system was at the heart of linking and mediating between different social forces and in securing the social basis of the state. And AD emerged as the fundamental party.[21] This party system may be characterized as a polity where processes of contestation, conflict resolution, and corruption have been accommodated through populist clientelism. It was a populist polity in the sense that political parties mobilized the masses, in particular, peasant leagues, informal and formal workers, and middle-class professional groups. It was a clientelist polity in the sense that party leaders provides state subsidies and employment to groups in the coalition, who returned the favor by voting for the party responsible for dispensing state patronage

The mechanisms of accommodating rent-seeking groups were achieved through a centralized and well-disciplined party system in this period. Political pacts were the mechanism of compromise and containment.[22] The Pact of Punto Fijo, signed in 1958 and refined during the Betancourt administration (1959–63), was the cornerstone of the pact-making process. In this pact, the major noncommunist parties (AD, COPEI, and URD) pledged that each party would respect the electoral results, whatever the outcome.[23] The parties maintained a "prolonged political truce" by depersonalizing debate and ensuring interparty consultation on policy issues, and by ensuring there would be a proportional distribution of state benefits. The Pact of Punto Fijo also reaffirmed the central role of parties in channeling interests and distributing resources.

Fourth, the exchange rate, after thirty years of stability, was devalued by 36 percent in 1961. Finally, the financial bailouts of the banking system averaged over 5 percent of GDP and reached 10 percent of GDP in 1961.

21. A "fundamental party" (Gutiérrez Sanín 2003) can be viewed as the natural governing party (in terms of electoral success) as well as the party whose mobilizations and strategies were central to regime founding. Acción Democrática is referred to as a fundamental party in the period 1958–93 because it never yielded its position in this period as the single biggest party in the either La Cámara de Diputados or El Senado and because it won five of the seven presidential elections.

22. On the historical role political pacts have played in transitions to democracy in different contexts, see O'Donnell and Schmitter 1986 and Higley and Gunther 1992.

23. The first agreement made among the leaders of AD, COPEI, and URD (the so-called Pacto de Nueva York) was made abroad while each of the leaders was still in exile. The general thrust of the agreement was that containing rivalries among the three main parties would improve the prospects of the democratic struggle (Ellner 1997, 204 n. 1).

The curtailment of radicalization and partisan conflict was further embodied in the second major pact signed in 1958, known as the Minimum Program of Government. All parties agreed to accept a development model based on a centralized state–led deployment of oil rents accruing to finance social welfare spending, state production in strategic industries such as steel and petrochemicals, foreign investment, and import protection and subsidies for local private industry.

The legitimacy of the AD was also enhanced by the implementation of a popular if modest land reform program in the early 1960s. Article 105 of the 1961 Venezuelan Constitution states: "The existence of *latifundia* is contrary to the interests of society" (quoted in Neuhoser 1992, 128). After 1958, AD used this authorization to carry out a land reform program, even though opposition to the original Agrarian Reform Law of 1948 had helped provoke the overthrow of the *trienio* democratic government (ibid.; Martz and Myers 1986).

The viability of the populist and clientelist pact also depended on balancing co-optation with the accommodation of middle- and working-class demands. The ability of AD to penetrate and co-opt the labor movement was a critical factor in the viability of the pact.[24] AD leaders were able to persuade the main unions to avoid strikes and limit wage demands. This was important in allaying the fears of business association leaders, who were expecting a return to the polarization of the 1940s. Business elites conditioned their support for democracy on the willingness and ability of AD to control strike activity. As indicated in table 8.1, a comparison of indicators of labor conflict (including dispute petitions and strikes) in different periods indicates the degree of relatively peaceful labor relations achieved in the decade following the signing of the Punto Fijo Pact in 1958.

In the period 1958–68 (under two AD administrations), annual strikes averaged 19.5 percent per year and dispute petitions averaged 45 per year in comparison to the much higher level of 70 strikes per year and 203 dispute petitions per year in the more radical populist AD program during the *trienio* (1945–48). While there is evidence of restraint in the period 1958–69, the number of strikes and petitions was still much higher than in all the years of military rule when labor unions were repressed harshly.[25] This was telling of the level of political

24. Party organizers, starting in the late 1930s, founded most labor unions. Since its second National Congress in 1944, AD has had the largest representation in the national labor federation, La Conferderación de Trabajadores de Venezuela (CTV) in the democratic era (Coppedge 1994, 30–35). In the period 1959–69, 50 to 70 percent of the members of the National Congress and Executive Committee of the CTV were affiliated with AD (31).

25. The general trends in labor strike activity are the same even after controlling for increases in the number of workers, as indicated in the level of disputes/petitions, strikes, and total worker hours lost per one million workers (see table 8.1).

TABLE 8.1 Indicators of labor conflict in Venezuela, 1935–1989 (all indicators are annual averages; numbers in parentheses are per 1 million workers)

President	Term of office	Party affiliation	Disputes and petitions	Strikes (legal and illegal)	Total worker hours lost (in thousands per year)
Authoritarian					
J. V. Gómez	1908–35		n.a.		
E. López Contreras	1936–41		4[a] (3.5)	1[a] (1.0)	n.a.
I. Medina Angarita	1943–45		11[b] (7.6)	3[b] (2.1)	n.a.
Trienio of democracy (1945–48)					
R. Gallegos	1947–48	AD	203 (135)	70 (46.7)	n.a.
Authoritarian					
Military junta	1948–50		5 (3.1)	5 (3.1)	n.a.
M. Pérez Jiménez	1950–58		5 (2.5)	6 (3.0)	n.a.
Democratic era					
R. Betancourt	1959–64	AD	52 (22.6)	20 (8.7)	265 (115.2)
R. Leoni	1964–69	AD	37 (13.7)	19 (7.0)	81 (30.0)
R. Caldera	1969–74	COPEI	222 (65.3)	166 (48.8)	2,116 (622.4)
C. Andrés Pérez	1974–79	AD	190 (45.2)	150 (35.7)	742 (176.7)
L Herrera Campins	1979–84	COPEI	181 (36.2)	155 (31.0)	1,624 (324.8)
J. Lusinchi	1984–89	AD	95 (15.6)	30 (4.9)	255 (41.8)

SOURCE: Banco Central de Venezuela, *Estadísticas socio-laborales de Venezuela*.
[a] For 1939 only.
[b] For 1943–44 only.

demands that populist clientelism was generating even during the era of pact-making. The increase in strike activity was to increase even further in subsequent periods as factionalism and interparty rivalry increased.

Significant increases in political accommodation also took place in other areas. One telling indication of change in the balance of political power can be seen in the patterns of social spending in the three decades following 1958. One of the key strategies of the political parties was to build middle- and working-class clientele. Increase in social spending became one of the main ways to deliver jobs and services to the middle- and lower-income groups, and was important for preempting more radical demands for distribution. Shifts in the composition of public spending were dramatic when compared with the era of authoritarian rule in the twentieth century. As Miriam Kornblith and Thais Maignon (1985, 205) demonstrate, social spending (which includes health, education, water and sanitation, and housing) as a percentage of total state spending grew from an annual average of 11.4 percent under Pérez Jiménez to 28.1 percent in the period 1958 to 1973. Between 1969 and 1973, the years immediately preceding the oil boom of the 1973, social spending averaged 31.4 percent of total spending.

A second telling indication of the scale of accommodation can be seen in the extent to which the dominant political parties permitted the appropriation of land by low-income groups in Caracas. The Pérez Jiménez dictatorship looked unfavorably on the spread of informal housing in urban centers. "[His] government's solution to the *barrios* was the bulldozer" (Karst et al. 1973, 7). After the overthrow of Pérez Jiménez, and before the election of Betancourt, the provisional military junta suspended evictions in urban shantytowns and offered public relief to the unemployed. This resulted in more than 400,000 mostly rural, poor people moving to Caracas in just over a year (Davis 2006, 59). In the next decade, both main political parties, AD and COPEI, built much of their support base among the poor by allowing the expansion of informal housing in the hills around Caracas (ibid; Pérez Perdomo and Bolívar 1998, 125–26). Policies of promoting appropriation of urban land contributed to rapid rates of urbanization in Venezuela in the 1960s and 1970s. The legacy of these policies was that, by 2005, Caracas contained one of the largest "mega-slums" in the developing world (Pérez Perdomo and Bolívar, 28).[26]

---

26. A decidedly negative aspect of these polices was that many of the shantytown houses were built on unstable hillsides and in deep gorges surrounding the seismically active terrain of Caracas. The massive inflow of mostly Venezuelan rural poor (and Colombian immigrants in the 1970s and 1980s) led to cut-and-fill construction, which destroyed much of the vegetation that had hitherto

It is generally accepted that there was a significant and rising trend in corruption in the three decades after the Pact of Punto Fijo. This suggests that the political structures and practices that generated corruption were embedded in larger political strategies and institutional structures for maintaining social order and a minimum degree of capital accumulation. While most Venezuelan analysts see corruption as a problem of institutional design, Ruth Capriles (1991) makes the perceptive argument that administrative corruption in Venezuela has also been the result of a political project with a deliberate objective, namely to generate a political clientele for political party cadres. Accordingly, the middle class was developed as a related product of the clientelism that accompanied the dominant political project (ibid.). As long as a minimum degree of economic and political stability was maintained, political and economic corruption that sustained the populist clientelism was viewed as a legitimate cost. To the extent that patronage was centralized and predictable, it had less of a negative effect on the effectiveness of economic policy and implementation than it was to have during subsequent decades when political party fragmentation and polarization emerged.

The main legacy of the Punto Fijo Pact was to institutionalize a centralized form of political clientelism in which the political parties were the main channels of patronage. In terms of rent deployment and management, the pacts reinforced the central role of the executive in agenda-setting, policymaking, and implementation. In the period 1958–93, Venezuela was to remain a presidentialist system bonded to political parties. A key component of the pact that developed was that regardless of which party won the elections, each of the main parties was guaranteed some access to state jobs and contracts. This included a partitioning of the ministries, access to intermediate political posts in the bureaucracy, and more generally, a spoils system that would ensure the political and economic survival of all signatories, which included the main labor union federation (CTV) and the main umbrella business association (FEDECAMERAS).[27] The procedures and substance of the Pact of Punto

---

prevented major mudslides (Davis 2006, 122–23). As a result, major landslides and slope failures increased from less than one per decade before 1950 to an average of two or more per month at the beginning of the twenty-first century (123). In mid-December 1999, northern Venezuela (and particularly the Avila mountain area surrounding northern Caracas) was hit by torrential rains, which caused massive mudslides and debris flow which killed an estimated 32,000 people and left 140,000 homeless and another 200,000 jobless (123).

27. Levine (1973), Karl (1986), and Myers (2004) discuss the formal and informal arrangements of the Punto Fijo Pact in detail. Pact making was indeed a feature of many Andean countries in the 1950s and 1960s. Political pacts were a particularly prominent feature in Colombia and, to a lesser extent, Peru. Far from being a uniquely Venezuelan phenomenon, the division of ministries along

Fijo, and particularly the ethos of cooperation between political parties and the avoidance of conflict, were to be influential factors in political patronage throughout the democratic era (Naím and Piñango 1984).[28]

In effect, the decision to divide up ministerial posts and deploy resources to meet political criteria laid the seeds of politicizing the public administration, and allocating rents on the basis of political rather than economic criteria. The consolidation of populist support was to depend on tangible patronage, which AD (and eventually COPEI, its main rival after 1958) was to provide. Clients of the political parties (especially labor unions and the urban and rural poor) generally offered their votes and political loyalty in exchange for material benefits, while entrepreneurs offered substantial campaign financing in exchange for subsidies of various types. The patronage took many forms, both legal and corrupt. These included cheap investment credit, tariff protection and import licenses, employment opportunities in the public sector, housing and mortgage credits, and price controls on basic consumer goods.

It is important to highlight that the era of pacted democracy did *not* eliminate regime fragility.[29] Many left-wing groups, including the Communist Party and radical elements within AD, were demanding a more radical social program than the one AD offered. However, the lessons of the demise of radical populism meant that the leaders of AD, COPEI, and URD and leading business group members had agreed in clandestine meetings to exclude the Communist party and radical left-wing groups from sharing power in any future democratic government.

Ideological disputes over the exclusionary nature of the pacts had two major political costs. One was that AD suffered three damaging splits—in 1960, 1962, and 1967—which cut deeply into its electoral strength (Neuhoser 1992, 123–24; Coppedge 1994, 54–56, 98–103)[30] and was to underlie the factionalism

---

clientelist lines was a feature of different party systems throughout Latin America in this period (Kaufman 1977).

28. While AD was the dominant party in electoral terms in the period 1958–68, the victory of Rafael Caldera (founder of COPEI) in the 1968 presidential election (primarily because of damaging splits within AD) converted the Social Christian Party into an "alternative electoral pole" to AD and thus created the sense that there was genuine electoral competition.

29. The fragility of the Betancourt regime was particularly noticeable in urban and industrial areas (Ellner 2008, 60). While Betancourt defeated transitional president Wolfgang Larrazábal in the first elections with a comfortable national margin of 49 percent to 36 percent, most of this margin was due to AD's support in rural areas. The URD and PCV came first and second in Caracas, where AD came last. Larrazábal also won in the central industrial states of Carabobo, Aragua, and Miranda, where the greatest concentration of manufacturing workers resided.

30. The conciliatory attitude of the U.S. government toward AD also played a role in inducing splits in the center-left coalition. At the height of the Cold War, President John F. Kennedy praised

and divisions that were to plague AD in the three decades after 1958 (Ellner 1999, 104).[31] Second, the left turned to armed insurgency in the period 1960–67. While the exclusion of the left reassured business and the military, "the success of the Pacto de Punto Fijo cost Venezuela the largest guerrilla movement in Latin America" (Przeworski 1991, 91). The fragility of the regime meant that rent deployment based on political rather than economic considerations was required to accommodate the core constituency of big business, national labor unions, the military, and the Catholic Church.

In this context, it is important to highlight that the construction of privileges to core political constituents and elites is not an unusual mechanism for maintaining political order or even preserving democracy. North and colleagues (2007) argue that models of state-building make two assumptions that lead to misunderstanding how and why polities form. The first is that the state is modeled as a single actor. The second is that the state has a monopoly on violence.[32] Following the insights of Thomas Hobbes, they argue that a more realistic place to begin is to assume that the potential for violence is prevalent throughout society rather than concentrated or monopolized. That is, it makes sense to explain rather than assume that a particular state has a monopoly on violence. The establishment of political order and peace in the model requires the creation of incentives for groups to compete for resources through nonviolent mechanisms.

The principal solution through history to the classic Hobbesian problem of endemic violence is the creation of what North and colleagues (2007) call

---

Venezuelan democracy as a true alternative to communism and authoritarianism in the Western Hemisphere (Myers 2004, 11). The U.S. position undercut much of the old guard, who had equated opposition to AD with anticommunism. This rapprochement between the U.S. government and AD infuriated PCV, the leftist factions in AD and URD, and university students (Myers 2004, 19).

31. The early post-1958 period was indeed characterized by the rapid proliferation of parties across the political spectrum: from Marxist and socialist parties supporting armed revolution, to parties, such as the National Civic Crusade, which promoted the return to power of the former dictator, Marcos Pérez Jiménez (Neuhoser 1992, 123). In 1958, four parties had significant electoral support (at least 5 percent), but by 1963 the number had increased to five parties, and in 1968 there were six parties (Gil Yepes 1981, 55). In the presidential elections of 1963 and 1968, four candidates captured at least 15 percent of the vote; in 1968 less than 10 percentage points separated the top four candidates—29.1 percent to 19.3 percent. In the first three congressional elections (1958, 1963, 1968), parties of the left steadily increased their share of congressional votes from 7.5 percent in 1958 to 24.4 percent in 1968, which was a higher share than that of COPEI and only 1.3 percentage points behind AD's share. AD's congressional representation fell rapidly: from 54.9 percent in 1958 to 29.6 percent in 1968 (Neuhoser 1992, 123).

32. Well-known examples include Olson's (1993) stationary bandit model and North's (1981) and Levi's (1988) revenue-maximizing monarch, as well as standard theories of rent-seeking (Buchanan et. al. 1980).

limited access orders (as opposed to the much rarer open access orders, which characterize advanced market economies). The limited access order creates limits on the access to valuable political and economic functions as a way to generate rents. The dominant coalition creates opportunities and order by limiting the access to valuable resources or valuable activities to elite groups. When powerful individuals and groups become privileged insiders and thus possess rents relative to those individuals and groups excluded (and since violence threatens or reduces those rents), the existence of rents makes it in the interest of the privileged insiders to cooperate with the coalition in power rather than to fight. In effect, limited access orders create a credible commitment among elites that they will not fight each other. Of course, the maintenance of rents depends on the stability of the coalition in power. Thus, through limited access orders the political system creates rents as a means of solving the Hobbesian problem of endemic violence and political disorder.[33]

While pacts were the foundation of political clientelism and the dominance of political criteria in distributing rent deployment in Venezuela, threats from both the right and the left reinforced (in the first decade after 1958) the need to use resource rents to build political clientele through centralized and disciplined party structures. The threats to regime survival were, according to political analysts, the main factor behind the *avoidance* of factionalism and tension that were inherent in the alliance of populist clientelism. Threats to regime survival also limited the proliferation of subsidies and rules that were to plague the polity after 1968 (Karl 1986; Kornblith and Levine 1995). As Levine (1978) points out: "More than any other single factor, the development of a leftist strategy of insurrection in the early 1960s consolidated democracy by unifying centre and right around AD in response to a common threat" (98). The maintenance of stable macroeconomic rules (e.g., stable fixed exchange rates, low inflation, fiscal balance) and generally rapid, though slowing, economic growth (Hausmann 1990, 333–59) was a manifestation of the stable and legitimate central public authority in the early years of democratic rule

While oil largesse did facilitate the level of patronage required to secure the consolidation of pact-making, it did not guarantee it. If mineral resource abundance explains why these pacts were successful, as, for example, Kevin

---

33. The model in North et al. 2007 generates two important research challenges. The first would be to examine the conditions under which coalitions that provide peace become fragile. Second, the North et al. (2007) model does not explain *why* limited access orders (which the authors posit is the relevant political form for all developing countries) perform so differently or why the same limited access order performs so differently over time.

Neuhoser (1992) claims, then it would be difficult to explain why pacted settlements were achieved in petroleum-poor settings such as Mexico in the 1930s, Costa Rica in 1948, or Colombia in the late 1950s (Burton and Higley 1998, 68), or why no other oil-exporting developing countries have made a transition to democratic rule (pacted or otherwise). Pact-making more generally is the result of purposeful institutional design that depends on contingent combinations of circumstances: bold and unpredictable elite choices, delicate balances of elite autonomy, and mass pressures (ibid.). Settlements tame politics by establishing a basis for restrained and peaceful competitions between major elite camps. This was the main consequence of the Venezuelan settlements: an unbroken string of peaceful and binding electoral contests in a country that had no prior experience of such contests since the first post-settlement election in December 1958 (ibid.).

Such settlements did not preclude a resurgence of radical political forces. Rather, settlements made it likely that the main elite camps that forged them would cooperate to defeat or stifle resurgent radical forces. In Venezuela, governments dominated by AD between 1959 and 1969—despite internal AD splits and the loss to other centrist and center-right groups and parties—successfully resisted and increasingly marginalized Cuban-supported guerrilla movements, in large measure because AD had strong support from most business, trade union, and peasant organizations. In sum, both perceived and real threats to democratic rule and the construction of political pacts were instrumental factors behind the consolidation of a polity characterized by centralized political organizations and substantial consensus-building and cooperation between political parties in this period.

8.2.3 Growing Factionalism in the Era of Two-Party Electoral Rivalry, 1968–1993

The nature of populist clientelism changes gradually after 1968. The *pactismo* of the first decade gave way to electoral party rivalry and factionalism between and within parties, especially between 1973 and 1993. Several political observers note that the very consolidation of the regime, the defeat of the guerrilla movement, and thus the decline in threats to the regime reduced the urgency for reaching consensus through the pact-making process (Rey 1991, 557–67; Kornblith and Levine 1995, 48–58; Levine and Crisp 1995, 227–32). The growing importance of factions and factionalism between and within political parties is well documented in Venezuela from 1973 to the mid-1990s (Coppedge

1994; Ellner 1996), though there were seeds of declining cooperation between and within political parties as early as the mid-1960s (Ellner 1985; Rey 1991; Coppedge 1994).[34]

In comparative terms, the development of factions and factionalism is not an uncommon feature of economic and political conflict in societies undergoing either state-led or primitive accumulation (Sandbrook [1972] 1998). Following the standard definition in political science, a faction is a conflict unit operating within, and struggling for, control of a formal organization, which, in practice, is usually a political party (70). Factionalism can be a defined as a form of conflict (over wealth, power, and status) that unites people from different social categories with the aim of advancing the participants' individual (usually material) interests (64). The general structure of the faction in Venezuela is a pyramid organization headed by a high-level political party member allied with one of more big business conglomerates, small and medium-sized capitalists, labor union leaders, and lower-level political party cadres (Capriles 1991). Generally, factions are fluid multiclass alliances that are constantly undergoing changes in composition and loyalty given the inherent difficulties of defining the boundaries of membership. Politics in the context of primitive accumulation is the main avenue of upward economic mobility; yet the faction is usually unable to absorb the large number of ambitious individuals who emerge demanding privileges.

Several distributive conflicts accompanied the growing factionalism of the pacted democracy. The first concerned increases in labor conflicts. The AD-dominated Congress initiated more radical labor laws and encouraged a more combative labor movement. As a result, both COPEI administrations (1969–74 and 1979–84) were plagued by large-scale labor conflict, which increased the polarization and political instability in the country (Gelb and Associates 1988, 289–325; Neuhoser 1992; Coppedge 1994, 34). As can be seen from table 8.1, the number of labor disputes, strikes, and man-hours lost increased dramatically in the period 1969–84, compared with the first decade of democratic rule (1958–68), when centralized pact-making was still intact. The period 1984–89 saw a fall in labor strife, but this was due to the return of AD to power, and the attempts by the government to return to a social pact with labor during the AD-led Lusinchi administration (1984–88).

The second important distributive conflict occurred within business conglomerate groups tied to different political factions. The splits in AD in the 1970s

---

34. See also Monaldi and Penfold 2006 for an excellent discussion of the gradual decline in political party cooperation and growing executive-legislative tensions over the period 1958–2005.

were based on the growing rift between the old guard of AD, led by its founder, Rómulo Betancourt, who favored links with the traditional family conglomerates and a more limited state role in production, and the faction led by Carlos Andres Pérez, which demanded a more radical state program based on more public enterprise production. Moreover (and of considerable significance), the Pérez-led faction was committed to financing an emerging set of smaller-scale entrepreneurs so they could challenge the economic dominance of the more established family conglomerates. These rifts within AD can be traced back to some of the same groups that led the split from AD in the early 1960s (Ellner 1999). Gumersindo Rodríguez, for example, who was a member of the left splinter group called Los Muchachos, became Carlos Andres Pérez's powerful planning minister during the period 1975–79 (Coppedge 1994, 98).

Rapid changes in patronage and policy and contestations to those changes were most evident in the first administration of Carlos Andres Pérez (1974–79). In this period, growing factionalism within AD between the old guard (headed by former president Betancourt) and the new guard (headed by Pérez) came to the fore as Pérez used the massive state resources to provide subsidies to and procurement projects for an emerging set of family groups who had contributed to his campaign for the presidency (Coppedge 2000).[35] The economic groups close to the Pérez administration were popularly known as the "Twelve Apostles," many of whom came, like Pérez from the Andean region.[36]

The ties between President Pérez and the Twelve Apostles were forged largely during his acrimonious power struggle to assume the AD presidential candidacy; his isolation in AD and lack of control over the party hierarchy convinced him of the necessity of establishing a power and financial base outside the party machinery. The names of these businessmen appear in many of the financially most lucrative contracts awarded in the period 1974–78, including the Guri Dam, Cementos Caribe (the licensing of a new cement factory), the new Zulia steel mill, the Pentacom petrochemical project, and the construction of Parque Central (the largest shopping mall/office complex in South America at the time), among many others (see Karl 1982, 463–512). For Pérez and his faction

---

35. Whereas the average cost per vote in developed country elections averaged from two to three dollars in 1978, in Venezuela the cost approached fifty dollars (Myers 1980).

36. There were more than twelve individuals involved, but the nickname became part of Venezuelan political parlance. The term was coined by Duno (1975) in a political tract critical of the Pérez regime. Some of the main family groups included the following: Tinoco, Pérez Briceno, Asis Espejo, Febres Cordero, Di Mase, and Cisneros. The Cisneros group became the most successful by the 1990s, and was one of the few groups to develop extensive commercial holdings internationally.

of AD, the Twelve Apostles represented an attempt to democratize capital by breaking the hegemony of traditional large family business groups.

The rise of emerging business groups also profoundly affected the main national business chamber, FEDECAMERAS (Ortiz 2004, 79–80). In 1979, amid bitter conflict, control passed from traditional economic groups to Ciro Añez Fonseca, a former general manager of FEDECAMERAS, who represented an alliance of emerging groups and long-marginalized business elites from the interior of the country. The large traditional business groups would never regain their former power and influence within FEDECAMERAS. More important, the political basis for collective action and consensus-building within the large business sector was severely curtailed.

The third dramatic conflict that characterized the post-1968 era concerned the increasing resistance from important factions of the big business community and the rival political party, COPEI, to the growing role of state production in the economy. As noted in Chapter 7, the Pérez administration sought to make state-owned enterprises the focal point of the big push, natural-resource-based industrialization strategy. The increased role of the state in the so-called strategic sectors of the economy (oil, iron ore, steel, petrochemicals, hydroelectric power, bauxite, and aluminum) represented an important disruption one of the main political rules of the game, that the state should facilitate private sector investment in industry through tariff protection and subsidized long-term credit. The post-1973 era saw the emergence of the state challenging the private sector in industrial production. As the economic historian Mauricio García Araujo (1975) noted: "This generation of Venezuelans is witness to the rise of a special new oligarchy ... fifty-six people generally designated by political or party criteria—who decide how to spend, how to employ, how to administer 19,298 million bolívares a year, each year. The economic power that is accumulating in this bureaucratic oligarchy ... has no parallel in the rest of the economy" (quoted in Karl 1997, 142).

Overall, factionalism led to an increased contestation over patronage, which in turn, fueled a greater degree of politicization and political polarization in the era of two-party electoral rivalry. While the Pérez faction of AD wanted to redress the dominance of the old oligarchy, many other factions within AD and COPEI and within the business associations challenged the rise of new oligarchies in the private sector (such as the Twelve Apostles) and the rise of the state enterprise oligarchy within the state. As such, it is not surprising that Ellner (1985, 38–66) finds a breakdown of interparty agreement in the period 1976–80 compared with the period 1967–71. From 1976 to 1980 there was a significant increase, according to Ellner, in the questioning of the motives of political rivals, accusations of violations of the political rules of the game, and constant warnings regarding the gravity of the political situation. Francisco Monaldi and Michael Penfold (2006) similarly argue that interparty cooperation and executive-parliament consensus increasingly declined in the period 1973–93. Other political analysts have argued that the increase in political factionalism was accompanied by an increase in whistle-blowing and the use of the corruption scandal in the 1980s and 1990s as a weapon of political competition (Capriles 1991; Pérez Perdomo 1995; Karl 1997, 138–85). Such accusations were further fueled by the divisions between and within political parties and between rival big business groups that were the result of the manner in which economic liberalization was introduced (as discussed in Chapter 5).

It is worth noting that the coincidence of oil nationalization in 1976 and the very large increase in fiscal resources due to the oil booms undoubtedly upset the political balance in the country. In particular, it could be argued that this coincidence upset the balance of interparty relations, and thus fueled increasing factionalism as competing interest groups vied to capture power in an increasingly state-centered economy. The political pacts were formed and consolidated in the 1960s, a period of relatively stable (and even declining) oil-export earnings. As a result, the state was more dependent on domestic and foreign private sector investment to achieve economic growth. The oil nationalization likely lessened the counterweights (that is, the checks and balances) to the state, and thus perhaps lessened the sense that there should be a limit to centralized discretionary authority. The reduction in this sense of limits probably increased the abuses of power and made corruption less predictable and more contentious as a result.[37] In any case, this period was characterized by the fragmentation of the polity and a decided reduction in cooperation both within and between political parties.

8.2.4 Further Political Fragmentation: The Emergence of Multiparty Electoral Contestation, 1993–1998

The period 1993–98 represents a decline in the two-party hegemony as economic decline and the divisiveness of economic liberalization led to dissatisfaction with the two main parties by the electorate and growing factionalism within the

---

37. While corruption is likely to increase transaction costs in an economy because such transactions are secret and unenforceable in the legal system, the growing unpredictability of corruption (that is, a greater uncertainty about whether a bribe "gets what a briber seeks") will increase

two main political parties. Two main factors led to this growing factionalism.[38] The first (and perhaps most important) factor was the decisions of the two most popular and influential leaders of Venezuela's two main parties, Carlos Andres Pérez (AD) in 1989 and Rafael Caldera (COPEI) in 1993, to distance themselves from their parties. Both leaders seized upon crisis situations to reinvent themselves as political outsiders. They did so with political messages and platforms that were the opposite of what they and their respective parties had established over the previous forty years. Dramatic policy switches have been shown to be a destabilizing event for fragile democracies (Stokes 1999). The decision of Pérez, leader of AD, to implement neoliberal reforms through the use of nonparty technocrats was detrimental in two ways. First, Pérez's party-neglecting strategy (Corrales 2002) accentuated factionalism within AD and made implementing reforms politically contentious. Many AD party members blocked reforms in Congress and ultimately supported the impeachment of Pérez. They considered his strategy a betrayal on two fronts: one for implementing neoliberal policies, and two for naming very few AD party members to the Cabinet. Second, the launching of a neoliberal economic program went against the policies that had defined AD's legitimacy for decades. AD became a fundamental party by championing the working class and peasants and had built its reputation (however tarnished it had become) by advocating and implementing state-led developmentalism, anti-imperialist struggles, and economic nationalism. Neoliberal reforms launched by AD's most established politician divided what AD stood for in the minds of their militants and sympathizers. The loss of AD's party identity most likely contributed to the significant decline in party identification through the 1990s.

In 1993, Rafael Caldera lost the nomination of the party he founded, COPEI, to Oswaldo Alvarez Paz, one of the emerging regional politicians that decentralization and direct state elections (established in 1989) had created. Caldera won the presidential election with a loose coalition of small left-wing parties under the umbrella of the new "party" he founded, Convergencia (Convergence). Convergencia's main ally in government was the Movimiento al Socialismo (MAS, the Movement Toward Socialism), which was an established, but small

---

transaction cost uncertainty in the economy, which is analogous to a growing insecurity of the property rights purchased as a result of corrupt/illegal transactions. The political conditions under which some forms of corruption produce greater security of property rights (as, for example, in China) and others produce less security of property rights (as, for example, in Venezuela, particularly in the post-1973 period, or in Pakistan) remains an important area for further research.

38. Chapter 5 also examines the political economy underlying the breakdown in cooperation within and between the main political parties in the period 1989–98.

left-wing party. The rapid but short-lived rise of Convergencia had serious consequences for the cohesion and legitimacy of the party system, hitherto controlled by AD and COPEI. First, Caldera's victory was proof that it was possible to win the presidency by running outside traditional party affiliation. Second, Caldera split the COPEI vote, and thus divided what was a solid center-right nationally organized *alternative* to AD and civil society. COPEI did not survive this fracture of its middle-class and business support.[39] Third, this period sees a growing proliferation of regionally based and personality-centered political parties competing for the presidency and Congress in the period 1993–98 (Molina 2004, 169).[40] With the rise of Convergencia and La Causa R (a party that successfully competed for the labor union support AD traditionally received), representation of the center-left vote became divided between these two parties, AD, and MAS, at both national and regional elections.

Evidence of the growing fragmentation of the party system can be seen in the increase in the number of effective parties (ENP) contesting elections.[41] In the period 1973–88, the ENP averaged 2.5 for the presidency and 3.3 for the Congress (Crisp 1997, table 4.1). In 1993, the number of effective parties competing for the presidency rose to 5.6, and the number of effective parties in Congress rose to 5.6, though AD and COPEI remain the two largest parties in both chambers (ibid.). In the 1998 elections, the ENP reached 6.1 for the legislative elections (Monaldi and Penfold 2006, 27).[42]

The very negative and disappointing experience of government in the Caldera administration (1994–98) further undermined the legitimacy of the party

---

39. As Molina (2004) notes: "The event that seems to have undone Venezuela's two-party system was the decision of COPEI's founding leader, Rafael Caldera, to abandon the party and run for the presidency as an independent.... Had Caldera not divided the party he founded and nurtured, it is probable that in 1993 the pattern of alternation of power between the two Punto Fijo regime-sustaining political parties would have withstood the challenge of . . . anti-establishment forces" (163–64).

40. Several dissident AD regional organizations have been the source of new parties: Alianza Bravo Pueblo and Un Nuevo Tiempo (Molina 2004, 169). COPEI sustained even more damaging defections: three new parties have defected from its one-time clientele: Convergenica (as mentioned above), Proyecto Venezuela, and Primera Justicia (169).

41. The effective number of parties is N = ? vi 2, where vi is the fractional share of votes for the *i*th party.

42. Further evidence of growing political instability and party fragmentation was frequency with which electoral rules were changed. While electoral rules were stable in the period 1958–93, Monaldi and Penfold (2006, 29) note that, in the period 1993–2000, electoral rules for the legislature were changed four times and the elections of 1993, 1998 and 2000 had different electoral formulas. The authors argue that the proliferation of institutional reforms eroded the strict control that party leaders had over nomination procedures, which, in turn, weakened party discipline in the legislature.

system. Caldera, was after all trying to govern with political party input (including a rapprochement with AD), as opposed to Pérez, who was convinced that AD, and the party system more generally, were moribund. Three manifestations of this disappointing performance were noteworthy. First, Caldera inherited one of the worst banking crises in Venezuelan history (see Chapter 5) and exacerbated the situation by shutting down the largest bank, Banco Latino, which was owned by an economic group close to the previous Pérez administration (M. Rodríguez 2002). Second, there was a growing incoherence in state ministries as Caldera tried to accommodate the fractious coalition. Finally, there was no clear and coherent economic strategy; and as a result, there were no fewer than four economic plans initiated in Caldera's government and a large turnover of ministers (De Krivoy 2002).

The introduction of political decentralization and fiscal federalism in the early 1990s was the second main factor that contributed to the fragmentation and loss of party discipline in the two main parties (AD and COPEI) in the democratic pact. According to Michael Penfold-Becerra (2002), the post-1989 reforms, which initiated the direct election of mayors and governors and led to the devolution of state spending to states and municipalities, lowered the barrier to entry of marginal and emerging parties and encouraged politicians within the two main parties to develop local alliances and assert autonomy from national party bosses. Decentralization, in the context of rapid economic reforms and economic crisis, along with relentless media coverage of corruption scandals concerning the state and political parties, provided opportunities for marginal but strong parties such as MAS, but more important, embryonic and structurally weak political "parties," such as La Causa R and Proyecto Venezuela, and later Movimiento Quinta República (MVR, Fifth Republican Movement, headed by Hugo Chávez Frías) to compete in elections at the state level.

The emergence of federalism drastically changed the alliance strategies followed by political parties. AD, COPEI, and MAS all developed alliance-bloc systems with other (often smaller) political parties as a strategy to protect their regional leaderships (Penfold-Becerra 2002). In 1989, AD established alliances with an average of 2.18 parties per state for the twenty-two gubernatorial elections. By 1998, AD allied with an average of 7.5 parties per state. COPEI was allied with an average of 5.57 parties per state in 1989 and with an average of 9.0 parties per state by 1998. The electoral premium COPEI obtained from these alliances increased from an average of 7 percent in 1989 to 20.6 percent in 1998. In 1998, Chávez's electoral organization, MVR, was gaining strength at the regional level and by 1998, on the coattails of Chavez's victory, won 17.7 percent of the governorships. What is telling about these regionally based parties is that they never organized at the national level. In sum, this period is characterized by a growth of multiparty competition and the fragmentation of the party system, which reduced the possibility of effective coordination of government policies.

8.2.5 The Rise of Antiparty Politics and the Growing Polarization of Politics, 1999–2005

The failure of the political parties to meet economic challenges along with the growing polarization that neoliberal reforms opened the space for the emergence of a political outsider. During the electoral campaign of 1998, Hugo Chávez and his electoral organization, Movimiento Quinta República (MVR), campaigned on an anticorruption, anti-neoliberal, anti–political establishment discourse that called for the transformation of the political system and the Constitution.[43] The focus of his platform was the promise of a constituent assembly. Growing levels of poverty and the policy switch to a neoliberal agenda (in the form of the Agenda Venezuela) during the Caldera administration severely reduced the popularity and legitimacy of the traditional political parties.

Chávez became the leader of the Bolivarian movement long before he rose to power. He joined the Movimiento Bolívariano Revolucionario 200 (MBR-200, Bolivarian Revolutionary Movement 200) in the early 1980s. The MBR-200 began as a clandestine military organization composed of junior military officers (including Hugo Chávez) who were critical of the established political and economic order. The group discussed politics, made plans for revolutionary change, and formed alliances with other leftist activist groups and intellectuals (López Maya 2003, 74–75).

The MBR-200 created the MVR as an electoral structure. The MVR "was not conceived to be a party, but rather an electoral front for the MBR-200" (83). The MBR-200 was designed to protect the fragile structure of the Bolivarian movement from the unpredictability of the electoral process (82–83). The MVR refused to make any alliances with traditional parties; instead, the MVR constructed a broad alliance with new and alternative movements, which together became known as the Polo Patriótica (PP, Patriotic Pole). MBR-200 leaders did

---

43. For a discussion of the political economy factors that contributed to the rise of Chávez, see Roberts 2003 and Di John 2004. On the ideological roots of the *chavista* movement, see Gott 2000 and Garrido 2000, 2002, and 2003.

not want their ideological orientation compromised by the real politics of constructing electoral alliances.

The rise and survival of Chávez administration owes much to the effectiveness of his radical discourse, which gained him a loyal following of lower-income groups in both urban and rural regions.[44] Chávez directly attacked not only the political class but also the economic oligarchy as enemies of the people—referring frequently to the "horrór a la oligarquía." He reversed the privatization of the oil industry, gradually eliminated the independence of the Central Bank, and introduced capital controls in 2003 in the wake of a two-month oil strike led by oil industry executives, the business sector, middle- and upper-income groups, and traditional labor unions. A further aspect of the Chávez administration discourse has been the politicization of social and economic inequality (Roberts 2003). While Chávez has had little support among traditional labor unions, the prominence of agrarian reform in his electoral platform and during the period 2000–2005 represented a radical stance toward agrarian property rights not seen since AD's proposals for land reform in the mid-1940s.[45]

The rise of this radical antiparty politics has generated several important tendencies that have weakened the capacity of the state to revive economic growth (except when large oil windfalls permit large-scale public spending, which occurred in the period 2004–5). First, the radical nature of the political discourse has led to a growing polarization of politics (Roberts 2003). The period 2002–3 saw numerous massive street demonstrations both supporting and resisting the Chávez administration, highlighted by a two-month national strike, which included the nearly complete shutdown of the oil industry.[46]

---

44. Support from the urban poor has been particularly important in face of the fierce opposition to Chavez's radical politics from the private media, middle-class, and business groups, and in face of opposition in parts of the military and tacit U.S. backing for a failed coup attempt on April 11, 2002. See Valencia Ramírez 2005 for a more general discussion of the government's support base.

45. It should be noted, however, that evidence available suggests that the share of income of the poorest segments of the population declined in the period 2000–2005. According to a 2005 Venezuelan Central Bank survey of households, the share of income of poorest 20 percent of the population declined from 6.3 percent of total household income in 2000 to 4.1 percent in 2005. At the same time, the share of the richest 10 percent also declined from 41.6 percent to 37.6 percent. The groups that have improved their share in total household income most have been those households in the 6th, 7th, 8th, and 9th income deciles (the so-called middle classes) (Banco Central de Venezuela 2007, 91).

46. Indeed, much of the growth collapse in the period 1998–2003 (see table 2.1) was caused by the two-month national strike (which included the suspension of oil production) in December 2002 called by the opposition to Chávez. In the first trimester of 2003, GDP fell 25 percent compared to the previous year. For a discussion of the continuities and changes in the level and motivations of popular protests over the period 1958–99, see López Maya and Lander 2005.

Moreover, relations between the state and big business have been more antagonistic than in any time in the democratic era.

Second, there has been a deinstitutionalization of political organizations. The 1999 Constitution banned the financing of political parties, which limits the organizational strength of the opposition. In the absence of nationally based political party opposition (the result largely of the collapse of AD and COPEI), the private television and print media have become the most institutionalized force of opposition (Hellinger 2005, 17). Moreover, the administration has limited the organizational development of its own party, which is itself subject to intense factionalism and numerous defections of one-time supporters (including high-level military officers).[47] Many of the defections within the government and the growing intensity of the opposition are reminiscent of the *trienio* (1945–48), when the then dominant political party, AD, was overthrown because it initiated radical social reforms without consulting opposition parties, the business sectors, and other important actors such as the Catholic Church. Finally, both opposition and *chavista* labor unions have been subject to internal faction-fighting (Ellner 2005). These factors lessened the possibility of consensus-building and cooperation within the state and between the state and business groups (Monaldi et al. 2004; Monaldi and Penfold 2006).[48]

Third, there has been a purposeful strategy to circumvent state institutions in the delivery of social services. One of the main political reasons for this was the purposeful strategy of the administration not to draw on incumbent bureaucratic personnel, who historically were associated with AD and COPEI. This resulted in the underutilization of some talented pools of labor.[49] The

---

47. There are several examples of this. Luis Miquilena, one of Chávez's leading political strategists and interior minister, resigned from the MVR and the government in 2002. Miquilena favored using traditional political structures and reaching out to the political opposition, both of which met with disfavor among leading MVR and government leaders (Alvarez 2003, 158–60). In 2002, MAS, traditionally the most popular left-wing party in Venezuela (and one of the main political allies of the MVR), joined the opposition, and some of Chávez's most dedicated supporters defected and formed a party called PODEMOS (Molina 2004, 166). The most recent example is the defection in 2007 of former general Raùl Baduel, who was formerly defense minister in the Chávez administration.

48. The factionalization of Chávez's own support base can be seen in the high degree of cabinet instability. In the period 1958–88, cabinet members lasted an average of 2.13 years in their position (in a five-year term). In 1989–93, ministers lasted only 1.4 years, in 1994–99, their average tenure increased to 1.8 years, but in 1999–2004, average ministerial tenure declined to 1.3 years (Monaldi et al. 2004, 34).

49. As such, the armed forces and immigration of Cuban health and education workers fill many of the gaps needed. One of the biggest losses of talent occurred by the ill-advised attempt by oil sector executives and workers to strike in December 2002. The Chávez administration fired 18,000 employees associated with this revolt. However, by 2005, the government accepted that many would be allowed to be employed by contractor firms (Ellner 2008, 161).

administration set up several government missions to improve education, health, and housing in shantytowns. However, these missions, funded by resources from the state oil company, are executive-led and bypass state ministries in their planning and implementation phases, as even the head of the state oil company, Ali Rodríguez Araque, a Chávez loyalist, acknowledged (Wilpert 2004).[50] This has created a dual state structure that has further led to the fragmentation of the state despite an increase in the centralization of power.[51] As a result, there has been little attempt to build the state's organizational capacity or build political links between ministries.

Fourth, there has been a lack of any coherent production or export strategy.[52] This is due to several factors. The first is the antagonistic relationship Chávez has with many big business groups.[53] The second is the much reduced capacity of the private sector to act as an engine of growth (Ortiz 2004). Few of the dominant economic groups of the early 1990s were major players a decade later. According to Nelson Ortiz, the 1994 financial crisis wiped out more than two-thirds of the private sector and few of the economic groups were engaged in mineral or manufacturing export activities (ibid.).[54] A third important factor is the ideology of *chavismo*, which is focused on supporting small-scale businesses and cooperatives through micro credit schemes and, most important, the emphasizing of pro-poor social programs as the cornerstone of government policy. What economic cooperatives the state has promoted have taken

---

50. There are divergent views on the extent to which the government missions have improved the social welfare of the poorest groups. For a positive view, see Weisbrot and Sandoval 2007; for a decidely more negative assessment, see Moreno and F. Rodríguez 2005, 96–103, and F. Rodríguez 2008. Crime rates, however, have unambiguously deteriorated in the 2000s. From an already high national level of 25 murders per 100,000 inhabitants in the mid-1990s, the murder rate increased to nearly 32 per 100,000 inhabitants by 2003, the fourth highest in the world behind Colombia, South Africa, and Jamaica (United Nations Survey of Crime Trends, various years). By 2007, the murder rate in Venezuela reached 48 per 100,000 inhabitants, the second highest in the world after El Salvador ("Deadly Massage" 2008, 56).

51. On the factionalism among important *chavista* supporters with respect to labor and business organizations, the oil industry, and the extent to which traditional state structures should be circumvented in the implementation of social policy, see Ellner 2008, 139–74.

52. See table 7.4 for evidence of stagnation in non-oil export revenues in the period 1999–2003.

53. Since the resignation of Maritza Izaguirre as finance minister in mid-1999, FEDECAMERAS, the main national business association, has gone unrepresented in important ministerial posts (Ellner 2008, 126).

54. Ortiz (2004) examines the extent of this reduced private sector productive capacity by looking at the dramatic decline in market capitalization of the Caracas Stock Exchange, which includes most of the country's largest companies. Whereas it exceeded US$13 billion in 1991, twelve years later it had fallen to approximately US$3 billion. These figures do not reflect the control and management of a large number of agricultural, commercial, industrial, and financial companies that passed to foreign ownership. The value of companies actually controlled by Venezuelan investors at the end of June 2003 was less than US$1 billion.

the form of social programs generating employment, and have not, in general, generated economically dynamic and productive small businesses (Ellner 2008, 190–91). The government has prioritized social policies in housing, education, and health through the creation of presidential-led Social Missions.[55] Such policies lie at the center of the loyal political support the government has among lower-income groups. Central government social spending as a percentage of GDP increased from 8.2 percent of GDP in 1998 to an annual average of 11.5 percent of GDP in the period 2000–2005 (Weisbrot and Sandoval 2007, table 2).[56] The state oil company, PDVSA, has been instrumental in financing the various Social Missions. PDVSA funding of social investment was US$249 million in 2003, but subsequently rose dramatically to US$1.24 billion in 2004 and US$6.91 billion in 2005 and was set to rise to US$13.26 billion in 2006.[57] The use of oil revenues to fund massive social programs may be one of the reasons that that PDVSA has experienced significant production capacity problems in the period 2002–7.[58]

In effect, the period 1999–2005 saw, not only a growing polarization of politics, but also Venezuela's dependence on oil revenues reach historically high levels. The decline of the political party system in this period has meant that the links between the state and interest groups have become more deinstitutionalized than at any point since 1958. Moreover, there has been considerable faction-fighting within the ruling coalition, as witnessed by defections to the opposition and the high turnover of government ministers. The fragmentation of the *chavista* movement has contributed to the mismanagement of state resources, even those directed toward the poor. This can be seen by also examining events beyond 2005. Despite rapid rates of economic growth (averaging 9.5 percent in the period 2006–7), and historically high oil revenues, inflation increased from 14.4 percent in 2005 to 22.5 percent in 2007, a year that also

---

55. See www.missionesbolivarianas.gov.ve for a description of the social missions in Venezuela.

56. This figure rose to 13.6 percent of GDP in 2006 (Weisbrot and Sandoval 2007, table 2).

57. From official figures provided by the Venezuelan Information Office at http://www.rethinkvenezuela.com/downloads/Social%20Missions.htm. It is clear that low-income groups tend to benefit more from the government social missions. According to the Venezuelan Central Bank, among the poorest 40 percent of households, over 60 percent of those, on average, claimed they benefited from at least one government mission. Among the richest 30 percent of households, less than 30 percent of those claimed they benefited from at least one government mission (Banco Central de Venezuela 2007, 35).

58. See Padgett and Gould 2007, 51–53. Before the oil strike in 2002, Venezuela exported more than 3 million barrels per day. By 2007, OPEC calculates that oil production was approximately 2.4 million barrels per day, though the Venezuelan government claimed production was back to 3.2 million barrels per day (Padgett and Gould 2007, 53).

saw massive food shortages of basic consumer items. The mismanagement of the economy contributed to the first referendum defeat of the government in December 2007.[59]

While it is true that there has been a growing centralization and power of the executive in this period, the prospects of implementing an effective production strategy have been limited by the antagonistic stance of the regime toward many sections of the private sector (domestic and foreign), the fragmentation of state ministries, and the reduced collective action capacity of the domestic private sector as witnessed in the dilapidated state of business associations. Even if one argues that effective industrial policy is not so much about targeting sectors but providing the institutional milieu within which state and private business can cooperate and coordinate actions to facilitate promising new sectors (Hausmann and Rodrik 2006), the political and institutional foundations necessary for such action would appear to be remote in a polity characterized by fragmented political organizations and state ministries and a polarization of politics more generally.

8.3 Conclusion

In sum, this chapter, drawing on an extensive political science and political economy literature, documented the changing nature of the Venezuelan polity in the period 1920–2005. Table 8.2 summarizes the main trends in the nature of the polity and political settlements that evolved over the period 1920–2005 and includes what many historians would concur was characteristic of most of the nineteenth century. The period 1920–58 was largely authoritarian, but the polity resembled (to recall the framework employed in section 6.4) a consolidated state with centralized political organizations. Political pacts and compromises were dominant in the period 1958–68, and the polity maintained its characteristic of a consolidated state with centralized political organizations though the populist and clientelist nature of these pacts were underlined. The growing

---

59. In 2007, the government proposed a series of initiatives to develop laws that would institute a more "socialist" state in Venezuela. The referendum was narrowly defeated, mostly as a result of low voter turnout among supporters of the government (Carlson 2007). The defeat has convinced leaders that the development of the United Socialist Party of Venezuela (PSUV) is essential to revive political support for the government (Janicke 2008), a decided switch in the long-standing rhetoric of antiparty politics. See Ellner 2008, 175–94, on the tensions within the *chavista* movement between those who support a grassroots/social movement approach to politics and those who support political party–building.

factionalism in the Venezuelan polity in the period 1968–2005 was also examined. The fading threats to democratic regime survival in the late 1960s were identified as crucial to the transformation of the polity from a consolidated state with centralized political organizations into a state that was still consolidated but with increasingly fragmented political organizations. In effect, centralized populist clientelism in the period 1958–68 turned into more competitive and contentious populist clientelism characterized by more intense electoral rivalry, increasing factionalism within the dominant party, AD, and less interparty cooperation in the period 1973–93. The growth in factionalism was also documented as an important outcome of the increase in multiparty electoral rivalry over the period 1993–98 where the fragmentation of political organizations worsened further. The period 1998–2005 saw the emergence of radical antiparty politics, which created a polarization of political contestation, increased factionalism, and a deinstitutionalization of political organizations.

Of course, it could be contended that the categorization of the entire period 1968–2005 into one category, namely "strong state with fragmented political institutions," downplays important nuances and differences in the evolution of

TABLE 8.2 *Periodization of political structures and political settlements in Venezuela*

Type of polity/political settlement	Period
Unconsolidated, weak/fragile state	1820–1900
Consolidated state with centralized political organizations (mostly under authoritarian rule)	1900–1958
Consolidated state with centralized political organizations (transition to pacted democracy underpinned by populist clientelist support base)	1958–68
Transition to growing fragmentation of political organizations and decline in interparty cooperation (political pacts and consensus begin to break down as the threats to democracy fade)	1968–73
Consolidated state with fragmented political organizations (AD-COPEI party hegemony, growing factionalism within AD, declines in interparty cooperation, conflicts between traditional and emerging business groups, business groups critical of growing power of state managers in heavy industry)	1973–93
Consolidated state with fragmented political organizations (decline in hegemony of AD-COPEI, emergence of multiparty electoral democracy; increase in anti-political party, anticorruption discourse)	1993–98
Consolidated state with fragmented political organizations (rise in antiparty politics, growing polarization of politics, decimation of organized political opposition parties, increase in massive street demonstrations, increase in executive power)	1999–2005

the party system that political analysts (referred to in this chapter) have identified.[60] While there has indeed been substantial variation in the nature of the Venezuelan party system, the evidence reviewed does suggest that that inter- and intraparty collaboration was declining over time, which is the main point of the classification. I have grouped the party system into this one category for the sake of parsimony in order to assess evidence in a systematic way. Further subdivisions of the party system would not have added any further insights with respect to the main proposition being evaluated.

It also could be contended that such a categorization—"strong state with fragmented political institutions"—could be applicable to any democracy. While tensions and factionalism within party systems undoubtedly exist in all democracies, the evidence reviewed in this chapter suggests that even in the Venezuelan case, the party system was significantly less cohesive and disciplined in the post-1993 period, for example, than it was in the period 1958–73. In comparative terms, the Venezuelan party system in the 1990s was certainly more fragmented than its counterparts in many democracies such as South Africa, Chile, or Costa Rica, to name just a few. While more systematic research is required to assess the degree of party factionalism, there is no reason to assume that this variable is the same in all democracies in all contexts.

Finally, the argument put forward does not suggest that only authoritarian regimes are capable of implementing advanced ISI and big-push industrialization strategies. I argue only that centralization of political organizations may be necessary for the effective implementation of such strategies. However, such centralization has occurred in competitive-party systems such as in France and Israel, in less than authoritarian one-party dominant states such as Malaysia, and in more obviously authoritarian contexts such as South Korea and Taiwan.

In any case, the evolution of political settlements in the period 1968–2005 represents a long trajectory of an increasingly *reduced* political and institutional capacity to formulate and implement economic strategies that require coordination of investment; the cooperation of state, labor unions, and big business over economic strategies; and a state capable of monitoring and disciplining infant industries. The Venezuelan polity in the post-1968 era increasingly resembled a consolidated state with fragmented political organizations in a conjuncture where development strategies required more centralized political and bureaucratic organizations.

---

60. I thank Javier Corrales for bringing this point and the ones in the subsequent two paragraphs to my attention.

9

A NEW VIEW ON THE POLITICAL ECONOMY OF GROWTH IN VENEZUELA

The inadequacy of reigning explanations for the growth collapse in Venezuela (examined in Chapters 3 and 4) is that these analyses assume that the growing centralization of the state- and state–led big-push strategies *necessarily* produce dysfunctional rent-seeking, cronyism, and corruption. Venezuelan policymakers in the 1970s can be accused of switching development strategies too early by adopting an unrealistically complicated and ambitious industrial strategy in the context of a relatively underdeveloped manufacturing sector in 1970. While this may explain the increases in capital intensity, and while there is no doubt that the suddenness with which Venezuela switched strategies in 1973 contributed to pressure placed on state and private sector capacities, the prolonged stagnation of the Venezuelan economy still requires explanation.[1] The coincidence of a change in development strategy with a decline in growth was thus not inevitable, but a historically specific feature of the Venezuelan political economy in the period 1968–2005.

Convincing explanations of relative performance have to identify which features distinguish Venezuela from more rapidly growing latecomers with similar institutional interventions and levels of corruption. What needs to be explained is why apparently similar levels of oil abundance, rent-seeking, and corruption throughout the period 1920–2005 were associated with significant

---

1. The challenges of Venezuela's development strategy were particularly great because the strategy of big-push, natural-resource-based industrialization was *combined* with the second stage of ISI. Many Latin American economies introduced more advanced industries such as steel during the first stage of ISI.

A New View on the Political Economy of Growth....................................227

changes and ultimately decline in the economic performance of the non-oil and particularly the manufacturing sector over the course of the twentieth century. The most salient differences in economic performance were rapid non-oil and manufacturing output growth in the period 1920–80 and moderately high non-oil and manufacturing productivity growth in the period 1950–68, but much lower non-oil and manufacturing productivity growth between 1968 and 2005 and a collapse in non-oil and manufacturing output growth between 1980 and 2003. In addition, those same explanations need to identify why productivity and eventually output growth collapsed within Venezuela as it switched development strategies from generally small-scale, consumer goods import substitution to a more capital-intensive big-push industrialization. Such big-push strategies were not very different from those adopted by more successful industrial policy states such as Korea, Taiwan, and Malaysia.

9.1 Mapping Development Strategies and Political Settlements

It is now possible, following the framework developed in sections 6.3 and 6.4, to map the extent to which development strategies and political settlements were sufficiently compatible to generate rapid output and productivity growth. The evidence compiled thus far enables us to suggest an alternative explanation for the significant changes in output and productivity growth in Venezuela that is more compatible with comparative and historical evidence than those provided by the reigning explanations explored in Chapters 3 and 4.

Let us return to the main propositions established in section 6.4. At the end of the nineteenth century Venezuela was one of the poorest countries in Latin America. The historical evidence suggests that the weakness of the central state, rampant regional warlordism, and numerous episodes of civil war generated insecure property rights for most asset holders and thus low investment throughout the century. This sufficiently explains why there was very little economic development and is something Thomas Hobbes could have told us centuries ago.

Over the course of the twentieth century the Venezuelan state did indeed become a leviathan in the sense that it consolidated its monopoly control over the means of violence. If this process was not legitimate in the eyes of all interest groups, it was nevertheless sufficient to prevent large-scale political violence (not to mention civil war) and permitted the relatively peaceful transition from authoritarian rule to competitive democratic politics. However, a Hobbesian framework cannot explain why a consolidated state and its support base can generate policies and institutions that produce very different records of economic growth.

The first proposition is that relatively high non-oil and manufacturing growth in the period 1920–68 and relatively respectable rates of non-oil total factor and labor productivity growth in the in the period 1950–68 were due to a broad *compatibility* between the development strategies undertaken and the nature of political settlements. Table 9.1 presents a more detailed account of this compatibility based on the historical analysis undertaken. The political settlement that reigned for most of the period 1900–1958 (discussed in section 8.1) facilitated rapid industrial growth for the following reasons. First, the consolidation of the central government that ended regional warlordism provided a measure of predictability to the polity. The consolidation of the Venezuelan polity was, however, *not* built through democratic and transparent bargaining processes. State formation occurred with significant levels of authoritarianism, cronyism, and networks of corruption. Second, the formation of state institutions such as the central bank and state development banks was an important institutional means through which macroeconomic stability was maintained and through which capital accumulation in the non-oil economy, including manufacturing, was financed. Third, the absence of organized and powerful interest groups, especially among the intermediate or middle-class groups, which had not yet appropriated power, gave the state a relative power that it did not obtain in the post-1958 political settlement. Fourth, the influence of the United States government in imposing a trade agreement prevented the creation of blanket, or indiscriminate, protection and created a contingent selectivity to infant industry protection. Selectivity was further enhanced by the absence of effective organized pressure from the family conglomerates, which, in any case, were presented with important accumulation opportunities in real estate, construction, banking, and commerce. The rise of political party and labor union power was to change the urgency and purposefulness of state-led import-substitution strategies, as discussed in Chapter 8.

By 1958, Venezuela was still in the relatively easy stage of ISI, and behind other countries in South America (table 7.1). Thus, there was no urgent need for the state to be selective in rent deployment and or to discipline rent recipients. Because there were still ample substitution possibilities in small-scale and consumer goods industries, selectivity and discipline would not have greatly affected economic performance.

Most analyses on Venezuela do not discuss the mechanisms through which oil abundance can translate into rapid long-run growth. As discussed in section

TABLE 9.1 An overview of evolution of regime type, development policies, and economic outcomes in Venezuela, 1920–1958

Period	Regime type	Main political trends and settlements	Industrial policy orientation	Stage of ISI/ dominant technologies	Main economic results
1920–45	personalist military	centralized political organizations dominate	liberal trade regime with little state-led industrial financing; most industry is private.	easy state of ISI; artisan manufacturing production dominates.	rapid non-oil economic growth; low inflation
1945–48	democratic	radical populist clientelism; increasing political polarization	liberal trade policy maintained, though state-led industrial financing increases significantly.	easy stage of ISI	rapid non-oil and manufacturing growth; low inflation
1948–58	personalist military	centralized political organization dominate	liberal trade policy with increases in state enterprises in steel and electricity	easy stage of ISI with significant growth in consumer durables	rapid non-oil and manufacturing growth; low inflation; non-oil total factor productivity growth and manufacturing productivity growth relatively high

6.4, rapid growth during the early stage of ISI is a common feature of many low-income economies. In the period 1920–58 Venezuela was characterized by a relatively consolidated state with centralized political organizations that was pursuing a development strategy to promote easy ISI. However, as discussed in the previous chapter, the nature of the political settlement changed significantly with the transition to and consolidation of democracy in 1958. Such changes were to make the management of the first stage less efficient and would also lead to a "premature" adoption of more advanced development strategies.

The second proposition is that the slowdown in non-oil and manufacturing growth in the period 1980–2005 and the substantial declines in non-oil productivity growth from 1968 on were the result of a broad *incompatibility* between the development strategies undertaken and the nature of political settlements. In this period, Venezuela was a relatively consolidated state with increasingly *fragmented* political organizations that was pursuing a more advanced stage of ISI and a big-push, natural-resource-based heavy industrialization policy. The nature of the incompatibility was a growing political fragmentation and discontinuity in populist clientelist coalitions underpinning the state precisely when the development strategy called for more centralized state control and continuity in policy orientation.

The maintenance of high output growth in the 1970s was, as suggested, due to high levels of public investments in manufacturing industries and subsidies and protectionism to the private sector infant industries, both of which were financed by oil boom revenues. This growth became unsustainable in the context of increasingly unproductive investments and the massive coordination failures in macroeconomic policy that resulted in the capital flight cum debt crisis of the 1980s.

9.2 The Economic Effects of Political Fragmentation and Populist Clientelism in Venezuela, 1968–2005

The implications of the historical patterns of state-society relations characterized by populist clientelism, pact-making, and increasing factionalism (discussed in the previous chapter) have been profound. The priority of preserving democracy meant that political party and state leaders needed to accommodate the growing factions of the populist coalition. They accomplished this by providing economic rents to those groups in return for political support. In essence, centralized rent deployment and management came to be based on political rather than economic criteria.

While pacts protected embryonic democratic institutions from pressures of more open political and economic contestation, the feasibility of such pacts depends on partners extracting private benefits, or rents, from democracy (Przeworski 1991, 87–94). In effect, pacts conferred monopoly privileges on the recipients of rents. Extensive rents were awarded to family conglomerate capitalists in the form of subsidies and protection (Naím and Francés 1995). The strategy of "excessive avoidance of conflict" and the need to keep the "insider" groups content meant that selectivity in rent deployment and the discipline of rent recipients declined significantly.

The decline in threats to the regime and an increase in the intensity of electoral rivalry, in turn, led to growing levels of factionalism between and within political party alliances. Maintaining centralized and coherent populist clientelist alliances in the context of primitive accumulation and state-led industrialization is difficult because such coalitions are fragile in the face of the large-scale levels of political mobilization of groups and classes with contradictory demands. On the one hand, the state and political parties and organizations that mediate demands become the site of politicized struggle over resource allocation. On the other hand, the inevitable excess of demand for rights relative to resource availability creates shortages. Because parties were the road to state patronage, powerful incentives existed to form more particularistic and even personalist factions linked to different leaders.

What is more, where clientelism and factionalism are endemic, agents inevitably adapt their strategies in order to survive. If the state is subjected to clientelist pressure and organized groups systematically get payoffs, then not developing clientelist links with the state results in economic "extinction." Here, eventually all classes and groups become clients within different sections of the fractionalized political parties and the fiefdoms they maintain in the state bureaucracy. Overall, in post-1968 Venezuela factionalism led to general lack of continuity in policymaking; greater political polarization and numerous changes in policy; and the proliferation of rules, agencies, and subsidies (Naím and Piñango 1984; Karl 1997, 116–85; Coppedge 2000). The period 1999–2005 saw a growing polarization of politics and a deinstitutionalization of political organizations that further exacerbated the trend of political fragmentation.

After 1968 the Venezuelan polity ever more closely approximated the category of a consolidated state with fragmented political organizations. At the same time, in the period 1960–73, the technological characteristics of the economy switched from the easy stage of ISI to a more advanced stage of ISI. Moreover, the decision of the government in the early 1970s further accentuated this transformation by initiating a big-push, natural-resource-based industrialization strategy. This meant that the economic and political challenges facing the state and the demands on state-business relations increased significantly, since the dominant development stages and strategies in the post-1968 period required greater investment coordination, greater selectivity in rent deployment and industrial subsidies, greater capacity to discipline poor infant industry performers, and a high degree of continuity in economic policymaking. The subsequent decline in non-oil and manufacturing productivity growth in the post-1968 period and the growth collapse in the post-1980 period can be interpreted as a mismatch between the development strategy and the nature of the polity.

9.2.1 Coordination Failures in Macroeconomic Policy in Post-1968 Venezuela

The growing factionalism of populist clientelism in post-1968 Venezuela resulted in a series of distributive conflicts that led to instability in macroeconomic management and coordination failures among competing political factions and economic ministries. The signs of growing economic instability were evident in several spheres. The negative consequences of increasing factionalism and interparty rivalry produced dramatic coordination failures in macroeconomic policy within the state in the period 1976–85. First, consider the dramatic episode of the capital flight cum debt crisis that occurred in the period 1976 to 1985. The big-push, natural-resource-based industrialization drive, initiated in 1974, was compromised by increasing external debt, which, in the Venezuelan case, perversely financed capital flight of roughly equal proportions. The state-owned enterprises borrowed dollars abroad and converted them to bolívares so that they could complete the massive investment projects planned. At the same time, the government purposefully sold dollars to the private sector. This extraordinary sequence of events was intimately related to the growing factionalism and polarization within and between political parties.

Between 1976 and 1985, the net public sector external debt rose from nil to over US$30 billion in Venezuela, an outcome that was indeed paradoxical since Venezuela ran a cumulative current account *surplus* of nearly $10 billion in the period 1974–85. In most developing countries, debt is normally incurred to finance insufficient internal savings, current account deficits, and fiscal shortfalls, none of which applied to Venezuela over this period. The origin of net external public debt that accumulated is accounted for, as indicated in table 9.2, completely by capital flight (M. Rodríguez 1991). Miguel Rodríguez refers to this juncture of economic mismanagement as one approaching the terrain of "magical realism" (ibid.).[2] Indeed, only Mexico and Argentina approached the ratio of capital flight to change in external debt that Venezuela experienced in the period 1976–85.

There were several factors that led to the capital flight cum debt crisis.[3] First, the Pérez administration attempted to undertake an ambitious development plan of big-push, natural-resource-based industrialization (see Chapter 7) through decree, not consultation. This upset factions within AD and COPEI. Second, the executive branch of the government realized that the financing needed to fulfill the plan would outrun oil revenues during *its own* administration, and thus encouraged state managers to undertake a major foreign borrowing program. This was to extend to the next administration, which (as it turned out) was headed by the rival party COPEI. Third, the isolated manner in which the megaprojects were initiated meant that the main goal was to initiate foreign borrowing, which was the main way to avoid political battles over appropriating state oil revenues domestically. Since Venezuela was running current account surpluses, the extent and level of foreign borrowing was unnecessary. Nevertheless, the political expediency and possibility of foreign

TABLE 9.2 Capital flight in Latin America, 1976–1985

Country	Capital flight (annual average, billions of U.S. dollars)			Capital flight/change in external debt (%)
	1976–82	1983–85	1976–85	
Argentina	27	−1	26	62.7
Brazil	3	7	10	12.0
Chile	0	1	1	6.4
Mexico	36	17	53	64.8
Venezuela	25	6	31	101.3

SOURCE: Sachs and Larrain 1993, 705.

NOTE: Capital flight is defined as the difference between capital inflows (the sum of the change in foreign debt and direct investment and the sum of the current account deficit plus increases in official reserves and other public dollar assets.

2. Capital flight figures may underestimate or overestimate the phenomenon. Such figures do not take into account the accumulation of interest and profits that foreign assets produce abroad, since balance of payments usually do not include these items in the national income accounts. Smuggling, nonreporting of imports, and over- and under-invoicing may further distort the data.

3. This rest of this section draws substantially on M. Rodríguez 1991 and Frieden 1991, 199–206.

borrowing meant that, in the period 1974–78, external public debt, driven by the non-financial public enterprises, increased by US$10.5 billion, while external public assets (of the state oil company and a sterilization fund, the Venezuelan Investment Fund) increased by US$9.6 billion over the same period. Fourth, Venezuelan borrowing by both state managers and private business conglomerates was not closely monitored.

Not only state but also *private* borrowers were free to borrow abroad without regulatory scrutiny. While medium-term and long-term public external borrowing needed congressional approval, the AD-dominated Congress throughout the period 1974–85 did not require that public enterprise managers (or private sector managers) seek approval for short-term borrowing.[4] This strategically allowed the state managers to continue borrowing even if the next administration curtailed flows to the public enterprises. Essentially, the Pérez faction of AD facilitated the possibility of massive primitive accumulation of close political allies, essentially state managers and emerging family conglomerates (such as the Twelve Apostles), who were awarded contracts and licenses in industry and construction.[5] The free convertibility of the currency did allow the traditional business groups to access international capital markets, an important indicator that the pact to allow all established conglomerates access to international capital markets was still operative.

The tensions created within AD, and between AD and COPEI, were to have devastating macroeconomic consequences in the subsequent administration of Luis Herrera Campins of COPEI. Herrera won largely because factions and splits within AD and corruption scandals surrounding the Pérez administration, which reduced AD's legitimacy in the eyes of the voting public. This resulted in a compromise AD candidate who did not receive the same backing of the AD party that Pérez had.

Herrera was opposed to the state-led industrial policies of Pérez and had argued that the foreign borrowing of the public and private sector had "mortgaged the economy." On the heels of a second oil hike in 1979–81, the economy remarkably fell into recession.[6] Herrera's administration moved to cool down what his economic team mistakenly believed was an overheated economy. The

---

4. See Chapter 7 for a discussion of the major development projects initiated in the period 1974–78.

5. The Pérez administration also secured the AD labor base with mandated wage increases and expanded government and employment programs.

6. Overall, GDP rose only 4 percent in the period 1979–83 after rising 55 percent in the Pérez administration. Unemployment went from 4.7 percent in 1978 to 11.9 percent in 1983, as real wages dropped 22 percent (after a 39 percent rise in the Pérez administration).

measures taken included the intention to slow or halt the state natural-resource-based industrialization projects, the removal of tariff protection for over 150 products, the removal of food subsidies, and the tightening of monetary policy, the latter leading to a massive overvaluation of the exchange rate in the period 1979–83 (see table 3.2).[7] The unexpected increase in oil revenues from $9.2 billion in 1978 to $19.3 billion in 1980 did not elicit a reevaluation of policies.[8] This was, in part, because of the political desire of Herrera's administration to reign in the primitive accumulation possibilities of new AD "oligarchs" in the state and private sector.

Despite the intentions of the COPEI administration to cool the economy, the executive could not control short-term borrowing by the state enterprises or the private sector. The AD-dominated Congress did not attempt to change the borrowing law, which meant that short-term borrowing remained exempt from congressional review. By the end of 1982, the gross external debt of Venezuela had reached US$33 billion, despite the fact that the economy did not need to indebt itself at all to meet its investment effort since it was running current account surpluses. Until 1982, when the debt crisis in the country became evident, there was little knowledge of the amount of external debt accumulated in the country. Medium- and long-term debt amounted to $18.8 billion. The public sector accounted for $15.8 billion, or 84 percent of medium- and long-term debt. Of much more concern was the accumulation of $14.3 billion of unregistered short-term debt, of which public enterprises accounted for $10.8 billion and the private sector $3.4 billion (Frieden 1991, table 6.2). When short-term interest rates rose in 1982, Venezuela, like the rest of debt-ridden Latin America, was to experience declines in investment and growth.

Under Herrera, the net external debt situation of the public sector changed dramatically for the worse. From 1978 to 1983, the external public debt grew from $12 billion to $27 billion, while external public assets remained stagnant at $11.5 billion (204). At the same time, external private debt increased from US$3.2 billion to US$8.0 billion, and private external assets—in the form of capital flight—grew by $29.2 billion. Remarkably, a $20 billion increase in foreign debt, three-quarters of it public, was offset by nearly $30 billion in capital flight (204).[9]

7. The overvaluation may have coincided with a tightening of monetary policy, but it is not clear that the latter caused the former: reducing M2 money supply should reduce prices and increase interest rates. The former reduces overvaluation, the latter increases it, so the effect is indeterminate.
8. All monetary figures in this section are in U.S. dollars.
9. Manzano and Rigobón (2007) argue that the low growth in the post-1970 period in mineral-abundant less developed countries was due to large-scale borrowing when mineral prices were

The COPEI administration, by opposing AD policy, faced a series of contradictory pressures, which were exacerbated by the hostile AD-majority Congress. On the one hand, the traditional family conglomerates, shut out of many new industrial and construction contracts, wanted conservative fiscal and monetary policies. The Herrera administration obliged in this regard. On the other hand, there were many more groups who opposed monetary and fiscal austerity. The growing public sector meant that there were an increasing number of employees and businesses dependent on ongoing public spending and foreign borrowing. Moreover, the currency began to appreciate sharply from 1980–82 as internal inflation grew faster than U.S. inflation rates. By mid-1981, the bolívar had appreciated 16 percent since 1979, and devaluation expectations were growing rapidly. Despite this real appreciation, there was no attempt to devalue since devaluation was thought to be inflationary.

In the context of deflationary macroeconomic policies, the private sector was faced with a diminishing set of investment opportunities and thus began to accumulate foreign savings at the same time the public enterprises were borrowing in an uncontrolled fashion. The key reason behind the continued foreign borrowing by public enterprises was that capital transfers from the central government earmarked to finance public enterprise investment declined by 50 percent in the Herrera administration, compared with the previous Pérez administration. The private investment climate also worsened when the labor movement, controlled by AD, organized massive strikes against the threat of further austerity and the withdrawal of food and housing subsidies.[10] The final important and contradictory policy was the maintenance of negative fixed interest rates in the context of free convertibility and declining business confidence. The maintenance of free convertibility was part of the pact to allow the Venezuelan middle and upper classes to invest and travel abroad. The maintenance of negative real interest rates was in part the result of political pressures to keep mortgage and consumer loans affordable.

---

buoyant in the 1970s and the subsequent inability of these economies to pay off these debts when mineral prices fell substantially in the 1980s. The growth curse in such economies, according to the authors, was ultimately due not to resource abundance per se, but to a debt overhang problem. During the slowdown and fall in mineral prices, such economies were unable to continue borrowing and had to pay off their debts. The problem with this argument is that it does not explain why the external debts were contracted or why the large-scale investments did not yield more robust export revenues in non-mineral sectors. See section 3.3 for a comparison between Malaysia and Venezuela of the differing developmental outcomes of large-scale debt build-up.

10. Evidence from table 8.1 indicates that during the Herrera Campins administration, the number of worker hours lost due to labor strikes was more than double the worker hours lost during the previous Pérez administration.

The distributive conflicts that factionalism and electoral party rivalry caused were behind the massive capital flight crisis. In 1981, an estimated $8 billion to $13 billion fled the country. In 1982 and 1983, again depending on the estimate, between $8 billion to $15 billion fled the country; in the first six weeks of 1983, capital flight averaged $500 million per week. What is more, the Herrera administration, while unable to control state enterprise short-term foreign borrowing, raided the investment funds of the state oil company, PDVSA.

The purpose of this (ultimately destructive) strategy was to sell dollars in order to reduce liquidity and support an increasingly unviable exchange rate! In September and December of 1982 respectively, the executive ordered at least $8 billion of PDVSA's external assets converted into public bonds and another $1.8 billion of external assets used to underwrite public sector deficits (Ascher 1999, 213–17). This appropriation of funds, just six years after nationalization, broke the trust and ended cooperation between the central government and PDVSA, and was another telling indicator of the increasing breakdown in the continuity and coordination of macro and sectoral policies. Over the period 1975–86, 85 percent of PDVSA's investment fund had been raided by the executive (213–17). In February 1983, on a day that became known as "Black Friday," the government could no longer sustain a rate of 4.2 bolívares to the dollar, a rate that had remained fixed since 1961. The currency was devalued by 40 percent, capital controls were imposed, and a multiple exchange rate system was installed, which lasted throughout the AD-led Lusinchi administration (1984–88).

Finally, the extent to which the private sector was permitted to accumulate wealth through unproductive means during the capital flight cum debt crisis is telling of the weakness of the state and the costliness of this weakness in the context of big-push development strategies. There were several indicators pointing to the extent of massive state subsidization of the private sector, which included subsidized loans and access to foreign exchange. First, in the period 1974–78, the current account deficit of private sector activity averaged 8 percent of GDP, which represented the private sector debt to the public sector (M. Rodríguez 1991, 255). Total private sector debt, which the government guaranteed, amounted to $8 billion over the period 1974–83. Second, with the devaluation in 1983, the government did not apply capitalist rules of the game by forcing bankruptcy proceedings on firms unable to service their debt; rather the government made private liabilities payable at the old parity (255). Capital flight also allowed for a massive level of tax evasion, since Venezuelan residents do not pay tax on assets held in the United States, which received the bulk of the flows. Finally, the state in the following Lusinchi administration (1984–88), in order to accommodate the business sector, which supported the democratic regime and its political pacts, de-dollarized public and private sector debt after the devaluation, which meant that few of the agents paid for the debt incurred—a further indication of the inability of the state to discipline primitive accumulators or impose the capitalist rules of the game. Despite the large increase in funds to the private sector, the productivity of the manufacturing sectors, which were mostly privately owned, fell considerably in the 1980s (see tables 3.4a and 3.4c).

The consequences of the capital flight cum debt crisis were devastating for the Venezuelan economy. A country with a positive current account surplus plus the increase in $33 billion in gross external debt during the period 1974–83 financed, in the aggregate, the accumulation of $10 billion in foreign exchange reserves in addition to more than $30 billion in capital flight—a syndrome Miguel Rodríguez (1991) referred to as "the Venezuelan Disease." The transfer of assets to the private sector not only favored capital owners over workers but also saddled the state with debt payments for the remainder of the 1980s that reduced the resources available for investment. As a result of the massive debt it had assumed, the government was forced to increase savings to pay back its liabilities. The debt increased from 41 percent of GNP in 1980 to 79.9 percent of GNP by 1989. Debt repayment led to reductions in social spending and public investment, which in turn contributed to declines in economic growth and real wages. The claim that macro coordination failures in this period were the result of misguided policies (M. Rodríguez 2002) misses the point that these "policy failures" were the result of the vital distributive conflicts generated by factionalism, interparty rivalry, and the demands of populist clientelism.

The second important macroeconomic manifestation of the distributive conflicts of populist clientelism and factionalism was the increase in inflation rates. It is well known that rising inflation reflects increases in the intensity of distributive struggles, and the increasing inability of the state to manage such conflicts.[11] Bob Rowthorn (1971), for instance, demonstrates how conflicts over the distribution of income and power between organized workers and capitalists can generate inflation in capitalist economies. In particular, he argues that organized workers and capitalists will struggle, if possible, to defend their income shares in the face of exogenous shocks such as a rise in real import costs that result from devaluations of the currency. Increases in real import

---

11. Given the growing weakness of labor unions in Venezuela throughout the 1990s, it is plausible to assume that producers and import merchants were more successful in shifting the burden of rising import costs onto workers by raising prices.

costs negatively affect real wages (to the extent that workers consume imported products) and negatively affect profits (to the extent that there is an imported component of inputs into domestic production). Each group will respectively attempt to compensate for the loss in disposable income by bargaining for higher wages or increasing producer prices.

While Venezuela's inflation rates have historically been relatively low by Latin American standards, the increases in inflation rates in the 1980s and 1990s have been significant in terms of the country's own record of low inflation, as indicated in table 9.3. In the periods 1950–59 and 1960–69, inflation averaged 1.4 percent and 1.0 percent respectively. In the 1970s, average inflation rates rose to 6.6 percent. As a result of the distributive conflicts, the average annual inflation rates increased dramatically to 23.0 percent in the 1980s, only to rise further in the liberalization period of 1990–98, to 50.1 percent. Clearly, economic liberalization did not generate a constellation of political constituents capable of imposing stable macroeconomic management on the state. Under the decidedly less neoliberal policies pursued in the Chávez administration (in the period 1999–2005), the inflation rate declined to an annual average of 20.5 percent, which was still very high compared to that of the period 1960–80.

The combination of big-push development strategies and factionalism within the populist clientelist coalition contributed to increasing inflation in ways that did not occur in polities attempting large-scale industrialization strategies such as Malaysia and South Korea. The main difference in the latter countries was the much lower middle-class and elite capture of the state apparatus (Khan 2000b). The fact that Venezuela maintained current account surpluses in the period 1968–2005 and did not suffer prolonged or severe negative external trade shocks in this period provides further support for the argument that the inability of the polity to settle *domestic* distributional conflicts was at the heart of economic management problems in the economy.

TABLE 9.3 Inflation rates in Latin America, 1960–1998 (average annual change, %)

	1960–70	1970–80	1980–90	1990–98
Average	21.6	74.0	197.6	260.8
Venezuela	1.0	6.6	23.0	50.1
Argentina	22.4	132.9	565.7	281.1
Brazil	46.1	36.7	354.5	936.4
Chile	26.6	174.6	21.4	12.7
Colombia	12.0	19.3	23.5	23.7

SOURCE: World Bank, *World Development Indicators*.

A third indication of failed macroeconomic policy can be seen in the increasingly frequent episodes of exchange rate crises and massive devaluations particularly after 1983 (Hausmann 2003, 265–66). From 1942 to 1961, the exchange rate remained fixed at 3.30 bolívares/dollar. Following a balance of payments crisis in the late 1950s, the bolívar was devalued to 4.20 in 1961 and remained fixed between 4.20 and 4.40 from 1961 to 1982. In the period 1983–99, the country experienced four different currency regimes and six currency crises. By 1988, the free market rate had fallen to 28.2/dollar during the multiple exchange rate regime in the period 1985–88. Unification of the exchange rate as part of the economic liberalization package brought a further devaluation to 50.5/dollar by the end of 1990. The decline continued throughout the 1990s and into this century, with the bolívar falling to below 600/dollar by the end of 1999, and 2,090/dollar by 2005. The increase in the frequency of currency crises and the episodes of massive devaluations, particularly after 1983, reflects not only distributive conflicts but also the high costs of macroeconomic failures in the context of more ambitious big-push development strategies.

A fourth negative effect of factionalism and the fragmentation of political organizations was the increase in the perception of risk by industrial investors. The increasing risk that accompanied factionalism saw a long-run reduction within the private banking system of commercial loans to industry. For instance, from 1963 to 1973, industry borrowed an annual average of 29.6 percent of all loans made by the Venezuelan commercial banks. This fell to 19.0 percent from 1974 to 1988, and to 16.8 percent from 1989 through 1998.[12] The longer gestation and thus higher risk of industrial ventures was becoming less attractive over time. This result is consistent with the general collapse of private sector investment in the non-oil economy in the period 1980–2003. Here again, economic liberalization (initiated in 1989) did not arrest the long-term increase in factionalism and related distributive conflicts. Although I could not find comparable data after this period, the output collapse of the manufacturing sector in the period 1998–2003 would suggest that commercial loans to industry continued to decline.

Another indicator of substantial declines in investor confidence can be seen in the evolution of real money balances, which provides a useful proxy for depositor confidence (Blavy 2006). The ratio of broad money to GDP declined steadily from 40 percent of GDP in 1985 to 25 percent by 2005 (6). Over the same period, deposits with commercial banks declined from 60 percent of GDP to

---

12. These data are calculated from Minsterio de Finanzas, *Base de datos*.

25 percent of GDP and domestic credit from 52 percent of GDP to 14 percent of GDP (6). This has resulted in low and declining levels of financial intermediation, which limited the role of the banking sector in financing real economic activity. Given that heavy and more sophisticated industries (which were dominant after 1973 period) require substantially more investment than light manufacturing firms to be competitive, the reduction in loans imposed much larger economic costs than would have been the case if the development strategy had concentrated on light manufacturing (as in the Thai and Colombian cases discussed in sections 6.3 and 6.4).

Disintermediation of the banking system was accompanied by a substantial reduction of lending operations and a growing share of other types of operations—including cash transactions, off-balance sheet operations, and offshore operations.[13] Checking accounts constituted more than 50 percent of total deposits at the end of 2005, and approximately two-thirds of time deposits matured in less than sixty days. Off-balance sheet operations were even larger than total deposits in the period 2002–5.[14] By 2005, off-balance sheet operations began to rise at an accelerating pace, reaching more than 100 percent of assets and 130 percent of deposits.

Perhaps the most damaging aspect of the decline in investor confidence has been the high and increasing levels of capital flight. Earlier, it was noted that there was substantial capital flight in the period 1976–85 (see table 9.2). This trend continued in the period 1994–2005, as indicated in table 9.4. It should be noted that these high levels of capital flight persisted despite the institution of capital controls in the periods 1994–96 (as a result of the banking crisis) and in the periods 2003–5 (in the wake of the oil sector strike). Moreover, the estimates provided in table 9.4, taken from data provided by the Bank of International Settlements (BIS), provide a *conservative* estimate of the total private assets held abroad by Venezuelan citizens, since they account for only a subset of offshore operations (Blavy 2006, 10). Assets abroad reached US$50.5 billion in December 2005, more than doubling in the period 2003–5. This conservative estimate of assets held abroad by Venezuelan citizens represented 38 percent of GDP in 2005 (10). The most recent acceleration of capital flight was due

13. This paragraph and the following footnote draw on Blavy 2006, 9–11.
14. In Venezuela, banks are allowed to book two types of financial operations as off-balance sheet items: trust funds (*fideicomisos*) and ceded investments (*inversiones cedidas*). Trust finds are investment funds administered by banks. Ceded investments are shares in a specific investment vehicle where clients receive a claim on a portfolio, mostly government paper. The absence of reserve requirements on off-balance sheet operations is an incentive for banks to expand their off-balance sheet operations.

TABLE 9.4 Assets of Venezuelan residents abroad, 1994–2005 (in billions of U.S. dollars)

	Assets*
1994–98	23.1
1999–2003	29.3
2004	33.7
2005	50.5

SOURCE: Blavy 2006, table 5.
* External assets of banks located in bis reporting countries.

to the polarization of politics (and resulting political instability and fragmentation) and the inability of the state regulatory structure to effectively limit the purchase of dollars.

It seems then that the macroeconomic policy failures between 1968 and 2005 were the result of the failure of the state to coordinate macroeconomic policies while undertaking a big-push development strategy. The reasons behind such failures are numerous but one neglected interpretation is that the increasingly fragmented and contested nature of the polity generated distributive conflicts that made the *macroeconomic costs* of such failure particularly *acute* because the state was attempting such a demanding and investment-intensive strategy. If the development strategy and polity had been more compatible, it is likely the growth collapse in post-1980 Venezuela would not have been so severe.

9.2.2 Microeconomic Failures: Post-1968 Industrial Strategy and Performance in Venezuela

Factionalism and electoral party rivalry have microeconomic effects as well. The first effect of clientelist and populist spending, and the growing fragmentation and contestation of political organizations, can be seen in the dramatic increase in state employment. The capital-intensive nature of industrialization in Venezuela meant that, in the context of rapid urbanization, high urban unemployment could have been destabilizing for the regime. The number of nonfinancial public enterprises grew from fewer than thirty in 1958 to more than four hundred by 1985 (Segarra 1985, 132). Beginning in 1941, when the political parties began to establish their power, the use of employment patronage in the public sectors became an important vehicle for building political clientele, accommodating populist and clientelist demands, and thus maintaining political stability.

State-salaried employment grew in absolute terms throughout the period 1941–94, as indicated in table 9.5. As well, the share of state-salaried employees in total salaried employment increased from just 7.1 percent in 1941 to 31.1 percent in 1971 and peaked at 39.2 percent in 1981. The strength of the clientelist patterns of patronage was also evident in the first four years of the economic liberalization period (1989–94), when, ironically, retrenchment of a bloated state bureaucracy was one of the goals of the economic reforms. This period did not see any retrenchment at all. Rather, there was an increase in the absolute number of state employees from 1.14 million in 1998 to 1.17 million in 1994, and the share of state-salaried employees increased from 30.4 percent to 32.0 in 1994.

The development of patronage networks to build middle-class clientele can also be seen in the significant increase in real current public spending. As indicated in table 9.6, the share of public investment in total public spending declined steadily in the period 1950–98. The decline in the share of public investment in total public spending went from a peak of 46.8 percent in the period 1950–57 to a low of 17.7 percent in the period 1989–98.[15] This decline was due principally to the growth in personnel expenditure in state enterprises (Karl 1997, 104) and to interest payments on the external debt. In the period 1958–73, the share of interest payments averaged less than one percent of total current spending. In the period 1974–88 that share increased to nearly 20 percent, and rose to an average of 25 percent in the period 1988–98. The main negative economic consequence of the increase in current spending was that

TABLE 9.5 Growth and composition of salaried employment in Venezuela, 1941–1994 (number of positions in thousands; share as a percentage of total)

	Total	Private	Share (%)	Public	Share (%)
1941	587.1	545.4	92.9	41.7	7.1
1971	1,789.5	1,233.2	68.9	556.3	31.1
1981	2,785.6	1,694.6	60.8	1,091.0	39.2
1988	3,769.7	2,624.7	69.6	1,145.0	30.4
1994	3,663.2	2,491.9	68.0	1,171.3	32.0

SOURCE: Banco Central de Venezuela 1993, tables 11-1, 11-4, 11-5; Betancourt et al. 1995, table A.2-a.

NOTE: Salaried employment refers to manual workers, professionals, technicians, and clerical and administrative staff. Workers in the informal sector, independent professionals, and owners are excluded.

15. This percentage is unusually high but owes to the heavy emphasis on infrastructural and construction projects in the Pérez Jiménez administration.

the public investment rate was much lower than it might have been given the increase in real public spending over the period 1950–98. In the period 1999–2005, the annual average of public gross capital formation as a share of total public spending declined further to 16.3 percent.[16]

The decline in public investment was particularly pronounced in infrastructure investment, which is an important component of economic growth (Calderón et al. 2003; Calderón and Servén 2003). Venezuelan investment in infrastructure (including telecommunications, electricity, and transport), most of which comes from the public sector, declined from 0.49 percent of GDP in the period 1981–85 to 0.10 percent of GDP in the period 1996–2000, as indicated in table 9.7. While the percentage decline in Venezuelan infrastructure investment was similar to Latin American trends, the level of Venezuelan infrastructure investment was considerably lower throughout the 1980s and 1990s (Calderón et al. 2003).

The clientelist nature of employment patronage in the state sector also negatively affected the economic performance of public enterprises. This was mainly due to the appointment of managers for political rather than professional reasons, weak monitoring of enterprise finances, and the purposeful creation of unnecessary public employment. The latter occurred because state enterprises were used as conduits for political patronage (Kelly de Escobar 1984). For instance, Antonio Francés (1993) traces how clientelism has affected the national telephone company, Compañía Anónima Nacional Teléfonos de Venezuela (CANTV). From the late 1960s, there were several clashes between

TABLE 9.6 Growth and composition of real public spending in Venezuela, 1950–1998

	Total real public spending[a]	Share of public investment (%)[b]
1950–57	22.7	46.8
1958–73	54.5	34.5
1974–88	113.5	21.3
1988–98	119.5	17.7

SOURCE: OCEPRE, *Presupuesto fiscal anual*, various years; Banco Central de Venezuela, *Informe anual*, various years.

[a] Annual averages in billions of 1984 bolívares.
[b] Public investment as a percentage of total public spending.

16. Author's calculation based on data from the Office of Public Finance Statistics (http://www.mf.gov.ve/archivos/2000040000/oefp-01-98-06-I.xls; accessed January 2008).

TABLE 9.7 Infrastructure investment in Venezuela and Latin America, 1981–2000 (as a percentage of GDP)

	Venezuela			Latin America		
	1981–85	1996–2000	% change	1981–85	1996–2000	% change
Telecommunications	0.19	0.01	−94.7	0.29	0.19	−34.5
Electricity	0.11	0.03	−72.7	1.38	0.2	−85.5
Transport	0.19	0.06	−68.4	0.81	0.15	−81.4
Total	0.49	0.10	−79.6	0.56	0.56	−80.1

SOURCE: Moreno and Rodríguez 2005, table 28; based on data from Calderón et al. 2003.

political appointees in management and the permanent technical staff, which led to the exodus of many of the company's best engineers. In the 1980s, two-thirds of the CANTV employees held administrative posts and one-third held technical posts; the reverse of what one finds in the most efficient telecommunications companies (4–5). The general overstaffing was reflected in the number of phone lines per employee, which in 1990 was 74.9 for CANTV compared with the largest national phone companies in the United States, France, and Japan, where phone lines per employee exceeded 170 (97).

A third important economic effect of populist clientelist politics on industrial performance was the lack of state disciplining of either private or public sector rent recipients after 1968. While immediate threats to the democratic regime induced consensus politics and interparty cooperation from 1958 to 1968, the consolidation of democracy and the end of guerrilla warfare by the late 1960s marked a decline in concerns about the return to authoritarian rule. With the waning of threats to the regime, and with the demands of populist clientelism creating ever more demands and opportunities for political leaders, consensus politics gradually declined and interparty factionalism increased, particularly within AD. The splits within AD were, in fact, to have a long-lasting negative effect on the ability of the state to deploy and manage rents and subsidies efficiently. The splintering of AD, which began in the mid-1960s, weakened the support base of the natural governing party and its allies in the state. As a result, state decision-makers could not afford to antagonize the core supporters of the regime, which were the political party cadres, the main business groups, and the labor unions. It is thus not surprising that there is little evidence of selectivity and targeting or disciplining in the protection or subsidization process (World Bank 1990; Naím and Francés 1995). ISI was as much a political project of buying the support of business groups and creating employment for the urban middle class as it was an economic project to diversify the production and export composition of the economy.

There were also some contingent factors in the post-1968 period that increased the politicization of industrial policy and thus exacerbated the state's inability to impose conditionality on rent/subsidy recipients. First, Venezuela joined the Andean Pact in 1973 and actively embraced Decision 24, which severely restricted foreign investment. During the Caldera administration (1968–73), the state also adopted the Andean Pact's policy of industrial programming by sector. Under this policy, the state planners of the different member countries determined which industries would be allocated to each country, thus giving the country exclusive rights for the development of that particular industry.

As it turned out, this policy severely restricted the degree of contestability and rivalry within the region, thus reducing competitive pressures (Naím and Francés 1995, 175). Second, the abrogation of the Reciprocal Trade Agreement in 1971 allowed for the possibility of more "blanket" protection than had been the case under the era of military rule or during the era of pacted democracy between 1958 and 1968.[17] Third, the legacy of export pessimism within postwar Latin America reduced the scope of emulating successful export strategies in the region (Fishlow 1991; Thorp 1992). The consequence of these historical contingencies for Venezuela (and its neighbors) was that levels of protection, through tariffs and non-tariff barriers, were much higher and distortional than first envisioned by the structuralist theorists of ECLA advocating ISI (Thorp 1992). Moreover, the inward-looking strategies of many countries meant that ISI became more of an employment and national production policy more than a policy to promote competitive manufacturing exports (Fishlow 1991). As a result, protectionism in Venezuela became significantly less selective in its coverage.

Finally, the fact that the state assumed control of all main export sectors (oil, iron ore, petrochemicals, aluminum, and steel) rapidly upset the traditional political economy equilibrium of the state financing of private infant industries. This probably reduced a certain "sense of limits" on state activity and led to dysfunctional cronyism within the state. As Hausmann and F. Rodríguez (2006) point out, the economies of Indonesia and Mexico (the two oil exporters that did not experience steep declines in economic growth in the face of downturns in the oil market in the mid-1980s) were dominated by non-oil exports under the control of the private sector. This may explain why these two economies, despite high levels of corruption, had a greater sense of limits within the state since their decision-makers had to bargain more productively with private sector exporters.

A fourth effect of populist clientelism was the lack of selectivity in the deployment of rents to emerging conglomerate groups. The proliferation of subsidies was a by-product of the politics of state-business relations. The division of business groups along shifting factional lines was to characterize the politics of big business from the early 1970s on (Naím 1989). This factionalism affected the dynamics of state-business relationships in several ways. First, there was a noticeable increase in the fragmentation of business associations, not only between large-scale and smaller firms, but also between the larger

---

17. See Arenas 1990 for an analysis of the gradual changes in the Reciprocal Trade Agreement of 1939 and the political economy surrounding its abrogation in 1971.

firms within the same sector (Corrales and Cisneros 1999; Coppedge 2000). As a result, state-business consultative groups became ineffective, which in turn made it difficult to design and implement industrial restructuring policies (Naím and Francés 1995).[18]

A corollary result of this fragmentation was a growth in *particularistic* bargaining between business leaders and political party leaders and ministers in charge of dispensing licenses and subsidies. One of the main channels of influencing was through campaign financing, which became decisive to electoral victory as Venezuelan elections in the era 1973–93 were among the most expensive in the developing world (Coppedge 2000). The growing reliance on campaign financing and personal favors, in the context of fluid factional changes, meant that there was little collective action among business groups, that no performance criteria were imposed by the state, that licenses and subsidies proliferated based on political rather than economic criteria, and that the business environment became less secure. In an environment where there was insecurity in government policy and with little export activity, rapid diversification of factories and products became the most effective means to spread risk among conglomerates in order to endure long-term survival (Naím 1989). Thus over-overdiversification was generated from both political party strategies to build clientele and from the defensive strategies of business groups to diversify risk in a rapidly changing and uncertain policy environment.

Evidence of the lack of selectivity in industrial policy can be seen in the level of firm entry in the Venezuelan manufacturing sector. Despite the widespread knowledge of the growing relative saturation of the internal market (see section 4.5), the period 1961–98 witnessed a relatively high level of (excessive) entry into manufacturing, as indicated in tables 9.8 and 9.9. Without an explicit export policy, the increase in the granting of protection licenses and subsidies led to overdiversification of products, which in turn generated suboptimal scale economies in plants. The number of large firms, which made up the vast majority of manufacturing investment assets, and which appropriated most of state credits, increased 131 percent from 1961 to 1971 and 68 percent from 1971 to 1982! The number of large firms increased 26 percent in the period 1982–88, despite declines in medium-sized firms over the same period. However, the number of large firms declined by 28 percent, the largest drop of any firm category, in the trade liberalization period, indicating the unviable nature

---

18. Pharmaceuticals, textiles, and auto parts are three sectors where factionalism hindered the development of coherent policies (Naím and Francés 1995, 178). See Coronil 1998, 237–85, for a detailed discussion of factionalism in the auto parts sector.

TABLE 9.8 Trends in the number and scale of firms in Venezuelan manufacturing, 1961–1998 (number of establishments)

	Total	Large	Medium-large	Medium-small	Small
1961	7,531	196	170	949	6,216
1971	6,401	453	386	1,138	4,424
1982	10,304	760	649	1,881	7,014
1988	10,238	961	612	1,897	6,768
1992	10,374	961	595	1,969	6,849
1998	11,198	693	486	1,832	8,187

SOURCE: OCEI, *Encuestra industrial*, various years.

NOTE: Large firms have more than 100 employees; medium-large, 51–100 employees; medium-small, 21–50 employees; and small, 5–20 employees.

TABLE 9.9 Growth in the number and scale of firms in Venezuelan manufacturing, 1961–1998 (percentage change in number of firms)

	Total	Large	Medium-large	Medium-small	Small
1961–71	−15	131	127	20	−29
1971–82	61	68	68	65	69
1982–88	−1	26	−6	1	−4
1988–98	9	−28	−21	−3	21

SOURCE: OCEI, *Encuestra industrial*, various years.

NOTE: Large firms have more than 100 employees; medium-large, 51–100 employees; medium-small, 21–50 employees; and small, 5–20 employees.

of many of these enterprises. This, in turn, suggests that credits and licenses were awarded more on political than economic criteria. The shock nature of the liberalization in 1989–92 also contributed to the inevitable decline in the number of firms as the Venezuelan manufacturing sector experienced significant decapitalization (Francés 2001).

The drop in the number of large firms in the era of economic liberalization is an indication of the inefficiency of state policies in the disbursement and management of rents as well as the weakness of the private sector to invest productively in the period 1971–88. Simply put, too many large firms with suboptimal scale and overdiversification of products were subsidized. Table 9.10 traces the evolution of the decline in the number of firms as classified by firm size and industrial sector in the period 1988–98. In this table, the sectors are ordered according to the percentage decline in number of large-scale firms in the period, beginning with the sector where that percentage decline is smallest.

TABLE 9.10 Evolution of number and scale of firms in selected Venezuelan manufacturing sectors, 1988–1998 (percentage change in number of firms, 1988–1996)

Sector	(isic)	Total	Large	Medium	Medium-small	Small
All manufacturing		9.4	−28.1	−20.6	−3.4	21.0
Iron and steel	(371)	34.9	−6.7	14.3	25.0	63.2
Nonferrous metals	(372)	−23.3	−14.3	57.1	−50.0	−28.6
Printing/publishing	(342)	−11.1	−25.7	−55.9	−4.3	−8.3
Chemical products	(352)	12.5	−26.2	11.1	−12.5	61.0
Plastics	(356)	−27.5	−29.3	−16.3	−28.8	−29.1
Transport equipment	(384)	33.3	−36.7	−33.3	−63.1	92.1
Metal products	(381)	−1.5	−40.9	−32.6	−5.6	4.2
Textiles	(321)	17.1	−44.8	0.0	8.3	65.6
Electrical machinery	(383)	1.7	−52.6	−25.8	6.1	46.9
Other manufacturing	(390)	−0.6	−56.3	0.0	−44.4	28.0
Wearing and apparel	(322)	−19.0	−64.9	−48.6	−33.8	−6.5
Wood products	(331)	66.0	−73.3	−62.5	−4.5	102.8

SOURCE: OCEI, *Encuestra industrial*, various years.

NOTE: Large firms have more than 100 employees; medium-large, 51–100 employees; medium-small, 21–50 employees; and small, 5–20 employees.

In the period 1988–96, the overall percentage fall in the number of large firms was 28 percent. The decline in the number of firms was *least* pronounced among the state-controlled capital-intensive sectors (steel, aluminum, industrial chemicals), which were also the sectors where productivity levels dropped the *least* in the 1990s (table 3.4c). This reflects the higher political costs of closing down state-run firms in a clientelist polity but also reflects the greater potential viability of these sectors. Aluminum and steel became the second- and fourth-largest exports in the country in the 1990s. However, there was a much larger drop in sectors controlled by private owners, reflecting the weakness of private-sector productive capacity. The sectors characterized by intermediate capital-intensive technology (transport equipment, metal products, and electrical machinery), which were also the sectors receiving the greatest subsidization and protection in the more advanced stage of ISI, dropped 36.7 percent, 40.9 percent, and 52.6 percent respectively. The clothing sector, a labor-intensive activity (which experienced among the most fractious firm relations within the textile business association), also experienced a dramatic decline of 64.9 percent in the number of firms. Such fractious firm relations in the textile sector was also an important reason behind the failure of the sector to restructure its production strategy in ways that would have enhanced its capacity to export, which was, in any case, weak throughout the whole of the post-1968 period.

It is plausible to argue that the decline in the number of firms could be due to other factors beyond those I have posited. The fact that many large firms failed after liberalization can be symptomatic of excessive protection or simply of the fact that they had been set up under expectations of economic growth that were not realized. The latter explanation is relevant but not inconsistent with my analysis. The creation of firms under very protectionist conditions in the period 1974–88 was the result of an explicit policy of patronage. The number of large firms actually increased (for reasons of political patronage) in the 1980s despite the slowdown in manufacturing growth in the 1980s compared to the 1970s. Second, the non-oil growth rate of the economy in the period 1990–98 was actually greater than in the 1980s (see table 2.1), so the dramatic reduction in the number of large firms cannot necessarily be attributed to declines in growth rates alone. What is true is that the reduction in tariff and non-tariff protection in the 1990s (a policy that undermined traditional patronage patterns) weeded out many of these unviable firms, though high real interest rates and the banking crisis of 1994 clearly affected the ability of many large firms to survive, particularly in the context of less protection.

The decline in the number of large firms is also consistent with the hypothesis advanced by Adriana Bermúdez and Omar Bello (2006) that labor legislation became significantly biased against large firms. While economic liberalization challenged traditional clientelist patronage patterns, not all of the influencing of the state was eliminated in the 1990s. It is well known that the banking sector was still influential in preventing financial deregulation (De Krivoy 2002). Similarly, labor unions and their allies (particularly the political parties of AD and of Convergencia) remained sufficiently influential to enact labor legislation that increased the labor costs of firms. It is no accident that these laws affected the large firms more in the 1990s because the bulk of the formal sector facing increased international competition resided in large firms. This resulted, as Bermúdez and Bello (2006) argue, in both downsizing and the closure of plants. What is not explained (and remains an interesting research topic) is why owners of industrial assets were unable to prevent this unfavorable legislation, or more generally, why the financial sector seemed to be able to protect its interests to a greater degree than the large-scale manufacturing sector. This is a trend that extended well into the first eight years of the Chávez administration (1998–2005).

One of the more damaging effects of the proliferation of unviable firms is the very high and increasing levels of excess capacity in industry maintained throughout the post-1968 period, as indicated in table 9.11. Sustained levels of excess capacity represent dynamic inefficiencies in capital allocation and use.

For the years when excess capacity data are available, there was a noticeable decline in aggregate capacity utilization from 67 percent in 1966 to 60 percent in the period 1985–88. Spare capacity reached a low of 44 percent in 1988 and increased to 55 percent in 1991, still very low considering that that year saw the fastest growth rate in manufacturing growth (12.2 percent) of any year in the period 1980–98.

The sectoral breakdown of excess capacity in large-scale firms is also very revealing of the greater costs of the failure of industrial policy in the more challenging periods of big-push heavy industrial strategies and the more advanced stages of ISI. Many of the intermediate and capital goods sectors had higher levels of excess capacity than the overall level in the manufacturing sector. For instance, in the period 1985–88, average manufacturing capacity utilization was 60 percent, whereas capacity utilization in many heavy and industrial sectors—such as chemical products (44 percent), metal products (51 percent), nonelectrical machinery (39 percent), electrical machinery (54 percent), and transport equipment (38 percent)—was considerably *lower* than the already low average capacity for the manufacturing sector as a whole. In the liberalization period, all of these sectors remained below overall capacity utilization rates, with transport equipment falling to 12 and 22 percent of capacity utilization in 1990 and 1991 respectively.

Given the much higher fixed capital costs of these sectors as compared to light industry (which dominates the easy phase of ISI), the economic costs of unutilized capacity is much greater than would be the case for unutilized capacity in lighter manufacturing sectors. The capacity utilization of the state-dominated sectors (iron and steel and aluminum), while nowhere near full capacity, were better than the overall average. In 1985–88, 1990, and 1991, capacity utilization in iron and steel was 71, 62, and 67 percent respectively, while capacity utilization in aluminum was 71, 70, and 69 percent respectively. Again, exporting in these sectors has been instrumental in keeping utilization rates higher than the average capacity utilization for the manufacturing sector as a whole.

The inefficiency of overdiversification can be seen in the number of unviable small and medium-sized firms that were created in the period 1971–88.[19]

---

19. The extent to which an entrepreneur or sector is overdiversified is difficult to identify beforehand. I define the term broadly to refer to a context in which a significant portion of plants across manufacturing sectors are persistently operating at suboptimal levels of scale economy during periods of both recession and full employment. The widespread occurrence of spare capacity across sectors during downturns in demand (i.e., during recessions) does not qualify as overdiversification.

TABLE 9.11 Capacity utilization: Venezuelan manufacturing, 1966–1991 (actual output as a percentage of maximum possible output)

Sector	(ISIC)	Size of firm							
		Large				Medium and small			
		1966	1985–88	1989	1990–91	1966	1985–88	1989	1990–91
All manufacturing (excl. oil refining)		67	60	44	55	48	45	34	38
Food	(311)		60	50	64		48	46	45
Textiles	(321)		87	70	66		60	50	61
Wearing and Apparel	(322)		57	50	45		51	36	36
Wood products	(331)		71	45	51		74	70	68
Printing/publishing	(342)		75	50	55		57	37	35
Chemicals	(351)		77	60	71		65	46	47
Chemical products	(352)		44	36	49		53	32	36
Oil Refining	(353)		81	60	88		56	46	54
Rubber products	(355)		76	74	73		42	41	43
Plastics	(356)		62	53	52		48	39	45
Iron and steel	(371)		71	62	67		35	18	35
Nonferrous metals	(372)		71	70	69		48	55	46
Metal products	(381)		51	41	31		35	22	32
Nonelectrical machinery	(382)		39	35	38		26	22	25
Electrical machinery	(383)		54	38	50		35	38	27
Transport equipment	(384)		38	12	22		27	26	32
Other manufacturing	(390)		47	45	58		53	33	22

SOURCE: OCEI, *Encuestra industrial*, various years.

NOTE: "Maximum possible output" refers to the output possible given current working levels of human and physical capital; calculations based on yearly surveys with plant managers. Large firms are those with more than one hundred employees; medium and small firms are those with between five and ninety-nine employees.

The average capacity utilization rates of the smaller and medium-sized firms were lower than of large firms for all the years under consideration: 48 percent in 1966, 45 percent in 1985–88, 34 percent in 1989, and 38 percent in 1990–91. Capacity utilization rates within many of the heavy industrial sectors—such as iron and steel (35 percent), nonferrous metals (48 percent), metal products (35 percent), nonelectrical machinery (26 percent), electrical machinery (35), and transport equipment (27 percent)—were either near or far below an already low small and medium-size firm capacity utilization average. In the liberalization period, all of these sectors, except for nonferrous metals, remained below overall capacity utilization rates in 1990 and 1991. Once again, given the much higher fixed capital costs of these sectors as compared to light industry (which dominates the easy phase of ISI), the economic costs of unutilized capacity is much greater than would be the case of unutilized capacity in lighter manufacturing sectors even for small and medium-sized firms.

Product overdiversification and excessive firm entry (in nearly all sectors) also played an important part in the weakness of technological capacity of firms. For an infant industry characterized by economies of scale, large volumes of sales and output are necessary to spread fixed costs and accumulate learning (Kim and Ma 1997, 122). As a result, unrestricted or excessive entry leads to the development of too many firms with suboptimal plant size. Overdiversification becomes a constraint on the development of research and development (R&D) activities since suboptimal size results in an annual turnover per firm too small to undertake the risks and costs involved in directing resources to R&D. For Venezuela, Horacio Viana (1994, 128–29) suggests that the period 1975–90 is marked by relatively low levels of R&D spending in manufacturing industry in comparison with other Latin American economies, which, on average, direct fewer resources to R&D than less developed countries in other regions.[20] In a survey of six hundred large-scale manufacturing firms (defined as having one hundred or more employees) conducted in 1992 and 1993, only 19 percent reported that they dedicated any resources to "innovative activity" (which in the survey includes R&D spending, production process assessment, changes in machine design, and product innovation) in the period 1980–92 compared to the Latin American average of 27.9 over the same period (129).

---

20. Amsden (2001, 277–80) argues that one of the main problems of manufacturing competitiveness in Latin America in comparison to East Asian economies and India is due to relatively low levels of scientific research and patenting, a relatively low share of gross national product accounted for by science and technology, a relatively low share of R&D spending by the private sector, and a relatively low share of R&D spending accounted for by the manufacturing sectors.

Viana also notes that the limited degree of endogenous technological capacity was due mainly to the small size of firms, which limited the possibility of such enterprises to assume the costs and risks of R&D spending (175).

Finally, the effect of political fragmentation and uncertainty inhibited the development of long-term technical cooperation, or cooperation on issues related to the "supply chain" between firms in the same sector. As a result, there was no evidence of significant cooperation between firms and state-run technology institutes (165–68).[21] It is reasonable to suppose that the economic costs of suboptimal R&D spending is greater in the more advanced stage of ISI and during big-push industrialization because technological upgrading is more central to firm competitiveness than it is for light industry, which dominates the initial stages of ISI.

In sum, the failure of the state to impose conditions and selectivity in rent deployment and coordinate investment efforts was more costly for big-push and advanced stage of ISI strategies compared to strategies to promote firms in early stage of ISI. A more detailed summary of the mismatch between development strategies and political settlements in the period 1958–2005 is presented in table 9.12. As discussed in Chapter 8, the growing fragmentation of political organizations (which resulted from the plethora of political demands that populist clientelism generated) made centralized rent deployment and management increasingly ineffective in implementing more advanced-stage ISI and big-push industrialization strategies.

The negative effect of populist clientelism on state efficiency in rent deployment was reflected in the productivity performance of the manufacturing sector over time. Productivity growth lies at the heart of dynamic and competitive growth. The incompatibility of the populist clientelism with more demanding development strategies ultimately were reflected in declines in manufacturing growth from 1980, significant declines in non-oil productivity growth in the period 1968–98, and varying percentage declines in the relative productivity levels of all Venezuelan manufacturing sectors from the mid-1980s (tables 3.4a–c). Consider also the historical and comparative evidence of labor productivity in table 9.13. There has been a general decline in aggregate labor productivity growth in Venezuelan manufacturing since 1950.[22] The higher rate of labor

---

21. Only ten out of six hundred firms reported maintaining long-term relations with state-run technology institutes (Viana 1994, 149).

22. Labor productivity is used in the first instance because it is a relatively simple concept and easy to measure. For a discussion of the problems of interpreting different measurements of labor inputs and their implications, see Baumol et al. 1989, 225–50.

TABLE 9.12 An overview of evolution of regime type, development policies, and economic outcomes in Venezuela, 1958–2005

Period	Regime type/main political trends and settlements	Industrial policy orientation	Stage of ISI/dominant technologies	Main economic results
1958–68	Pacted democracy; less radical form of populist clientelism Interparty cooperation high Centralized political organizations dominate	State-led industrialization: blanket protection of industry through import quotas and tariffs and substantial increase in industrial credit. Manufacturing investment still dominated by private sector conglomerates	Transition period to more advanced stage of ISI (1960–73) Scale economies and exports become decisive to manufacturing productivity and output growth.	Rapid, but slowing gowth in non-oil and manufacturing growth Low inflation
1968–73	Political pacts begin to break down in 1968	Same as previous period	Same as previous period	Same as previous period; however, non-oil total factor productivity growth begins long period of decline; manufacturing labor productivity growth slows substantially
1973–93	Two-party electoral rivalry within democratic pact Growth in factionalism and fragmentation of populist clientelism and the party system Political fragmentation becomes salient and lasts through 2005	Continued blanket protections; proliferation of subsidies. Public enterprises in heavy natural-resource-based industry dominate manufacturing investments Multiple exchange rate system (recadi, 1984–88) generates large	Advanced stage of ISI; "big push" state-led natural-resource-based industrialization strategy, 1974–85.	Non-oil and manufacturing growth rapid 1973–80; stagnates thereafter Manufacturing labor productivity growth falls Proliferation of public enterprises Excessive entry into manufacturing sectors

Period	Politics	Policies	Economic Structure	Economic Outcomes
		subsidies for firms with political contacts and/or import licenses.		Little discipline of state or private subsidy recipients; Capital flight cum debt crisis (1974–85), indication of massive macro co-ordination failures Moderate and growing inflation
1993–98	Multiparty electoral rivalry Decline of AD and COPEI Rise of political outsiders and increase in anti-political party radical populism	Radical trade and financial liberalization Partial privatization of state-owned steel and aluminum enterprises.	Advanced stage of ISI continues; capital-intensive natural resource-based industries remain most productive relative to U.S. level.	Manufacturing growth, productivity, and investment collapse Sharp decrease in number of large-scale manufacturing firms Banking crisis (1994) High inflation Capital flight
1999–2005	Rise in antiparty politics Centralization of executive power Polarization of politics	Little attention paid to industrial strategy; oil opening policy reversed Introduction of capital controls Data collection of industrial survey curtailed	Same as previous period	Collapse in non-oil and manufacturing growth in 1999–2003; oil boom facilitates rapid short-term growth in 2004–5. Unemployment rises Capital flight continues Financial intermediation declines

productivity growth in the period 1960–68 (3.6 percent) than in subsequent periods is consistent with the argument that the era of pacted democracy (1958–68) had more continuity and consensus in policymaking (and hence had a higher degree of centralized political organization). Moreover, the periodization of development stages and strategies (see Chapter 7) suggested that, in the period 1960–68, the big-push, natural-resource-based industrial strategy had not yet started, and the easy stage of ISI was still coming to a close. These two factors also enhanced the prospects of rapid manufacturing productivity growth, since the economic and political challenges of effectively implementing early ISI strategies are less demanding than more advanced ISI strategies.

In comparative terms, the available evidence on labor productivity growth indicates that, in the period 1970–90, Chile and Argentina maintained rates of manufacturing productivity growth similar to those of Venezuela, while Colombia maintained a higher annual average at 2.7 percent per year. The higher rates of productivity growth of the Colombian economy may possibly be explained by the absence of populism and weakness of labor unions and the small-scale, labor-intensive focus of its industrialization process (see section 6.4). However, Venezuela compares much less favorably (in the same period) with more rapidly growing manufacturing economies, such as Thailand, South Korea, and (natural-resource-rich) Malaysia, where average annual labor productivity growth rates were 4.1, 9.4, and 4.7 percent respectively. These figures corroborate my contention that Venezuela's pattern of growth makes it far more similar to the Latin American pattern of poor industrial growth than rentier theorists such as Karl (1997) maintain.

While labor productivity is a useful proxy for the standard of living or consumption potential of the working population, it is inadequate, on its own, for illuminating the reasons why a given productivity has been reached. For instance, in an economy where labor skills are fixed, labor productivity may still grow because of technological changes that improve the *quality* of the capital stock or increase the *quantity* of capital equipment per worker. In this case, it would be inappropriate to attribute expansion in output to labor.

According to William Baumol and colleagues (1989), "Total factor productivity (TFP) is undoubtedly the better index of *efficiency* in input use while labor productivity is the more illuminating measure of the *result* of the process for its human participants " (227). This is the main rationale behind the effort to measure total factor productivity. Total factor productivity is meant to measure the efficiency of resource allocation within a static general equilibrium framework. Specifically, it is meant to provide an index of the rate of

TABLE 9.13a Venezuelan manufacturing labor productivity growth, 1950–1998

Period	Average annual growth in gross output per employee (%)
1950–60	9.6
1960–68	3.6
1968–70	1.8
1970–90	1.2
1990–98	0.7

SOURCE: Baptista 1997; OCEI, *Encuestra industrial*, various years; Banco Central de Venezuela, *Serie estadística*, various years.

TABLE 9.13b Manufacturing labor productivity growth in comparative perspective, 1970–1990

Country	Average annual growth in gross output per employee (%)
Venezuela	1.2
Argentina	0.8
Chile	1.2
Colombia*	2.7
Malaysia	4.7
Thailand	4.1
South Korea	9.4

SOURCE: Baptista 1997; OCEI, *Encuestra industrial*, various years; Banco Central de Venezuela, *Serie estadística*, various years; Amsden 1997, table 1; Katz 2000, table 1.

* Refers to the period 1970–96.

expansion of some productive unit's capacity to produce over and above the portion attributable only to an expansion of its input quantities. The conventional, and most common, way to calculate total factor productivity is to differentiate a Cobb-Douglas production function with respect to time, also known as the "residual method." This "crude" measure of total factor productivity is valid only under the (unrealistic) assumptions of perfect competition, disembodied technical progress, and constant returns to scale.[23] The most serious problems with the orthodox theory are that technical progress is exogenous to the economic system and that economic growth is independent of the investment ratio. With these assumptions, it is possible to calculate the marginal contributions of labor, capital, and the residual to the growth process.

23. For a critical discussion of the residual method, see Scott 1989, 69–127.

Despite the well-known problems with total factor productivity measurements, recent calculations of total factor productivity growth in the Venezuelan economy seem to support the proposition that the non-oil Venezuelan economy (of which manufacturing is an important component) experienced increasingly lower growth rates in productivity growth in the post-1968 period, compared with the period 1950–68. Consider table 9.14. According to Francisco Rodríguez's (2006, table 10) calculations, non-oil total factor productivity declined from an annual average of 1.10 percent in the period 1950–68 to minus 1.45 percent in the period 1968–84. While non-oil total factor productivity average annual growth improved to 0.31 percent in the period 1984–98, this growth rate was less than one-third the growth rate achieved in the period 1950–68. These total factor productivity figures are calculated with homogenous factors, which refer to increases in the quantity of factor inputs (labor and capital) *without* distinguishing between differences in the type and quality of those inputs over time. Rodríguez does, however, calculate non-oil total factor productivity by accounting for the heterogeneous factors that go into the production function.[24] The trajectory of non-oil total factor productivity calculated with heterogeneous factors (which take into account increases in the quantity of factor inputs that take into account improvement in the quality of labor and capital inputs over time) generates the same long-run trends.[25]

TABLE 9.14 Venezuelan non-oil total factor productivity growth, 1950–1998 (average annual growth at 1984 prices, %)

	Homogenous factors	Heterogeneous factors
1950–68	1.10	1.59
1968–84	−1.45	−1.68
1984–98	0.31	0.36

NOTE: Homogeneous factors refer to increases in the quantity of factor inputs (labor and capital) *without* taking into account improvements in quality of those inputs over time. Heterogeneous factors refer to increases in the quantity of factor inputs that take into account improvement in the quality of labor and capital inputs over time.
SOURCE: Rodríguez 2006, table 10.

24. To account for heterogeneities in human and physical capital, F. Rodríguez (2006, 524) divides the contributions of the capital stock into three categories (residential, machinery and equipment, and nonresidential) and the human capital stock (i.e., the labor force) into four categories by educational level (no education, primary, secondary, and higher).

25. Moreover, the Venezuelan TFP figures in the period 1984–98 are in the range of the TFP collapses occurring during the debt crises in Latin America (F. Rodríguez 2006, 524–25). Loayza et al. (2002), for example, estimate an average rate of TFP growth for Latin America for the 1980s of *minus* 1.29 percent.

9.3 Revising Recent Interpretations of Venezuelan Economic Decline, 1968–2005

It is useful to examine how the compatibility approach developed throughout Part Three complements or advances our understanding of the crisis in Venezuelan economic performance compared with political science and (more recent) economic explanations. With declines in economic performance and political stability, attention has also turned to the weaknesses of the democratic institutions (McCoy et al. 1995; Goodman et al. 1995; McCoy and Myers 2004; Crisp 1994; 2000, 173–93). As economic performance worsened, the political system has increasingly been described as a bloated state system dominated by closed elites (*cogollocracia*) and a clientelist party system (*partidocracia*).

The essence of much of the late-twentieth-century political analysis on Venezuela emphasizes either a breakdown in cooperation among elites or the growing exclusionary nature of the regime. Rey (1991) focuses on the shift from a more consensual "pacted" democracy in the period 1958–68 toward more intense electoral rivalry and less cooperation among and between parties in the period 1973–2005. Michael Coppedge (1994) examines the institutional logic behind the increasing factionalism and faction-fighting between and within parties and economic groups in the Venezuelan party system and the growing role of campaign financing by big business groups as electoral rivalry increased. While these insights are valuable (and provided the basis for a periodization of the nature of the polity discussed in Chapter 7), there is no attempt to explain in either analysis *why* increased electoral rivalry or faction-fighting generated would generate less efficient state capabilities in undertaking industrial policy and macroeconomic management. Electoral rivalry and factionalism did not seem to diminish the effectiveness of industrial policy as much in, for instance, Italy, Thailand, or Colombia—all countries well known for party factionalism—as they did in Venezuela.

Other political institutional analyses focus on the way the political institutions of Venezuela's pacted democracy perpetuated social exclusion by mobilizing bias in favor of big business (Crisp 1994, 2000). According to this argument, effective influencing works in Venezuelan democracy through a vast array of advisory commissions in the decentralized public administration. In the period 1958–93, domestic capital interests were apparently favored more than any other group in the decentralized public administration, and their influence is largest in the very bodies that were supposed to regulate their activities (1494). In this respect, Brian Crisp argues, Venezuela is similar to other late-developing capitalist economies, particularly within Latin America. The rigid institutional exclusion also meant that many issues were left off the agenda, such as creating a more competitive industrial base or establishing a tax base beyond petroleum revenues (Crisp 1997, 193). Crisp proposes that an increase in the internal democracy of political parties, unions, and business associations will be required to improve the representation and effectiveness of Venezuelan democracy (Crisp 1994, 1507).[26]

The idea that Venezuelan democracy is similar to that of other late-developing capitalist democracies (in the sense that domestic capital is disproportionately favored) is subject to at least three important shortcomings when viewed in comparative and historical perspective. First, if domestic capitalists were similar everywhere in their influence, it would be difficult to explain why Venezuela has had among the least unequal income distributions in Latin America, or why indeed its income distribution is similar to that of Malaysia.[27] Second, if domestic capital can successfully capture the state, then the variation and change in the effectiveness of accumulation strategies in Venezuela over time are not explained. Moreover, the relatively poor performance of Venezuela in relation to other middle-income late developers—particularly more exclusionary or authoritarian ones such as Chile, Malaysia, Korea, Mexico, Thailand, and Taiwan—is also not possible to explain within the corporatist framework Crisp investigates.

The political science literature provides a rich analytical and historical account of the functioning and evolution of Venezuelan democracy. However, the inability of this literature to map the economic effects of the historically specific evolution of political contestation is an important lacuna in understanding the evolving sociopolitical crisis Venezuela faced in the period 1980–2005. Economic stagnation has surely been at the heart of political crisis, as most political analysts of Venezuela acknowledge (Goodman et al. 1995). The dramatic decline in economic performance and real wages in the period 1980–2003 has certainly had clear political effects, the most notable being the substantial decline in the legitimacy of the once dominant political parties AD and COPEI.

The political science literature does not consider how different stages of development and economic development strategies affect the prospects for both

26. Hillman (1994) also argues that the political crises of the 1980s and 1990s were caused by the undemocratic, and exclusionary nature of the Venezuelan party system.

27. It is important to note that income distribution figures are based on surveys that do not adequately measure capital income. Venezuelan capital shares (both national and in manufacturing) tend to be very high (as indicated in tables 3.4a–c). I thank Francisco Rodríguez for bringing this point to my attention.

the state and political parties to channel resources in a more egalitarian way. In the process of primitive accumulation, political party cadres and the state are actively engaged in divisive decisions. This is because the construction of capitalism in its early stages is inherently divisive and often rewards and deprives individuals in arbitrary ways. Central to the problem of variation and change in growth is the need to explain why privileged asset owners are compelled to accumulate capital by engaging in productivity-enhancing investments in some contexts but not in others. Moreover, the construction of capitalism in less developed economies is a very unjust and contentious process, since the drivers of and beneficiaries of growth are often limited to "privileged" asset owners and (to a lesser extent) workers in the few dynamic sectors of the economy. This creates substantial problems of politically legitimating the structure of property rights and other privileges that emerge in less developed economies, where far more subsistence agricultural workers and informal urban workers benefit much less even when growth rates are rapid.

The problem of legitimating a more advanced stage of ISI and big-push industrial strategies is even more demanding, since vast state resources need to be allocated more selectively (and thus in an even less egalitarian manner) than in early ISI strategies. If the construction of a viable capitalism in the context of late development were a mere technical problem, then many more less developed countries would be raising their income per capita levels to that enjoyed in advanced industrial economies. The fact is that this is occurring in very few less developed economies (whether mineral resource-rich or not).

Moreover, the comparative and historical evidence on late development suggests that the political compromises in most developing countries are even more fragile than those in advanced industrialized countries (Khan 2005). This is because establishing legitimate rule in the context of processes of primitive accumulation and late development is inherently difficult. While primitive accumulation and state-led development are inherently divisive, late development imparts a greater role to the state, and particularly state development banks, in socializing the risks of catching up. The construction and financing of capitalism in late developers has been more overtly political because the state has been more central to financing capital accumulation. The generally greater levels of political instability and political and economic corruption indicate that the legitimacy of institutional structures is under great contestation. The political science literature neglects not only the context in which late capitalist development is created but also fails to explain why such a process has differential outcomes both within economies over time and across (similarly corrupt) late-developing countries.

That some more successful industrial policy states have been significantly more authoritarian than Venezuela suggests that the focus placed in the Venezuelan political science literature on the level of political competition and whether demands are channeled through democratic procedures may be less important than the issue of who is capturing the state and for what ends. Political analyses of the party system do not examine whether maintaining political stability and legitimate rule is in any way affected by the changing nature of the economic development strategy. This problem is also neglected by institutional economists, who argue that democracy, *in general*, is conducive to economic development because it provides checks on potential state predatory behavior such as the arbitrary confiscation or attenuation of property rights (North 1981; Olson 1993).

Indeed, much of the recent literature on governance (North 1990; World Bank 1997a) argues that growth requires certain prerequisites such as competitive electoral democracy, a well-functioning bureaucracy, economic liberalization, and so on. In this perspective, a failing state contains a set of institutional structures that deviate from a modern Weberian bureaucracy. State failure is measured roughly in this view as the distance of a failing state from "best practice," that is, bureaucratic structures in advanced industrialized countries. This view is dominant among the international financial institutions and donors.

The problem with this view, what may be called the prerequisite view of development, is that there is little evidence that economic liberalization, democratization, low corruption, the absence of patrimonial rule, or even a modern Weberian state have been inputs into long-term economic development (Chang 2002; Rodrik 2004a; Khan 2005; North et al. 2007). The evidence also suggests that rapid growth is not a function of countries scoring well on the standard "good governance" indicators (such as voice and accountability, political stability and a lack of political violence, controlling corruption, rule of law, bureaucratic quality, and expropriation risk) (Svensson 2005; Khan, 2006). Almost all late-developing countries score significantly below average on OECD governance indicators. What is more, the fastest-growing late developers obtain governance scores *very similar* to the average score of slow-growing developing countries. This implies that the idea that countries require good governance to grow is ahistorical, a point stressed by Gerschenkron (1962) long ago.

Patrimonial politics, clientelism, corruption, limited representation, and even political violence are found in *all* late-developing states, particularly ones at low levels of development. The key issue is to explain why some of these states become more developmental over time, and why some find it difficult to maintain even a semblance of basic public authority, let alone developmental features. Explaining this variation and change in state capacity in the context of patrimonialism, corruption, and clientelism represents an important research challenge that the state failure literature has yet to adequately address.

As argued in Chapter 6, if different strategies require different levels of selectivity and concentration of economic and political power to be initiated and consolidated, then the problem of legitimacy and inclusiveness cannot be adequately examined as isolated from economic strategy. If economists often fail to incorporate politics when examining the state intervention, then political scientists often fail to examine the political challenges that different technologies and stages of development generate. The growing factionalism of the Venezuelan polity was particularly costly in terms of the efficiency of investment because it occurred in the context of the more advanced stage of development, which requires a greater centralization and coordination of investment. The growing proliferation of subsidies and licenses to Venezuelan firms in the period 1965–88 also suggested that the Venezuelan state, far from being exclusionary, was too inclusionary and unselective in its patronage patterns for it to produce an effective industrial strategy in the more advanced stage of ISI and particularly during its big-push industrialization strategy.

My explanation also attempts to move beyond simply identifying conflict as the source of growth decline. Recently, economic analysts have argued that growing distributive conflicts have played an important role in the growth collapse of the Venezuelan economy (Hausmann 2003). The analysis presented here tries to explain both the emergence of conflict in a specific historical conjuncture and its effects in comparative and historical perspective.

Hausmann argues that evidence of increased conflict can be seen in the rise in Venezuelan interest rates (an indicator of growing risk) in the period 1983–98. Evidence of Venezuela's increased risk is its consistently higher spreads in the Emerging Market Bonds Index and its weaker credit ratings vis-à-vis other Latin American countries such as Colombia, Brazil, Argentina, and Mexico. Moreover, this increased premium in Venezuela over this period has occurred despite the fact that Venezuela was a net creditor to the world (i.e., the country ran current account surpluses) and maintained lower debt-export ratios than its Latin American counterparts. Hausmann argues that the increased risk premium reflects not a problem with macroeconomic fundamentals, but the increase in distributive conflicts within the economy. As a result, "the mechanism through which this conflict affects growth is its impact on the cost of capital (i.e., the interest rates premium), leading to decline in output and capital per worker" (267).

While the identification of increased risk and conflict in the post-1983 Venezuelan economy is useful, these arguments have some important shortcomings. First, the long-run slowdown (if not collapse) in manufacturing productivity growth in Venezuela started in the late 1960s, well before there was a serious deterioration in Venezuelan interest rates. Second, there is no attempt to explain either the nature of the conflicts or why they were more salient after 1983 than before. In particular, there is no attempt to examine the extent to which changes in either the development strategy or changes in political contests might affect the probability of conflicts arising. In sum, explaining the growth slowdown as a mismatch of economic strategy and historically specific political strategies improves upon existing explanations that do emphasize political institutions and distributive struggles as the source of Venezuela's poor economic performance.

9.4 Conclusion

The long-run rapid growth in Venezuela from 1920 through 1980 and relatively respectable rates of manufacturing and non-oil productivity growth from 1950 through 1968 can be attributed, at least in part, to the compatibility of early stage ISI with centralized political organizations. The relatively rapid growth in the 1970s was the result of high investment rates facilitated by buoyant oil revenues and not by particularly high levels of productivity growth, which in fact was declining. Consequently, the significant declines in manufacturing productivity growth (1968–98) and a non-oil growth collapse (1980–2003) can be attributed, at least in part, to a growing incompatibility between more advanced ISI and big-push industrialization strategies and an increasingly fragmented and polarized polity.

Macroeconomic manifestations of this incompatibility included declining legitimacy of economic policies, leading to the massive capital flight crisis in the period 1974–85, growing inflation after 1980, several massive exchange rate devaluations after 1982, and a reduced confidence in and effectiveness of the banking system. These macroeconomic outcomes were important manifestations of the political failure of the party system to contain conflicts. The growing fragmentation of the state was shown to be primarily a result of dispersed political power and factionalism, and not due primarily to institutional design failure as suggested by mainstream models of corruption and rent-seeking.

After 1968 there were also several important microeconomic manifestations of the growing incompatibility between the economic strategies and political settlements. The political rationale of maintaining populist support negatively affected the ability of the state to be selective in the disbursement of rents and to discipline rent recipients. The populist and clientelist accommodation documented in this and the previous chapter has been manifested in Venezuela through well-known patterns of excessive entry into industry, excessive white-collar employment patronage in state enterprises, contradictory and rapid changes in regulations, and volatile and (politically) aggressive competition among factions of capital allied to different political patrons. The costliness of ineffective rent deployment was magnified in the context of the second stage of ISI and big-push strategies, since such strategies required massive investment outlays. These costs were manifested through significant productivity declines in the non-oil and, especially, the manufacturing sector in the period 1968–98.

Our understanding of the complex interactions between economics and politics is still far from satisfactory. Mainstream growth theories have not adequately explained long-run variation in growth across countries or variations in growth in one country over time. The framework presented here suggests that exploring the interactions of development strategies and political settlements promises to provide a more defensible explanation (at least in the Venezuela case) of growth and productivity trends that is more consistent with comparative and historical evidence. It is hoped that refining this approach will lead to testable hypotheses about growth trajectories across a wider range of middle-income economies.

PART 4

BEYOND THE VENEZUELAN CASE

10

THE POLITICAL ECONOMY OF GROWTH IN MALAYSIA AND VENEZUELA

Malaysia achieved one of the highest growth rates among middle-income countries in the period 1970–2000. The contrast between the Malaysian growth experience and the Venezuelan one illustrates the importance of how different political settlements and the nature of internal threats to regime survival can affect the efficiency of centralized rent deployment. The similarities between the two countries are as follows. First, both countries, in their regional contexts, possess abundant natural resource rents.[1] Second, both countries are medium-sized economies with populations just above 20 million. Third, both economies have had similar income distributions from 1970 on, though relative to their respective regions, income distribution is less unequal in Venezuela and more unequal in Malaysia (tables 2.12 and 2.13). Fourth, both countries undertook big-push, natural-resource-based industrialization strategies in the 1970s and 1980s (for evidence on Malaysia, see Jomo and Gomez 2000).[2] The investment rates for both economies were among the highest in their respective regions in this period, although Malaysia maintained a much higher rate after the early 1980s (table 3.8). As well, both economies were characterized by highly concentrated manufacturing sectors where big firms dominated. Fifth, both countries

---

1. While Malaysian natural resource exports were a considerably smaller percentage of total exports than was the case in Venezuela (see table 4.2), the Malaysian share of natural resources as a percentage of GDP in the period 1970–89 were greater than 30 percent, which was higher than Venezuela's share (25 percent) over the same period (Sachs and Warner 2001, fig. 1).

2. This makes the comparison with Malaysia more relevant than many other middle-income, mineral-abundant economies such as Botswana, Ecuador, or the Gulf states, where big-push heavy industrialization strategies were not salient features of their respective growth trajectories.

have attempted, through centralized and clientelist rent deployment, a populist redistribution toward middle-class and less privileged groups (on Malaysia, see Jomo and Gomez 1997). Finally, there was substantial amount of rent-seeking and corruption in the two countries in the 1980s. Historical accounts suggest that corruption in Malaysia was substantial through its period of state-led high growth in the 1970s (Jomo 1986, 243–72). There is also substantial evidence that political lobbying and cronyism were widespread through the privatization drive in Malaysia in the 1980s and early 1990s (Gomez and Jomo 1997, 75–165).

Why, then, did Malaysia enjoy rapid growth (see table 2.2 on comparative growth rates), while Venezuela has experienced growing ineffectiveness of industrial policy, a growing unpredictability of macroeconomic policy, and massive capital flight since the late 1970s? More generally: why were the outcomes of the rent-seeking process (that is, the types of rents created) in Malaysia more productivity-enhancing and growth-enhancing than in Venezuela despite substantial levels of inputs into rent-seeking in both countries? This chapter attempts to address these questions through a brief examination of the historical political economy of Malaysian economic development, especially since 1970.

10.1 The Political Origins of Centralized Political Organization in Malaysia

In the period 1970–2000, the Malaysian polity could reasonably be characterized as a consolidated state with centralized political organizations. While there has been electoral competition, Malaysia is essentially a one-party state. Nevertheless, it is also a polity characterized by significant degrees of clientelist and populist pressure to redistribute income and assets away from the dominant economic groups, comprised mainly of ethnic Chinese, toward the more economically marginalized, but majority ethnic Malay population. Most interesting are the factors that led to the formation and maintenance of centralized political organizations despite significant clientelist and populist demands that could easily have fragmented and even polarized political organizations and contestations, as has happened in Sri Lanka, for example, or as we have seen, in Venezuela.

In Malaysia, a centralized pattern of resource flows emerged in response to middle-class demands that proved to be compatible with rapid growth and structural transformation of the economy. Paradoxically, the well-known ethnic divide in Malaysia between its minority Chinese-Malaysian capitalists and the majority Malay population allowed this centralized redistribution strategy to emerge (Khan 2000b, 98–101). The political isolation of the Chinese capitalists in the Malaysian polity meant that the ruling coalition in the state could effectively "tax" them for the benefit of emerging intermediate Malay groups.[3] Moreover, since the bulk of the transfer could be legal, the need for illegal exactions was far less. This "affirmative action" policy gained multiclass support among Malays and is probably, according to Khan (2000b, 99), an important reason why Malaysia maintained one of the least corrupt regimes among middle-income economies in the 1990s (see table 4.2). The nature of political pacts and the growing costs of electoral rivalry in Venezuela made the state more vulnerable to capitalists' demands for protection, and for particularistic bonds to develop between capitalists and parts of the executive.

The centralized solution to the redistributive problem and the relatively "hard" democracy that emerged in Malaysia was an unintended consequence of the internal threats and opportunities that emerged from race riots between Chinese and Malays in 1969 (Putzel 1995). The political consolidation and restructuring, which took place after the riots, established the United Malays National Organization (UMNO), the Malay party in the ruling Barisan national coalition, as the dominant political organization in the country. The effect of this was to consolidate potentially competing Malay clientelist groups into a unified structure and, at the same time, to give the state leaders dominance over Chinese-Malay capitalists. Let us look briefly at the historical origin of this settlement.

Malaysia gained independence from the British in 1957. Independence was achieved in a relatively nonviolent way through a negotiation with the elite-dominated UMNO. Under Tunku Abdul Rahman, the UMNO constructed the National Alliance with the Malay Chinese Association (which was the main source of Alliance campaign financing) and the Malaysian Indian Congress.

---

3. One indication of the leverage the Malaysian state had with upper-income groups is the relatively high rates of personal and corporate income tax collected. In the period 1975–78, average annual personal income tax in Malaysia was 2.1 percent of GDP (above the East Asian average of 1.8 percent of GDP). In 1985–88 average annual personal income tax in Malaysia was 2.4 percent of GDP (above the East Asian average of 2.3 percent); in 1997–2002 it rose to 6.1 percent of GDP (which was the highest rate in East Asia, where the average was 3.9 percent of GDP) (Di John 2006, table 2). Average annual *corporate* tax collection as a percentage of GDP was 9.6 percent of GDP (above the East Asian average of 6.0 percent of GDP) in the period 1985–88, and was 8.4 percent of GDP (which was the highest rate in East Asia, where the average was 6.9 percent of GDP) in the period 1997–2002 (Di John 2006, table 3). In Venezuela, in the period 1975–2002, annual personal income taxes averaged a mere 1.0 percent of GDP (13–14), a telling indicator of the lack of state power over upper-income groups.

UMNO won an overwhelming majority in parliament in pre-independence elections. At the time, domestic business was almost entirely in the hand of the Chinese (who were well organized politically and professionally), and the Malays dominated the political elite.[4] UMNO, like AD in Venezuela, was able to channel and mobilize the large Malay peasantry to support the regime, rather than move into radical opposition. In contrast to Venezuela, independent trade unions were destroyed during battles with the Malaysian Communist Party before independence. Trade union power was further reduced by repressive labor legislation enacted in the decade after independence (Jomo 1986, 235–36).[5] The lower level of urbanization in Malaysia also meant that employment pressures in the cities and the possibility of mobilizing urban labor were of less concern than in Venezuela.[6]

As in Venezuela, the dominant political party in Malaysia, UMNO, faced redistributive demands from its support base, which included a significant part of the intermediate and middle classes. However, the relative power of the intermediate classes was historically less in Malaysia than in many developing countries (Khan 2000b, 98–101). As important, the relative power of labor unions and business groups was also less. Populism simply never reached the same levels as in Venezuela. As such, redistribution remained more coordinated and centrally controlled in Malaysia, since political party pressure and electoral rivalry were less of a concern for state leaders than their Venezuelan counterparts. Let us explore the nature of the power balances in more detail.

That labor unions in East and Southeast Asia were weaker than those in Latin America in the 1970s and 1980s is well documented (Banuri and Amadeo 1991). In the early 1980s, when both Venezuela and Malaysia were undertaking their respective big-push industrialization programs, comparative evidence suggests that labor unions were more powerful in Venezuela. According to table 10.1, the percentage of the labor force that was unionized was 44 percent, whereas in Malaysia the percentage was 8.7 percent. This difference in unionization rates

---

4. In the colonial period, Chinese capitalists were active in commerce, artisanal manufacturing, and tin mining. There was no explicit colonial policy to favor them over Malay capitalists. Rather, the Chinese exploited opportunities. The Chinese were able to control immigrant coolie-labor through secret societies, which employed extra-economic coercion (Jomo 1986, 168–77). This allowed Chinese capitalists to accumulate wealth in their ventures, including tin and rubber plantations. The Malay elites, according to Jomo (209–45), were less interested in pursuing business ventures, since they were offered relatively high salaries and benefits in the colonial administrative apparatus.

5. There is little evidence that Chinese capitalists had much influence in the enactment of repressive labor legislation.

6. In 1980, the urban population as a percentage of the total population was 42 percent in Malaysia and 80 percent in Venezuela (World Bank, *World development indicators*, 2007).

is generally representative of the difference between Latin America and Asia. Moreover, the right to strike was operative in Venezuela in the early 1980s, whereas striking was prohibited in Malaysia. Again, differences in the right to strike were also generally representative of the differences between the two regions, though Chile under Pinochet was the one exception in Latin America at that time.[7] Finally, Venezuela in the early 1980s was experiencing much more labor conflict than Malaysia. This also is indicative of the greater union strength in Venezuela. In the period 1980–84, there was an average of seventy

TABLE 10.1 Labor union power and labor conflict in less-developed countries in the early 1980s

	Labor union strength indicators, 1983			Labor conflict, 1980s	
	Percentage unionized	Closed shop	Right to strike	Period	Worker-days lost to labor disputes (average per year per 1,000 workers)
Polarized					
Brazil	48.6	no	yes	—	n.a.
Chile	28.5	no	limited	1980–81	158
Venezuela	44.5	no	yes	1980–84	70
Philippines	24.0	yes	yes	1980–84	60
Pluralist					
India	4.5	no	yes	1980–84	153
Pakistan	3.5	no	yes	1980–84	22
Bangladesh	3.0	no	yes	1980–84	28
Decentralized					
Indonesia	4.8	no	limited	1980–82	1
South Korea	7.0	yes	limited	1980–82	2
Malaysia	8.7	no	limited	1980–83	3
Thailand	1.1	no	limited	1980–83	4

SOURCE: Banuri and Amadeo 1991, tables 6.3 and 6.4.

NOTE: According to Banuri and Amadeo (1991, 176), *polarized* labor market institutions are those with a history of mobilization, organization, conflict, and success. They are capable of imposing real costs on the economy in defense of their interests but unable to impose cooperative solutions at the national level. The power of *pluralist* labor market institutions is based on alliances with other collective actors such as political parties or ethnic groups (as in the United States, Canada, Italy, and France). *Decentralized* labor market institutions are strongly circumscribed and divided and have reduced influence in national politics. Wage bargaining always occurs at the enterprise level and the right to strike is strongly limited (as in Japan and Switzerland).

7. Even here, Chile had the highest number of strikes and was still highly unionized.

worker-days per one thousand workers lost because of labor disputes, while in Malaysia, in the period 1980–83, an average of just three worker-days were lost. Venezuela's labor unions, strongly allied with the main political party AD, were again representative of more combative labor unions in Latin America than in East and Southeast Asia in that period.[8]

The nature of the difference that the political mobilization of urban middle-class clientelism played in Malaysia and Venezuela can also be seen by tracing employment profiles of workers in largely state-owned heavy industries. Table 10.2 indicates the percentage of manual workers in total employment in three heavy industrial sectors (industrial chemicals, iron and steel, and non-ferrous metals) in Korea, Malaysia, Venezuela, Chile, and Bangladesh in the period 1974–90. Following Venkataraman Bhasker and Mushtaq Khan (1995), one of the indicators of unproductive political clientelism is the overmanning of public sector enterprises with white-collar (administrative, secretarial, and managerial) workers. The proportion of manual workers steadily declined in Venezuela in all three sectors over the period. In fact, the employment profile in the heavy industrial sectors in Venezuela is generally much more similar to that of Bangladesh, where clientelism is prevalent in the public sector (see Khan 2000b) than to that of Malaysia, where the employment profile is more like that of a developmental state, such as South Korea. This, in part, indicates the ability of the Malaysian state to engage less in productivity-reducing political patronage. Also telling is the divergence between Venezuela and Chile. In the period 1974–76, the employment profile in the state-owned heavy industrial sectors was similar. This is not surprising, since clientelist and populist patronage characterized the Chilean polity in the 1960s and early 1970s. However, in the period 1988–90, the Chilean heavy industrial firms (many of which were privatized between 1980 and 1982) manifested an employment profile generally closer to the developmental state pattern of South Korea than the populist clientelist pattern of Venezuela.

10.2 The Political Economy of Rent Management and Capital Accumulation in Malaysia and Venezuela

It is still necessary to explain why Malaysian business groups, in the period 1970–2000, were relatively less able than their Venezuelan counterparts to capture

---

8. Interestingly, in South Asia, while unionization has been low, the strength of labor unions has been higher as indicated in their right to strike and in the higher indices of labor conflict. The

TABLE 10.2 Venezuelan employment profile in heavy industry: A comparative perspective, 1974–1990 (manual workers as percentage of total employment, by sector)

	Industrial chemicals (ISIC 351)			Iron and steel (ISIC 371)			Nonferrous metals (ISIC 372)		
	1974–76	1980–82	1988–90	1974–76	1980–82	1988–90	1974–76	1980–82	1988–90
South Korea	68.3	68.1	70.8	85.4	83.2	85.1	68.3	78.4	75.5
Malaysia	57.1	57.9	81.0	76.3	76.5	87.5	70.0	66.7	91.6
Chile	65.8	65.4	64.9	72.4	74.3	75.1	70.4	65.3	60.6
Bangladesh	66.5	50.0	54.0	78.6	49.2	63.5	n.a.	n.a.	57.6
Venezuela	62.3	51.3	44.0	71.3	64	59.3	77.6	65.3	51.6

SOURCE: UNIDO, *International yearbook of industrial statistics*, various years.

rents from the state and use such rents in unproductive ways. In order to address this issue, an examination of the relative political weakness of the Malaysian capitalist class would be required. Indeed, historical analyses of the political economy of Malaysia in this period strongly suggest that the political weakness of the Chinese and Chinese capitalists was the by-product of political mobilization (Putzel 1995; Jesudasen 1989).

A consensus emerged among the Malay elite that the economic dominance of the Chinese needed to be reversed. The Chinese make up one-third of the population and thus are not negligible in the electoral political system. Nevertheless, the political legitimacy of UMNO was largely based on effectively making the Chinese second-class political citizens while, at the same time, providing them with opportunities to continue investing in the domestic economy, thus creating job opportunities for working-class Malays. The integration of Singapore, Sarawak, and Sabah in 1963 led to a threat to both Malay dominance and UMNO supremacy within the nascent democracy. The threat to Malay political supremacy, once Singapore was incorporated into Malaysia, ignited a sense of collective resentment among Malays (Putzel 1995, 246). The tensions that ethnic politics generated led to the forced expulsion of Singapore in 1965.

The legacy of the threat to Malay supremacy had several important consequences for the political economy of Malaysia. First, the main capitalists (the ethnic Chinese) in the country lost the political power to legitimately buy off politicians in return for monopoly rents. Maintaining privileged positions for Chinese capitalists would have undermined the New Economic Policy, the cornerstone of UMNO's legitimacy. Second, increases in collective resentment and distrust among the Malays for the Chinese made it more difficult for Chinese capitalists to form clientelist arrangements with intermediate classes of Malays and the political factions representing them. Third, distrust and resentment of the Chinese among the politically dominant Malays brought the issue of income and asset distribution to the fore of the political agenda (Malaysia has among the most unequal distribution of income in Southeast Asia and an income distribution similar to Venezuela's).[9] Chinese privilege was associated with this inequality. Finally, although the first prime minister, Tunku Abdul Rahman, was a strong advocate of parliamentary democracy, the "ethnic question" in Malaysia meant that ethnicity took primacy over the democratic principle of universal suffrage. As a result, regime legitimacy depended more on effective

---

strength of factionalized middle-class clientelist factions, as described in Khan 2000b, may be the reason behind the disproportionate power of unions, especially public sector unions, in South Asia.

9. See tables 2.12 and 2.13.

"affirmative action" programs and asset redistribution benefiting Malays than in instituting democratic procedures and electoral rivalry. In this respect, populist clientelism in Malaysia was both more authoritarian and more centralized than in Venezuela. The fact that electoral competition was not as important a source of legitimacy in Malaysia as it was in Venezuela may be why Malaysia, unlike Venezuela, avoided an increase in factionalism.[10]

The cornerstone of Malaysia's centralized "ethno-populism" was the New Economic Policy (NEP) begun in 1970. The NEP sought to create socioeconomic conditions for "national unity" through massive economic redistributions programs aimed at achieving the twin goals of "poverty eradication" and the "restructuring of society." The main shifts in policy were to give the state a more active role in the economy and give *bumiputras* (Malay and indigenous groups) preferential access to education, professional training, state employment, business opportunities, and state credit. The increase in oil and natural gas rents (from the mid-1970s through the mid-1980s) provided the state the resources to finance much of its expansion in production and credit allocation. The NEP sought to reduce the incidence of poverty from nearly 50 percent in 1970 to 16 percent by 1990 (Jomo and Gomez 2000, 287). The NEP also envisioned raising the *bumiputra* share of corporate equity from 2.4 percent in 1970 to 30 percent by 1990.

After 1970, the role of the state-owned enterprises (SOEs) in Malaysia increased substantially for the purpose of increasing professional and employment opportunities for Malays, a vital part of the NEP's ethnic affirmative action program. The number of SOEs increased from 10 in 1957 and 22 in 1960 to 109 by 1970, to 656 by 1980, and to 1014 by 1985 (Jomo and Gomez 2000, 288–89). In line with the growth of SOEs, the public sector's share of the GNP rose from 29 percent in 1970 to a peak of 58.4 percent in 1981 (288).[11] As in Venezuela, the growth of SOEs, especially in heavy industry in the early 1980s, was accompanied by declining capital productivity in the economy. The average incremental capital-output ratio (ICOR) rose from 2–3 in the 1970s to 5–6

10. The factionalism of many other poor economies with ethnic rivalry, such as Sri Lanka and many polities in sub-Saharan Africa, suggests that the maintenance of a centralized regime in Malaysia can be attributed to a historically specific political strategy and organization. The possibility of controlling ethnically defined capitalist groups is also more feasible when they are dispersed in the territory, which is the case in Malaysia. When the dominant capitalist group is from one ethnic group and is concentrated in a particular region, their collective action capacity is likely to increase and thus they are less likely to accept a subordinate position in politics (at least not without substantial and even violent political resistance).

11. According to Jomo and Gomez (1997, table 12.4), the share of general government expenditure in GDP rose from an annual average of 33 percent in the 1970s to 45 percent in the 1980s.

in the early 1980s, while public sector ICOR rose from 6–7 in the 1970s to 15–16 in the first half of the 1980s (Jomo and Gomez 2000, 289).

While most of these SOEs were loss-making enterprises (289), they were an important conduit for interethnic redistribution. This produced political stability and security of property rights for domestic and multinational capitalists, which underpinned high levels of investment, growth, and ultimately structural change in the Malaysian economy. Under the leadership of Mahathir bin Mohamad,[12] the UMNO maintained firm centralized control over a rapidly growing economy. The growing role of the state gave the prime minister and party cadres access to large resources for patronage.

The basic structure of the centralized patron-client flows that resulted from the NEP in Malaysia, as summarized by Khan (2000b, 98–101), worked as follows. The principal rent flow was from Chinese capitalists to the central party leadership of UMNO in the form of legal taxes, illegal extractions, and campaign financing. Malaysia's abundant natural resources allowed the state to access the largest natural resource rents per capita of any country in Southeast Asia. Collectively, these resources were used to create transfers to intermediate classes in many forms: including government-funded education programs, employment in public enterprises, subsidized loans from the banking system, and preferential access to business opportunities. The growth in the state in the period 1970–85 is indicative of the massive increases in patronage. Not only that, the massive privatization program of the mid-1980s and early 1990s was also used as a means to transfer rents to a series of Malay political constituents (Gomez and Jomo 1997, 75–165).

A second type of rent-seeking outcome involved the transfer of relatively small learning rents to domestic capitalists in order to induce their move into higher-technology sectors and the capture of natural resource rents by companies appropriating the right to exploit Malaysia's abundant natural resources (Khan 2000b, 98–101). While important, these types of learning rents were not as large as in more dirigiste industrial policy states like South Korea in the 1970s or Japan in the 1950s.

There is substantial evidence that the use of rents for affirmative action purposes generated large-scale rent-seeking costs in Malaysia (Gomez and Jomo 1997). The rent-seeking costs are, however, the *inputs* into the rights appropriation process (Hirshleifer 1994). These need to be compared with the outputs of this very same rights appropriation process (Khan 2000b). The rapid growth

---

12. Mahathir was president of Malaysia from 1981 to 2003.

of the Malaysian manufacturing sector suggests that the net benefits of political stability more than offset the costs of producing the rights generated through centralized state rent deployment.

In comparison with the more fraction-ridden form of centralized rent deployment in Venezuela, the Malaysian system maintained a centralized system of rent redistribution. Another important difference with Venezuelan populist clientelism is that there were fewer direct links between particular family conglomerate groups and political factions. In Venezuela, the regime relied on the capitalist groups to fund increasingly expensive and competitive elections. Big business groups in Venezuela were, in fact, central to the viability of political pacts that helped preserve democratic rule. This made it difficult for the Venezuelan state to create significant leverage vis-à-vis family conglomerate groups.[13]

UMNO was able to dominate the Chinese capitalist groups because the base of their political legitimacy was their alliance with the Malays. And because they had no effective electoral rival, it was unlikely that campaign financing could "capture" political decisions in dispensing privileges.[14] Thus, while Chinese capitalists were given opportunities to make money, they were not given licenses to appropriate state-created rents, regardless of how well they performed. The one-party dominance of the UMNO meant that the Malaysian regime, while no means immune from cronyism, was less vulnerable to capture by family conglomerate groups.[15] The relatively low tariff rates in Malaysia in the 1970s and 1980s (Bruton and Associates 1992, 261–64) were one indication of the relative weakness of business conglomerates to capture the state. The blanket and unselective protection provided for Venezuelan capitalists would never have gained legitimacy in Malaysia in this period.[16]

The political stability that the centralized rent distribution system generated was also instrumental in attracting multinational investment in high-technology

13. The very low personal income tax collection in Venezuela (see note 3) along with the inability of the state to effectively monitor rent recipients in infant industries (Naím and Francés 1995) are two telling indicators of the relatively weak leverage and power the state had with respect to upper-income groups.

14. This is not to say that there weren't significant amounts of cronyism in Malaysia. For an account of the patron-client links between UMNO leaders and business groups, see Gomez and Jomo 1997.

15. The Malay leadership under Mahathir bin Mohammad promoted much of economy's export boom in electronics and textiles by providing substantial incentives to multinational companies. In this way, the regime was able to use the presence of multinational as a mechanism of limiting the economic power of Chinese capitalists in the polity. In Venezuela, there was no analogous mechanism for preventing unproductive rent-seeking capitalists from capturing the state.

16. The influence of structuralist ideas in Latin America also influenced the implementation of blanket protection by justifying "inward-looking" ISI "at any cost" (Fishlow 1991).

sectors. The centralized political transfers, funded largely with the abundant rents from natural resources, meant that domestic redistributive demands were met by the state. The ability of the state to separate economic strategy from redistributive transfers was an important factor in convincing multinationals that their proprietary assets and business operations would not be negatively affected by local political processes.[17] According to Khan (2000b), this was a contributing factor in Malaysia's success in attracting multinational investment.

Finally, an important difference in the nature of rent deployment was the class nature of the dominant political parties in the two countries. In Venezuela, the vast majority of the AD members were from lower-middle-class and working-class backgrounds throughout the democratic era. The relationship between the different factions of party members and conglomerate groups became more insecure over time. Electoral rivalry, factionalism within AD, and the party's more populist political strategy contributed to an increasingly contradictory proliferation of rights and rents in Venezuela. Overall, the business community in Venezuela never successfully penetrated the party system (Gil Yepes 1981; Coppedge 2000). The UMNO, on the other hand, was not only more centralized, but became increasingly comprised of capitalist-led political factions within the party (Gomez and Jomo 1997, 26–27; Jomo and Gomez 2000, 296). In 1981, although teachers made up 41 percent of the delegates to UMNO's annual General Assembly, this share dropped to 32 percent in 1984, and declined further to 19 percent in 1987 (Gomez and Jomo 1997, 26). Meanwhile, businessmen constituted 25 percent of the delegates in 1987; and by 1995, nearly 20 percent of the UMNO's 165 division chairmen were wealthy businessmen (26).

The growing presence of business-led factions makes the Malaysian political system resemble the dominance of business factions in Thai politics (Doner and Ramsey 1997). The significance of this difference is that rent deployment and management in capitalist-led, centralized clientelism is more likely to be related to production and accumulation rather than simply redistribution, which characterized the populist clientelist factions that dominated in Venezuela.[18] The generally higher rate of productivity growth in Malaysia is a clue

---

17. The Free Trade Zone Act of 1971 created attractive incentives to lure multinationals. The measures included exemptions from customs regulations, tax breaks, and "favorable" labor legislation. By the mid-1980s, the FTZ firms, particularly in electronics and textiles, dominated manufacturing exports, which overtook the resource-based industries processing raw materials as the main source of foreign exchange.

18. For evidence of the superior productivity growth performance of the Malaysia's manufacturing sector compared to Venezuela's, see tables 3.4a–c and table 9.13b.

that rent deployment was based more often on economic considerations rather than purely political criteria.

10.3 Conclusion

In sum, the difference in growth performance between Malaysia and Venezuela, at least in the period 1970–98, cannot be attributed simply to differences in the structure of rent deployment or in differences in rent-seeking costs. Both countries had relatively centralized rent-deployment systems that imparted to the central government, and particularly the executive, strong centralized discretionary authority. The main point of the comparison is that the dynamic efficiency consequences of the institution of centralized rent deployment *cannot* be known in theory, in contrast to what most rent-seeking theorists propose.

The actual functioning of the institutions of centralized rent deployment can only be known through an examination of the concrete historical political economy underlying the motivations of leaders and their power to impose, whether directly or through the market, performance criteria on the recipient of rents and thus on rent use. The shifting grounds of what constitutes legitimate rule and the contingent outcomes of threat, political strategy, and the balances that flow from such strategies are crucial in influencing the ability of the state to deploy and manage rents effectively. Such differences in the outcomes of rent-seeking make more of a contribution in explaining the long-term differences in economic performance in Venezuela and Malaysia than simply identifying rent-seeking costs that are generated by centralized forms of rent deployment and state intervention. Even if one were to argue (rather implausibly) that the Malaysian state created fewer rents than the Venezuelan state, and thus fewer rent-seeking costs, the political factors underlying such an outcome would still need to be explained.

Moreover, this brief comparison shows that the relative success of the Malaysian growth experience in the period 1970–98 was grounded in a broad compatibility of development strategy and the structure and dynamics of the political settlement. For the historical and political reasons discussed, the impressive rate of economic growth in Malaysia in this period can be attributed to the construction and maintenance of a consolidated polity with centralized political organizations capable of coordinating investments and capable of limiting the extent to which unproductive producers and managers could maintain subsidies and rents regardless of firm performance. Both of these features were central to implementing a big-push heavy industrialization strategy that was more effective than in the Venezuelan case.

This does not imply that everything the Malaysian state promoted was successful. While there were several infant industries that failed to grow up (most notably, the heavily subsidized national car, the Proton), the main issue is that in Malaysia industrial policies succeeded more often than they did in Venezuela. And Malaysian policies succeeded at a greater rate, in part, because the political settlements underlying state support in Malaysia favored the effective implementation of economically and politically challenging big-push industrialization strategies. The political fragmentation and growing polarization of politics in Venezuela, as we have seen, limited substantially the prospects of implementing such a big-push industrial strategy.

This brief chapter is an initial attempt to develop a comparative historical political economy of growth. While a recent collection of analytical growth narratives has included some comparative studies, the vast majority focus on one case study (e.g., Rodrik 2003), comparative work will help sharpen the lessons that historical political economy can provide to an understanding of variation and change in growth dynamics across countries and within countries over time. More detailed historical work would, of course, be needed to assess the validity of the basic comparison outlined here.

11

CONCLUSION: RETHINKING THE POLITICAL ECONOMY OF GROWTH

This study has provided an alternative view of the late development process in Venezuela over the period 1920–2005. It has made two contributions with respect to the interpretation of growth trends in Venezuela in this period. The first provides a critical intellectual history of the reigning economic and political economy explanations of economic slowdown in Venezuela in the period 1965–2005, namely economic versions of the "resource curse," such as Dutch Disease models, and political economy models of the rentier state. I argue that the inefficiency of centralized rent deployment in Venezuela in the period 1968–2005 owes less to natural resource abundance per se than to an incompatibility of economic and political strategies. The argument does not deny that sudden and large inflows of oil revenues in the 1970s had a negative effect on economic management in Venezuela; rather, it claims that a longer-run view of Venezuelan economic history suggests that oil abundance has been compatible with cycles of growth and stagnation.

Because of its abundant oil reserves, Venezuela was long considered an "exceptional" case in the context of twentieth-century Latin American development. This study revises that conclusion. First, while Venezuela did not manage the sudden inflows efficiently in the 1970s, almost no Latin American economy has been able to manage sudden capital inflows without undergoing substantial macroeconomic destabilization (Palma 1998).[1] Second, the slowdown in Venezuela's productivity and output growth in the period 1968–2003 is part of

---

1. Chile after 1982 is an important exception.

a wider slowdown in economic growth throughout Latin America in the same period. The Latin American experience, as well as the fact that a very small set of latecomers have actually sustained catch-up, suggests that patterns of failure in Venezuela may not be so "exceptional" after all.[2] Third, the current breakdown of the Venezuelan political system, and particularly the decline in the legitimacy of political parties, the rise of new forms of populism, and the increasing polarization of politics, is part of a larger pattern in the Andean region and beyond (Roberts 1996; Weyland 1999). Finally, the failure of the Venezuelan state to maintain convergence with more advanced countries, despite many favorable initial conditions, demonstrates that the combination of late big-push heavy industrialization projects with democratic politics is a very fragile and challenging process in countries both well endowed and less well endowed with mineral resource rents. Further case studies could be undertaken to test this finding.

The second major contribution concerns the idea of viewing growth performance as a contingent and interactive process of economic and political strategy. For the period 1968–2005 in Venezuela, the basic incompatibility I identify is that politics became increasingly more factionalized and accommodating precisely at a time when the development strategy and stage of import substitution required a more unified and exclusionary rent/subsidy deployment pattern. Chapters 6–9 identified the changing nature of development strategies, political settlements, and political organizations, and examined the economic effects of the incompatibility identified.

The historical political economy framework presented enables us to improve upon reigning economic and political economy explanations of increasingly poor economic performance in Venezuela after 1968. What distinguishes Venezuela from more successful cases of big-push industrialization in the period 1965–2005 is not principally the degree of state intervention, levels of public ownership, fiscal deficits, or the scale of corruption, but the nature of political

---

2. For instance, there is a substantial literature on Latin American politics that identifies strong clientelist and populist bases of support in many polities in the region since the 1940s (Kaufman 1977; Cardoso and Faletto 1979; Rueschemeyer et al. 1992, 155–225). At the same time, economic historians of Latin America have argued that all the large economies of the region introduced strategies to deepen import-substitution in intermediate and capital goods (characteristic of the more advances stage of ISI) in the 1960s and 1970s (Cárdenas et al. 2000; Fitzgerald 2000, 75). While there was variation in the timing of the transitions in development strategies across the region, the subsequent slowdown in manufacturing growth in most of Latin America in the period 1975–2000 may, in part, be due to growing incompatibilities of economic strategies and evolving political settlements. This is an area ripe for further research in the comparative historical political economy of development in Latin America.

strategies and settlements that underpinned the transition and consolidation of its democratic regime. In Chapter 10 (where the Venezuelan and Malaysian experiences with heavy industrial strategies were compared), I attempted to show how and why the nature of political organization and settlements, and the drivers of state legitimacy, matter for the effective implement of industrial policies in the context of mineral resource abundance.

My review of the political science literature found the nature of political strategies and settlements that underpinned the transition and consolidation of Venezuela's democratic regime ultimately led to the fragmentation and polarization of political organizations and contestations. And growing fragmentation became particularly pernicious to the prospects of economic growth because it occurred in the context of the more advanced stage of ISI and, especially, in the context of a big-push industrialization strategy.

Recent work on the types of economies that have experienced growth accelerations in the past fifty years suggests that the framework developed in this work may have a wider relevance. Hausmann, Pritchett, and Rodrik (2004) examine growth accelerations in the period 1957–92 and found that they occur overwhelmingly in countries at low-income or lower-middle-income levels of per capita income. There were few cases of growth acceleration among upper-middle-income countries. This is consistent with my argument that the earlier stages of industrialization present fewer economic and political challenges than strategies to upgrade into larger-scale and higher-technology sectors. It is for this reason that so few middle-income countries (many of which require more sophisticated production strategies to upgrade their manufacturing sectors) have been converging to the levels of income per capita found in more advanced industrialized economies. In recent years, the fastest-growing economies, such as China, India, and Vietnam, are all still well below the Latin American average income per capita level.

The major problem for sustaining manufacturing growth in Latin American economies (as well as other middle-income countries) is that they are caught between low-cost producers, such as in China and India, and high-tech producers within the OECD. The struggle to capture world markets in intermediate-technology sectors requires effective production strategies. This in turn requires degrees of coordination and cooperation between state, business, and labor groups that has proved difficult in a continent where populism, clientelism, political fragmentation, and more recently, anti–political party politics reign. Venezuela is hardly an exceptional case in this respect. Without the emergence of more effective centralized political, corporate, and labor institutions, the challenges of industrial upgrading and export diversification will remain formidable.

While the region has experienced more rapid growth since 2004, much of this growth is based on high world prices for agricultural commodities, minerals, and fuels. Historical evidence strongly suggests that this is not a sustainable growth path. Sustained growth depends on the ability of economies to diversify the production and export structure toward more sophisticated products with strong technological and production proximities (Hidalgo et al. 2007).

The idea of viewing economic growth as a function of the compatibility between development strategies and historically specific political settlements also has more general implications for understanding the political economy of institutional formation and change in the process of late development. The first concerns the idea that incompatibility is not the result of either cognitive failure or a knowledge gap. The simple identification of suboptimal policies and institutions (or what can be called Type I failure) is an inadequate basis for understanding the persistence of these problems in some countries as opposed to others. Type I failure, in the first instance, could be due to knowledge gaps or policy mistakes. For example, the state may provide protection to an entrepreneur (in an infant industry) who turns out to be incompetent. The initial decision to subsidize such an entrepreneur may be due to incorrect models on the part of decision-makers. However, Type I failure of this sort becomes less interesting as an explanation for suboptimal policies and institutions *over time*. As time passes, decision-makers become aware of these shortcomings either through their own observation or from the plethora of international experts who continually identify the problem over the years.

A more interesting and relevant issue is Type II failure—that is the failure of the government to correct or change suboptimal policies and institutions even when there is widespread knowledge of the problem. Type II failure is closely related to Type I failure, but involves analyzing the reasons behind the persistence of policy and institutional failures over time.[3] In explaining Type II failure it is necessary to incorporate historical political economy (and not merely technical) analysis, since understanding institutional change requires analysis of the interaction of economic and political processes. The ability of the state to withdraw subsidies from an inefficient producer may result in enterprise bankruptcy, which not only removes income from the entrepreneur but also will result in job losses. The failure of the state to withdraw subsidies

---

3. For a discussion of Type I and Type II failure, see Khan 1995.

often is due to political opposition rather than a lack of knowledge of the inefficient policy.

The fragmentary nature of economic policy and institutional formation and the growing ineffectiveness of industrial policy were widely known by Venezuelan decision-makers to be impediments to economic growth. There was knowledge of the problem. However, coordination of a fragmented power structure is not simply a technical issue of design, but involves power struggles and bargaining (Elster 1989, 183). Incompatibility refers not to some abstract mismatch between an economic structure and political structure, but to an inability to change institutions in a way that will not destabilize the basis of legitimate rule in a given historical conjuncture.

Second, the framework developed is a critical extension of the pioneering attempt to analyze the interaction of political and economic strategies in late developers as expounded by Guillermo O'Donnell (1973) in his work on the bureaucratic-authoritarian state in Latin America. O'Donnell distinguishes between the early stages of ISI and later industrial deepening, that is, developing intermediate and capital goods industry, which he considered more difficult. Based on a political economy framework, O'Donnell argues that the process of "deepening" the productive structure, that is, developing intermediate and capital goods industry, tends to be associated with the rise of authoritarian regimes in late-developing, or (what he refers to as) dependent economies.

The logic of O'Donnell's model is as follows. In many Latin American countries, the early phase of ISI expanded the working and urban middle classes and led to the emergence of powerful populist political parties and coalitions on the basis of redistributive strategies. In the early stage, rapid economic growth underwrote the costs of such social welfare strategies. However, once the easy stage was finished, the development of a technologically more dynamic growth path required capital goods production, which, in turn, required higher levels of investment. For O'Donnell, the growing import demand of capital goods created balance of payments crises, which increased the unpredictability of macroeconomic conditions through an increase in inflation. The stabilization policies that followed reduced the incomes of low-income groups, divided populist coalitions, and generated political crises.

The coalition that emerged from these crises, according to O'Donnell, was to include bureaucrats and the military. These groups, in turn, supported an authoritarian solution to the challenge of industrial deepening. Authoritarianism was seen to be compatible with the needs of repressing consumption and unsustainable social welfare spending that militant labor unions and populist politicians mobilized. Based on this analysis, O'Donnell suggests that democratic politics is not always favorable for sustained late development.[4]

The Venezuelan experience, along with that of other long-standing late-developing democracies such as Colombia, Costa Rica, and India, suggests that purposeful policies of industrial deepening do not necessarily generate, as O'Donnell suggested, shifts toward authoritarianism. The lack of any correlation between industrial deepening and authoritarianism in these countries suggests that development strategies do not determine the course of politics.[5] Political settlements (which are historically specific bargains over institutions) can have an independent effect on economic outcomes. What is telling, however, is that, in the period 1950–2000, there are practically no successful developmental states that have not had significant periods of authoritarianism. The value of O'Donnell's analysis was to draw attention to the conflictual nature of more advanced and technologically complex industrial challenges in late developers.

Third, a comparative, historical, and interactive approach to economic growth underlines the importance of incorporating politics into an understanding of the effectiveness of a given institution. While there are many theoretical justifications for state interventions, much of the developmental state literature and new institutional literature does not adequately explain why similarly designed interventions *fail* in many contexts. The brief comparison with the Malaysian case was illustrative of the importance of understanding how different political settlements, and the nature of threat facing a polity, can affect the efficiency of centralized rent deployment.

Work on the developmental state has focused on state bureaucratic capacity to the neglect of the ability of the state to penetrate and control the demands and relationships of the working class and other interest groups. The developmental state literature emphasizes *how* a state intervenes effectively, but less attention is paid to where the power to implement policies, and where the policy goals and motivations come from in the first place (Kohli 1999, 2004). The power to enforce institutions and coerce simply cannot be known from the distribution of property rights. One needs to include parametric information from history and politics to know the enforcement costs of a given institution (Mann 1993, 59).

The emphasis that the developmental state theorists place on the technical role of institutions in coordinating "incentives" is incomplete. The mechanisms

---

4. A point also made earlier by Barrington Moore (1966).
5. On the autonomy of political processes, see Gramsci 1971 and Przeworski 1985.

of control, selectivity, and discipline within the state are themselves a set of institutions, which are, in turn, a series of processes whereby groups and individuals bargain for material and political advantage. As Kenneth Arrow (1974) has pointed out, the use of authority is itself an exercise of authority. Moreover, the effective use of authority is sustainable only when centralized authority embodies a minimum level of legitimacy, what Arrow calls "convergent expectations" (72), or what Alexander Field (1981, 186) calls "reciprocal expectations." That the legitimacy of a central authority depends on convergent expectations implies its fragile nature (186); hence the formidable task industrial policy (particularly in the more advanced stages) presents.

To the extent that the participants perceive that particular policies and institutions are fair, the costs of enforcing rules and property rights are enormously reduced by the simple fact that the individuals will not disobey the rules or violate property rights even when a private cost/benefit calculus would make such a decision worthwhile.[6] The sources of state failure in big-push industrial strategy in Venezuela were that they became incompatible with the political demands and struggles within populist clientelism, where widespread distribution of patronage was expected by relevant political leaders and their support base. The integrative framework proposed here suggests that developmental state theorists neglect the greater conflict and consequent strains on state legitimacy that large-scale, big-push strategies generate in the transition from the "easier" or more artisanal stages of ISI.

The incorporation of politics in our framework also builds on the work of Mancur Olson (1982), who argued that poor economic performance (in some stable advanced industrialized democracies such as the United Kingdom and the United States in the period 1950–82) was the result of the accumulation of distributional, rent-seeking coalitions who had captured the state for their particularistic advantage. The big idea of his influential work, *The Rise and Decline of Nations,* is that narrow producer interest groups will form political lobbies to influence the state and ensure that regulatory policies are made that benefit them. A classic example would be textile producers lobbying the state for protection against foreign competition, which enables domestic producers to charge prices higher than in the case of import competition. As a result, state capture, for Olson, tends to produce anticompetitive regulations, which slows down the process of "creative destruction" and thus limits economic growth

---

6. Several authors have stressed the importance of restraint in contesting authority as a central characteristic of legitimacy. Simon (1991) refers to this restraint as "docility," Putzel (1995) as "passive acceptance," and Levi (1988) as "quasi-voluntary compliance."

and technological development. Furthermore, Olson argues that the benefits of such protectionist policies are concentrated among a small number of lobby members, while the costs of such policies are diffused throughout the whole population, where the collective action problems of organizing resistance are formidable.

According to his argument, the more successful economic performers among advanced industrial economies after the Second World War II were Japan and West Germany, and this was because the Second World War had eliminated the narrow special interest groups that had impeded growth in both countries. For Olson, special interest groups can be growth-enhancing when such groups are "encompassing." The relatively successful performance of the Swedish economy in the twentieth century was due to the encompassing nature of its interest groups, with its labor and management groups negotiating at the national level to achieve growth-enhancing policies.

The argument developed in this study differs from Olson's framework. The idea that particularistic interest groups capture the state is an abstract one; what matters is what groups are capturing the state and for what purposes. If successful entrepreneurs capture the state, the growth prospects of the state are different than if clientelist groups with no business skills capture the state. Also, Olson's analysis does not explain the dynamics through which clientelist capture generates further downward spirals of capture by clientelist groups. Many advanced capitalist states are subject to capture (as Olson has analyzed), but that does not mean that the groups capturing the state are as detrimental to the viability of the economic system as in fragmented late-developing polities. The increase in growth rates of the U.S. and British economies (the two economies Olson pointed to as being most vulnerable to narrow interest group capture) in the post-1980 period supports this point. Olson also does not provide an analysis of why states in late developers generally are more central to capital accumulation than in more advanced capitalist countries where already large corporations are capable financially and technically of undertaking their own industrial policy. Implicit in Olson's argument is the assumption that if state intervention were removed, the economy would grow faster, an idea that is not supported by the evidence.

Fourth, viewing growth as an interactive and historically specific process means that case study approaches and inductive analysis are essential methods for understanding the variation and change of state capacity and economic performance. There are various reasons for this. First, the historically situated nature of rationality, or bonded rationality, implies that the dynamic between agency and structure and what constitutes legitimate rule is contingent and cannot be known a priori. Second, many social patterns form in historically specific ways and tend to persist over time. Closely related, macrosocial phenomena are shaped by a constellation of factors—not just one in isolation. This means that patterns and sequencing of causal conditions matter.[7] Indeed, causation is a matter of sequence, and since causal explanation needs to be tested against evidence of sequences, the case study method is appropriate. Cross-sectional variable research does not allow us to establish either sequence or agency, both of which are essential in developing convincing analytical growth narratives.

While the case study approach is imperative in establishing historically situated agency, or bonded rationality, it needs to be complemented by references to policies and institutions elsewhere. In assessing the development of the non-oil economy and manufacturing industry in Venezuela, a relevant question surely is "compared to what?" The historical record *within* a country provides the canvas of analysis. However, the patterns of the canvas become meaningful in a comparative context. As Barrington Moore (1966) suggests, "In the effort to understand the history of a specific country a comparative perspective can lead to asking very useful and sometimes new questions. There are further advantages. Comparisons can serve as a rough negative check on accepted historical explanations" (xiii). Reference to other cases allows for a check on whether or not the types of policies, institutions, and politics identified are unique features in a particular case. It is only through references to and analyses of other cases that historical reductionism can be avoided. The comparative data provided throughout this study, including the brief comparison with Malaysia, were meant not only to assess the validity of competing theories but also to provide checks on the analytical narrative of the Venezuelan growth process.

Fifth, this work can also be embedded in a larger framework of contributions to the study of economic growth that have pointed toward the need to understand historical contingencies, interactions, and nonlinearities in the growth process (Myrdal 1957; Kenny and Williams 2000; F. Rodríguez 2005, 2007). In particular, empirical growth economics has started to move away from the emphasis on cross-country linear growth regressions and toward a

---

7. As Elster (1989) points out: "The variety of interacting motivations is simply too large for any equilibrium theorems to be provable. . . . If social scientists forgot their obsession with grand theory, and looked instead for small and medium-sized *mechanisms* that apply across a wide spectrum of social institutions, some mathematical economists and Parsonian sociologists (to name but a few) might go out of business but the world would be a better understood place" (205; emphasis added).

deeper understanding of national specificities in the growth process (Rodrik 2003). All of these contributions point to the need to study particular cases in detail in order to understand the relevant interactions between the political and economic systems.

A particularly relevant recent contribution is *Getting the Diagnostics Right: A New Approach to Economic Reform* by Ricardo Hausmann, Dani Rodrik, and Andrés Velasco (2006), who point out that economic theory (via the Theorem of Second Best) actually leads us to expect that the reduction of a particular distortion may have very different effects on welfare (and growth) depending on the initial levels of other distortions, and illustrate the potentially complex interactions that can arise even in relatively simple theoretical models, such as the Theorem of Second Best.[8] The authors point out that within Latin America, specific economies may not be growing because of a variety of distortions and constraints that may be quite different across countries.

In a comprehensive appraisal of the results of a decade of reforms in the 1990s, the World Bank, a leading actor in promoting economic liberalization, has more recently concluded that the role of complex interactions may play a central role in understanding economic performance:

> To sustain growth requires key functions to be fulfilled, but there is no unique combination of policies and institutions for fulfilling them . . . different policies can yield the same result, and the same policy can yield different results, depending on country institutional contexts and underlying growth strategies. . . . Countries with remarkably different policy and institutional frameworks—Bangladesh, Botswana, Chile, China, Egypt, India, Lao PDR, Mauritius, Sri Lanka, Tunisia, and Vietnam—have all sustained growth in GDP per capita incomes above the U.S. long-term growth rate of close to 2 percent a year. (World Bank 2005, 12)

The recent (re)discovery in the growth literature of the interaction of economic and political processes has surely been the by-product of the failure of

---

8. According to the "theory of second best" (Lipsey and Lancaster 1956), in the context of numerous distortions, removing only some of the distortions can have a welfare- and growth-reducing effect. This could occur if efficient allocation depends on the interactions of two or more institutions or policies. The reduction in one distortion can have a welfare-reducing effect. For example, removing an external "distortion" such as controls on capital inflows, without addressing distortions and poor regulation in the financial sector, can be welfare reducing by contributing to a banking crisis, as occurred in Chile in 1982 and in Thailand and Indonesia in 1997 (Zettelmeyer 2006, 28 n. 19).

universal theories to explain the variation and change in development paths. Indeed, "there is crucial and important mismatch between the actual economic world and the 'picture' or 'model' of the economic world which underpins the current search for causally significant variables" (Kenny and Williams 2000, 2). Current thinking about economic growth often fails to grasp the complex causal nature of the social world, assuming that the components and processes of the economy are the same across countries (2). Only if all economies are fundamentally the same, in their components and processes, does the search for fewer and fewer universal explanatory principles or "laws" make sense (3). The failure of cross-country regressions, based on models that assume universal economic laws across countries, suggests that growth is a more complex process than can be captured by such models (5). Charles Kenny and David Williams argue: "The universal failure to produce robust, causally secure relations predicted by models might suggest a broader problem than statistical methodological weaknesses. The evidence appears to suggest that growth country experiences have been extremely heterogeneous, and heterogeneous in a way that is difficult to explain by using any one model of economic growth" (12). They also remind us that Gunnar Myrdal (1957) argued long ago that economic growth may not only have a great number of causes, but also that these do not work in any "linear" manner.[9] Myrdal suggested that economic growth should be examined using the concept of "cumulative causation" where a change in one factor would affect a number of other factors, and these changes would in turn feedback on the first factor (16). The essence of economic growth is that "it concerns a complex of interlocking, circular, and cumulative changes" (14). For Myrdal, this had two implications. First, it was "useless to look for one predominant factor, a 'basic factor'—as everything is cause to everything else in an interlocking circular manner" (19). Second, viewing economic growth in these terms meant abandoning the search for neat economic models: "The relevant variables and the relations between them are too many to permit that sort of heroic simplification" (101).

While Myrdal's insights and the recent growth literature (which has revived Myrdal's ideas and which draws attention to the importance of understanding the interactions of economic and political processes) are useful, they often do not specify how or why such interactions might operate in concrete historical settings. In particular, these studies do not attempt to systematically analyze the extent to which development failures and successes are the result of the

---

9. The quotes from Myrdal in this paragraph are taken from Kenny and Williams 2000, 13–14.

compatibility (or lack of it) between a given development model and a country's political institutions. In this perspective, my interpretation of the Venezuelan growth process attempts to make a small contribution to historical political economy, a field that has long been marginalized and even endangered by the dominance of neoclassical growth and development theory.

Our understanding of the compatibilities between economic and political processes is still rudimentary. However, it is hoped that this work will encourage the revival of thinking about economic growth in a way that is more historically and politically informed. It is also hoped that it will open up a set of hypotheses that could have explanatory power beyond the Venezuelan case. Further research is needed to uncover the extent to which a similar type of incompatibility could be at work in the apparent inability of many middle-income, late developers to implement complex and demanding industrial strategies. Moreover, there is scope to examine the extent to which the (in)compatibilities between development strategies and political settlements affected the course of economic growth in numerous African nations after decolonization.

Sixth, it may be suggested that the compatibility argument developed here cannot explain the fact that there has been a substantial growth slowdown not only in middle-income economies such as those of South America and the Middle East but also in many low-income economies such as in sub-Saharan Africa and in Central America. Thus, it could be argued that there are international conditions such as the imposition of structural adjustment programs, commodity shocks, open capital markets, and debt and capital flight crises that explain the growth slowdown after 1980. And indeed, such factors were central to both low-income and middle-income economies.

However, there are several growth trends across countries that support a greater focus on internal factors, which the compatibility argument developed in this book attempts. First, there has been a differential capacity among middle-income economies to deal with so-called adverse international conditions. In general, East Asian economies have fared much better than Latin American economies in the period 1980–2005. International factors cannot explain such variations in economic performance across countries.[10] Second, the post-1980 period *has* seen significant growth episodes and acceleration in Asia (e.g., China, India, and Vietnam) and in sub-Saharan Africa (e.g., Uganda,

10. Moreover, even if financial deregulation has increased the vulnerability of less developed countries to financial crises, it is one thing for an economy to experience a short-run macroeconomic crisis because of an external shock; it is quite another thing for that same economy to stay mired in slow growth after the shock.

Mozambique, Botswana, and Tanzania) to mention a few. Third, growth collapses in middle-income countries are likely to have been caused by very different internal factors than in low-income countries. Weak state formation and civil wars were surely important factors behind growth collapses in sub-Saharan Africa and Central America. In this sense, state consolidation is the primary problem. No development strategy is compatible with growth if centralized public authority and the integrity of the state is under armed challenge. In the former, state institutions are much more consolidated and there are fewer challenges (including armed challenges) to state authority than in low-income countries. Even where there have been long-term armed challenges to state authority in middle-income countries (such as Colombia), they have been territorially limited and have not greatly affected the macroeconomic management capacity of the state.

The emphasis on international factors as a cause of growth slowdown misses the important point that the economic policies and political changes required to revive growth are very different in countries at different stages of development and income per capita levels. Identifying what the binding constraints to growth are in specific contexts is what lies behind the "growth diagnostic" approach (Hausmann, Rodrik, and Velasco 2006). Both the "growth diagnostic" approach and the compatibility argument developed in Chapters 6–9 in this book examine the *internal* obstacles to growth. A focus on international factors ignores such issues. Moreover, there is little less developed countries can do about the evolving nature of global challenges. There is, however, a great deal some countries have done to meet these challenges. Surely, diagnosing how and why a failing economy can meet global challenges "from within" is much more likely to increase growth and reduce poverty in such contexts.

Complaining about how the world's rich economies exploit the less developed world is not likely to amount to much more than rhetoric and rock concerts pleading for more aid. This is not say that the international financial architecture and the all too often used "one-size-fits-all" approach to international interventions cannot be improved upon. The development community can surely do better through more focused and more reticent revolutionary work that seeks to unravel the conundrums of growth for those nations faced with the political and economic challenges of late development.

Seventh, the political economy perspective employed also has importance for the "good governance" agenda. It is common in the mainstream view to focus on identifying the "right" institutions, and the "right" incentives (World Bank 1997a). What is missing in this view is that the specification and enforcement of property rights that are central for creating incentives simultaneously specify a distribution of assets and privileges. Such a distribution requires state enforcement and a wide acceptance that such a distribution is legitimate. The abstract introduction of issues of "good governance" (the set of institutions and rights in a polity) cannot explain variations and changes of growth without incorporating historically specific politics (especially the structure of political organizations and the shifting grounds of legitimate rule) and production (stage of development and development strategy). Neither democracy, transparency, centralization of authority, nor legitimacy per se assures the construction of a developmental state. One has to ask: authority or legitimacy to do what? In this sense, the relationship between governance and economic performance, in our framework, can only be understood through the historically grounded study of the political economy of institutions and property rights as an interactive process of establishing incentives and patterns of distribution in the context of a specific development strategy.

This study also has important implications for how state capacity is viewed within the "good governance" agenda. There are two central problems with the way capacity is measured and diagnosed. The first is the aggregate nature of the measures. While state capacity-building is central to governance reform initiatives throughout the developing world, most measures of state capacity are based on subjective surveys. Using these surveys, recent work has tried to "measure" seven key areas of "good governance": security of property rights, voice and accountability, political instability and violence, government effectiveness, regulatory quality, rule of law, and control of corruption (see Kaufmann et al. 2005) and identify their contribution to economic performance.

The usefulness of these types of measurements and explanation is limited in several ways. First, they provide only indirect measures of capacity. Second, the results are aggregate measures, that is, they do not provide any information on the extent to which capacity varies across different state functions or sectors within a polity. Third, as mentioned, there is little evidence that "good governance" is either a necessary or sufficient condition for rapid economic development in late developers.

The historical evidence suggests that state capacity varies substantially across functions and sectors within polities. There are numerous examples of this. For instance, Venezuela (as this study has examined) has long maintained a stable democratic system but has been unable to promote export diversification or maintain rapid growth in the post-1980 period. South African tax collection capacity (by far the best among middle-income countries) is much greater than its ability to undertake industrial policy or tackle HIV/AIDS. Botswana's democratic institutions are among the most robust in the developing world, yet it has also been very poor at controlling HIV/AIDS. Brazil has among the highest levels of tax take but is not (politically) capable of collecting personal income and property tax. Brazil's industrial policy is also very uneven: success stories in aerospace and autos stand out, while many other sectors have been less successful. The Colombian state is known for having among the best macroeconomic management but has among the lowest tax takes in Latin America, and is unable to contain decades of guerrilla and paramilitary political violence. Tanzania and Zambia have had relatively poor records on economic performance but have been able to prevent large-scale political violence, unlike most of their neighboring countries. This variation in capacity is not picked up by aggregate measures and thus our understanding of why capacity varies so much within polities is limited in such a framework. Detailed historical analyses of the political coalitions and settlements underpinning specific state capacities are essential to increase understanding of variable state capacity within a polity and provide a fruitful ground for research both within Venezuela and beyond.

The second problem concerns how policymakers should incorporate comparative historical political economy analyses that attempt to explain *why* state capacity differs across countries. A typical argument (and one that is the essence of my comparison of Venezuela and Malaysia in Chapter 9) runs as follows: country X has a more developmental state than country Y because X's power structure and political institutions provide incentives for asset holders to accumulate capital in ways that are growth-enhancing, while Y's power structure and political institutions provide incentives for asset holders to accumulate capital in ways that do not add social value and/or are growth-restricting. Often the identification of antidevelopmental rent-seeking, clientelist, or patrimonial coalitions is central to such explanations (e.g., Bardhan 1984; Sandbrook 1985; Khan and Jomo 2000; Kohli 2004; Khan 2006).

There are two limitations to these types of explanations. First, they tend to map the political settlements with developmental outcomes (e.g., growth rates) but do not provide much information on how capacity may vary within a country. Second, political settlements, including property rights assignment, are historically specific and cannot be easily changed except in extraordinary circumstances (war, revolution, and so on). This reflects the well-know problem of the transferability of institutions from one polity to another. The fact that the governing political coalition is more developmental in one country will not help policymakers in less successful states because changing property rights and institutions requires the construction of coalitions that are lacking in the less successful country. Moreover, international donors do not have either the mandate or the in-country leverage to change political settlements.

What can be studied, however, is why some state and private sector capacities work better than others in the same polity. Identifying relatively successful sectors, firms, and state functions, especially in generally weak state environments, provides a potentially more fruitful way to identifying drivers of change. This avoids the unrealistic proposition that policies would work better if (the biggest if) only politics were more compatible with the interventions attempted. For example, it makes more sense to study why and how the Brazilian state *could* nurture the development of Embraer (one of the most successful passenger jet companies in the world) in one of the most challenging sectors to develop, even in economies with high levels of technological capacity.

There are three advantages of this approach. First, the macro institutional context is constant, which means that the relative success of Embraer in Brazil is more convincingly explained. Cross-country analyses on why aggregate industrial performance in Brazil is less successful than in South Korea, while useful in identifying how differences in politics and governance matter, do not explain this case. Second, because politics is the historically specific struggle over what determines legitimate rule, it is unlikely that the identification of more developmentally compatible politics in South Korea will make any difference to political change in Brazil. Third, the fact that Embraer managed to succeed in the macro institutional context of Brazil shows what *is* possible in Brazil, and could inspire more confidence in the pathways forward than pointing to success stories in very different and distant polities. This is not to say that macro institutional changes are not needed in Brazil for more companies like Embraer to emerge. Because such path changes are rare (and at best, long-gestating) a more relevant space for policy is to investigate why and how Embraer works.

Finally, this study has highlighted the theoretical and empirical shortcomings of viewing factors endowments (such as mineral abundance) as determining the path of institutions, politics, and policy. "Resource curse" arguments have viewed mineral abundance as detrimental to the prospects of economic development. If, however, Venezuela is to embark on a renewed growth path, it is worth mentioning that mineral resources can play a progressive role in economic development (Wright and Czelusta 2003, 2007). Much of the "resource cure" literature has equated mineral resources with terms such as "windfalls" and "booms." Contrary to the view of mineral production as mere depletion of a fixed natural "endowment," Wright and Czelusta (2003) show that so-called nonrenewable resources have been progressively *extended* through exploration, technological progress, and advances in appropriate (often country-specific) knowledge. Indeed, minerals constitute a high-tech knowledge industry in many countries; and investment in such knowledge should be seen as a legitimate component of a forward-looking economic development program (ibid.).

One of the main implications of this study is that the poor economic performance of the Venezuelan economy, particularly since 1980, has not been due to an overemphasis on minerals and natural-resource-based industrialization, but has been the result of the failure to develop such industries more successfully. The "resource curse" models do not provide a viable alternative for mineral-abundant countries. Surely, *not* exploiting what is an obvious comparative and competitive advantage in both minerals and vast hydroelectric capacity would be antidevelopmental in the Venezuelan context. What this study has also highlighted is that the political obstacles to developing such a strategy remain formidable in the Venezuelan context. This is very different from arguing that minerals are the source of an inevitable developmental malaise.

The period 2005–7 has seen rapid economic growth of nearly 10 percent per year. However, this growth has come in the wake of the greatest growth collapse in Venezuelan history in the period 1998–2003, and has coincided with increases in inflation and rising food shortages. Recent growth has almost totally been the by-product of buoyant crude oil revenues. This simply reinforces Venezuela's dependence on oil. Oil windfalls have nevertheless provided the possibility of substantial government spending targeted toward lower-income groups. Such a pro-poor use of oil windfalls is a noble aim which has the potential of redressing social and economic inequalities. However, it does not address the problem of reducing the dependency of the welfare of the low-income earners in Venezuela on the vagaries of a highly volatile commodity.

A much more sustainable developmental approach would be to also analyze how the productive capacity of the Venezuelan economy can be developed in such a way as to make mineral discoveries both more frequent and more of an input into domestic manufacturing production. The sustainable empowerment of low-income groups cannot be supported forever by the dependence on oil exports; rather, sustainable empowerment and a more pluralist society (and therefore a more meaningful democracy) depend on the rising demand for employment in a more diversified set of economic sectors. The developmental prospects of the Venezuelan economy depend, in large part, on overcoming the formidable political and economic challenges of achieving a more robust and sophisticated natural-resource-based industrialization.

REFERENCES

Abramowitz, Moses. 1986. Catching up, forging ahead, and falling behind. *Journal of Economic History* 46, no. 2: 385–406.

Acedo de Machado, Clemy, Elena Plaza, and Emilio Pacheco. 1981. *Estado y grupos económicos en Venezuela*. Caracas: Editorial Ateneo.

Acemoglu, Daron, Simon Johnson, and James Robinson. 2001. The colonial origins of comparative development: An empirical investigation. *American Economic Review* 91, no. 5: 1369–1401.

Adelman, Jeremy. 2000. Institutions, property, and economic development in Latin America. In *The other mirror: Grand theory through the lens of Latin America*, ed. Miguel Angel Centeno and Fernando López Alves. Princeton: Princeton University Press.

Adriani, Alberto. [1931] 1990. La crisis, los cambios y nosotros. In *Ensayos escojidos*, ed. Hector Vallecillos and Omar Bello Rodríguez, vol. 1. Caracas: Banco Central de Venezuela.

Aguero, Felipe. 1995. Debilitating democracy: Political elites and military rebels. In *Lessons of the Venezuelan experience*, ed. Louis W. Goodman, Johanna Mendelson Forman, Moisés Náim, Joseph S. Tulchin, and Gary Bland. Baltimore: Johns Hopkins University Press.

Aitken, Brian, and Anne Harrison. 1999. Do domestic firms benefit from direct foreign investment? Evidence from Venezuela. *American Economic Review* 89, no. 3: 605–18.

Alchain, Armen, and Harold Demsetz. 1972. Production, information costs, and economic organization. *American Economic Review* 62, no. 5: 777–95.

Alvarez, Angel. 2003. State reform before and after Chávez's election. In *Venezuelan politics in the Chávez era: Class, polarization, and conflict*, ed. Stephen Ellner and Daniel Hellinger. Boulder, Colo.: Lynne Rienner Publishers.

Amin, Samir. 1976. *Unequal development*. Brighton: Harvester Press.

Amsden, Alice. 1989. *Asia's next giant*. Oxford: Oxford University Press.

———. 1997. Understanding the government's role in late industrialization. *World Development* 25, no. 3: 1555–61.

———. 2001. *The rise of "the rest."* Oxford: Oxford University Press.

———. 2005. Promoting industry under WTO law. In *Putting development first: The importance of policy space in the WTO and international financial institutions*, ed. Kevin Gallagher. London: Zed Books.

Amsden, Alice, and Takashi Hikino. 1994. Staying behind, stumbling back, sneaking up, and soaring ahead: Late industrialization in historical perspective. In *Convergence of productivity*, ed. William Baumol, Richard Nelson, and Edward Wolff. Oxford: Oxford University Press.

Anderson, Benedict. 1983. *Imagined communities*. London: Verso.

Aoki, Masahiko, Kevin Murdock, and Masahiro Okuno-Fujiwara. 1997. Beyond the East Asian miracle: Introducing the market-enhancing view. In *The role of government* in *East Asian economic development: Comparative institutional analysis*, ed. Masahiko Aoki, Hyung-Ki Kim, and Masahiro Okuno-Fujiwara. Oxford: Clarendon Press.

Araujo, Orlando. 1969. *Situación industrial en Venezuela*. Caracas: Universidad Central de Venezuela.

Archer, Ronald, and Matthew S. Shugart. 1997. The unrealized potential of presidential dominance in Colombia. In *Presidentialism and democracy in Latin America*, ed. Scott Mainwaring and Matthew S. Shugart. Cambridge: Cambridge University Press.

Arenas, Nelly. 1990. *La denuncia del tratado de reciprocidad comercial entre Venezuela y los Estados Unidos*. Caracas: Centro Venezolano Americano.

Ariff, Mohammed. 1991. *The Malaysian economy*. Singapore: Oxford University Press.

Arrow, Kenneth. 1962. The economic implications of learning-by-doing. *Review of Economic Studies* 29, no. 3: 155–73.

———. 1974. *The limits of organization*. New York: W. W. Norton.

Ascher, William. 1999. *Why governments waste natural resources: Policy failures in developing countries*. Baltimore: Johns Hopkins University Press.

Astorga, Pablo. 2000. Industrialization in Venezuela, 1936–1983: The problem of abundance. In *An economic history of twentieth-century Latin America*, vol. 3, *Industrialization and the state in Latin America: The postwar years*, ed. Enrique Cárdenas, José Antonio Ocampo, and Rosemary Thorp. London: Palgrave; Oxford: St. Anthony's College.

Auty, Richard. 1990. *Resource-based industrialization*. Oxford: Clarendon Press.

———. 2001. Introduction and overview. In *Resource abundance and economic development*, ed. Richard Auty. Oxford: Oxford University Press.

Auty, Richard, and Alan Gelb. 2000. Political economy of resource-abundant states. Paper prepared for the Annual World Bank Conference on Development Economics, Paris, June.

———. 2001. Political economy of resource-abundant states. In *Resource abundance and economic development*, ed. Richard Auty. Oxford: Oxford University Press.

Balassa, Bela. 1980. The process of industrial development and alternative development strategies. Princeton University International Finance Section, *Essays in International Finance*, no. 141.

Banco Central de Venezuela [Central Bank of Venezuela]. 1992. *Series estadísticas de los ultimos cincuenta años*. Caracas: Banco Central de Venezuela.

———. 1993. *Estadísticas socio-laborales de Venezuela: Series históricas, 1936–1990*. Vol. 1. Caracas: Banco Central de Venezuela.

———. 2007. *III encuesta nacional de presupuestos familiaries, 2005*. Caracas: Banco Central de Venezuela.

———. Various years. *Informes anuales*. Caracas: Banco Central de Venezuela.

———. Various years. *Serie estadísticas*. Caracas: Banco Central de Venezuela.

Banuri, Tariq, and Edward Amadeo. 1991. Worlds within the Third World: Labour market institutions in Asia and Latin America. In *Economic liberalization: No panacea*, ed. Tariq Banuri. Oxford: Clarendon Press.

Baptista, Asdrúbal. 1988. Más alla del optimismo y del pesimismo: Las transformaciones fundamentales del país. In *El caso Venezuela: Una illusión del armonía*, ed. Moisés Náim and Ramon Piñango. Caracas: Ediciones IESA.

———. 1995. *Teoría económica del capitalism rentístico*. Caracas: Ediciones IESA.

———. 1997. *Bases cuantitativas de la economía venezolana, 1830–1995*. Caracas: Fundación Polar.
———. 1999. Venezuela: Constitution and economic order. Paper presented at the conference "Venezuela at the Crossroads," Centre of Latin American Studies, University of Cambridge, December 3.
———. 2003. Las crisis económicas del siglo XX venezolano. In *En esta Venezuela: Realidades y nuevos caminos*, ed. Patricia Márquez and Ramon Piñango. Caracas: Ediciones IESA.
Baptista, Asdrúbal, and Bernard Mommer. 1987. *El petróleo en el pensamiento económico venezolano: Un ensayo*. Caracas: Ediciones IESA.
Baran, Paul, and Paul Sweezy. 1966. *Monopoly capital*. New York: Monthly Review Press.
Bardhan, Pranab. 1984. *The political economy of development in India*. Oxford: Basil Blackwell.
Barzel, Yoram. 1989. *Economic analysis of property rights*. Cambridge: Cambridge University Press.
Baumol, William J., Sue Anne Batey Blackman, and Edward N. Wolf. 1989. *Productivity and American leadership*. Cambridge: MIT Press.
Bentham, Jeremy. [1789] 1982. *An introduction to the principles of morals and legislation*, ed. J. H. Burns and H. L. A. Hart. London: Methuen.
Bermúdez, Adriana, and Omar Bello. 2006. Structural changes in the Venezuelan labor markets as a consequence of labor regulations. Paper presented at the conference "Venezuelan Economic Growth, 1970–2005," Kennedy School of Government, Harvard University, April 28–29.
Betancourt, Rómulo. 1978. *Venezuela: Oil and politics*. Boston: Houghton Mifflin. Originally published in Spanish under the title *Venezuela, política y petróleo* (Mexico City: Fondo de Cultura Economica, 1956).
Betancourt, Keila, Samuel Feije, and Gustavo Márquez. 1995. *Mercado laboral: Instituciones y regulaciones*. Caracas: Ediciones IESA.
Bhasker, Venkataraman, and Mushtaq Khan. 1995. Privatization and employment: A study of the jute industry in Bangladesh. *American Economic Review* 85, no. 1: 267–73.
Bitar, Sergio, and Eduardo Troncoso. 1983. *El desafío industrial de Venezuela*. Buenos Aires: Pomaire.
Blavy, Rodolphe. 2006. Assessing banking sector soundness in a long-term framework: The case of Venezuela. IMF Working Paper WP/06/225, International Monetary Fund, Washington, D.C.
Blomström, Magnus, and Ari Kokko 2007. From natural resources to high-tech production: The evolution of industrial competitiveness in Sweden and Finland. In *Natural resources: Neither curse nor destiny*, ed. Daniel Lederman and William F. Maloney. Washington, D.C.: The World Bank; Palo Alto: Stanford University Press.
Blomström, Magnus, and Patricio Meller, eds. 1991. *Diverging paths: Comparing a century of Scandinavian and Latin American development*. Washington, D.C.: Inter-American Development Bank. Boss, Helen. 1990. *Theories of surplus and transfer: Parasites and producers in economic thought*. Boston: Unwin Hyman.
Boué, Juan Carlos. 1993. *Venezuela: The political economy of oil*. Oxford: Oxford University Press.
Brenner, Robert. 1976. The origins of capitalist development: A critique of neo-Smithian Marxism. *New Left Review* 104 (July–August): 25–92.

Brewer-Carías, Allan. 1975. *Cambio politico y reforma del estado en Venezuela*. Madrid: Technos.
Brito Figueroa, Federico. 1966. *Historía económia y social de Venezuela*. Vol. 2. Caracas: Universidad Central de Venezuela.
Brossard, Emma. 1993. Petroleum research and Venezuela's Intevep: The clash of the giants. Mimeograph, Instituto de Tecnología Venezolana para el Petróleo, Houston.
Bruton, Henry. 1998. A reconsideration of import substitution. *Journal of Economic Literature* 36 (June): 903–36.
Bruton, Henry, Gamini Abeysekera, Nimal Sanderatne, and Zainal Aznam Yusof. 1992. *The political economy of poverty, equity, and growth: Sri Lanka and Malaysia*. Oxford: Oxford University Press.
Buchanan, James. 1980. Rent-seeking and profit-seeking. In *Towards a theory of the rent-seeking society*, ed. James M. Buchanan, Robert D. Tollison, and Gordon Tullock. College Station: Texas A&M University Press.
Buchanan, James M., Robert D. Tollison, and Gordon Tullock, eds. 1980. *Towards a theory of the rent-seeking society*. College Station: Texas A&M University Press.
Burton, Michael, and John Higley. 1998. Political crises and elite settlements. In *Elites, crises, and the origins of regimes*, ed. Mattei Dogan and John Higley. Lanham, Md.: Rowman & Littlefield Publishers.
Caballero, Manuel. 1993. *Gómez, el tirano liberal*. Caracas: Monte Ávila Editores.
Calderón, César, William Easterly, and Luis Servén. 2003. Latin America's infrastructure in the era of macroeconomic crisis. In *The limits of stabilization: Infrastructure, public deficits, and growth in Latin America*, ed. William Easterly and Luis Servén. Palo Alto: Stanford University Press; Washington, D.C.: The World Bank.
Calderón César, and Luis Servén. 2003. The output cost of Latin America's infrastructure gap. In *The limits of stabilization: Infrastructure, public deficits, and growth in Latin America*, ed. William Easterly and Luis Servén. Palo Alto: Stanford University Press; Washington, D.C.: The World Bank.
Canache, Damarys. 2002. *Venezuela: Public opinion and protest in a fragile democracy*. Coral Gables, Fla.: North-South Center.
Capriles, Ruth. 1991. La corrupción al servicio de un proyecto politico. In *Corrupción y control*, ed. Rogelio Pérez Perdomo and Ruth Capriles. Caracas: Ediciones IESA.
Cárdenas, Enrique, José Antonio Ocampo, and Rosemary Thorp. 2000. Introduction. In *An economic history of twentieth-century Latin America*, vol. 3, *Industrialization and the state in Latin America: The postwar years*, ed. Enrique Cárdenas, José Antonio Ocampo, and Rosemary Thorp. London: Palgrave; Oxford: St. Anthony's College.
Cardoso, Fernando Henrique, and Enzo Faletto. 1979. *Dependency and development in Latin America*. Berkeley and Los Angeles: University of California Press.
Carlson, Chris. 2007. Pro-Chávez leaders examine reasons for Venezuelan referendum loss. http://www.Venezuelananalysis.com. December 6.
Cavarozzi, Marcelo. 1994. Politics: A key for the long-term in South America. In *Latin American political economy in the age of neoliberal reform*, ed. William C. Smith, Carlos H. Acuña, and Eduardo A. Gamarra. Miami: North-South Center.
Centeno, Miguel Angel. 1997. Blood and debt: War and taxation in nineteenth-century Latin America. *American Journal of Sociology* 102, no. 6: 1565–1605.
Chandler, Alfred, Franco Amatori, and Takashi Hikino. 1997. Historical and comparative contours of big business. In *Big business and the wealth of nations*, ed. Alfred

Chandler, Franco Amatori, and Takashi Hikino. Cambridge: Cambridge University Press.
Chandler, Alfred, and Takashi Hikino. 1997. The large industrial enterprise and the dynamics of modern economic growth. In *Big business and the wealth of nations*, ed. Alfred Chandler, Franco Amatori, and Takashi Hikino. Cambridge: Cambridge University Press.
Chang, Ha-Joon. 1994. *The political economy of industrial policy.* London: Macmillan.
———. 1998. The "initial conditions" of economic development—Comparing the East Asian and the Sub-Saharan African experiences. A background paper prepared for the UNCTAD *Trade and Development Report, 1998*, April.
———. 2002. *Kicking away the ladder: Development strategy in historical perspective.* London: Anthem Press.
Chang, Ha-Joon, and Ajit Singh. 1992. Public enterprises in developing countries and economic efficiency. UNCTAD Discussion Paper No. 48, United Nations Commission on Trade and Development, Geneva.
Chaudhry, Kiren Aziz. 1989. The price of wealth: Business and state in labor remittance and oil economies. *International Organization* 43, no. 1: 101–45.
Chen Chi-Yi, Michel Picuoet, and José Urquijo. 1983. Los moviminetos migratorios internacionales en Venezuela. *Revista de Investigaciones sobre Relaciones Industriales y Laborales* 4, nos. 10–11: 34–70.
Chenery, Hollis. 1979. *Structural change and development policy.* New York: Oxford University Press.
Chenery, Hollis, Sherman Robinson, and Moises Syrquin. 1986. *Industrialization and growth.* Oxford: Oxford University Press.
Clague, Christopher, Phillip Keefer, Stephen Knack, and Mancur Olson. 1997. Democracy, autocracy, and the institutions supportive of economic growth. In *Institutions and economic development: Growth and governance in less-developed and post-socialist countries*, ed. Christopher Clague. Baltimore: Johns Hopkins University Press.
Coase, Ronald. 1960. The problem of social cost. *Journal of Law and Economics* 3, no. 1: 1–44.
Collier, Ruth Berins, and David Collier. 1991. *Shaping the political arena: Critical junctures, the labor movement, and regime dynamics in Latin America.* Princeton: Princeton University Press.
Commons, John. 1932. The problem of correlating law, economics, and ethics. *Wisconsin Law Review* 8:3–26.
Consejo Nacional Electoral (CNE). Various years. *Resultados electorales.* Caracas: Consejo Nacional Electoral.
Cook, Paul, and Colin Kirkpatrick. 1988. Privatisation in less developed countries: An overview. In *Privatisation in less developed countries*, ed. Paul Cook and Colin Kirkpatrick. London: Harvester Wheatsheaf.
Coppedge, Michael. 1994. *Strong parties and lame ducks: Presidential partyarchy and factionalism in Venezuela.* Stanford: Stanford University Press.
———. 2000. Venezuelan parties and the representation of elite interests. In *Conservative parties, the right, and democracy in Latin America*, ed. Kevin Middlebrook. Baltimore: Johns Hopkins University Press.
COPRE. 1989. *La reforma en síntesis: Proyecto de reforma integral del estado.* Caracas: Comisión Presidencial para la Reforma del Estado.

Corden, Max, and Peter Neary. 1982. Booming sector and Dutch Disease economics: A survey. *Economic Journal* 92:825–48.
CORDIPLAN. 1973. *Posibilidad de exportaciones de la industrial venezolana*. Caracas: Oficina Central de Coordinación y Planificación.
———. 1990. *El gran viraje: Lineamientos generales del VIII plan de la nación*. Caracas: Oficina Central de Coordinación y Planificación.
Coronil, Fernando. 1998. *The magical state*. Chicago: University of Chicago Press.
Corrales, Javier. 2002. *Presidents without parties: The politics of economic reform in Argentina and Venezuela in the 1990s*. University Park: Pennsylvania State University Press.
Corrales, Javier, and Imelda Cisneros. 1999. Corporatism, trade liberalization and sectoral responses: The case of Venezuela, 1989–1999. *World Development* 27, no. 2: 2099–2122.
Crisp, Brian. 1994. Limitations to democracy in developing capitalist societies. *World Development* 22, no. 10: 1491–1509.
———. 1997. Presidential behavior in a system of with strong political parties. In *Presidentialism and democracy in Latin America*, ed. Scott Mainwaring and Matthew Soberg Shugart. New York: Cambridge University Press.
———. 2000. *Democratic institutional design: The power and incentives of Venezuelan politicians and interest groups*. Stanford: Stanford University Press.
Dahlman, Carl. 1980. *The open field system and beyond: A property rights analysis of an economic institution*. Cambridge: Cambridge University Press.
Dasgupta, Sukti, and Ajit Singh. 2006. Manufacturing, services, and premature deindustrialization: A Kaldorian analysis. WIDER Research Paper No. 2006/49, World Institute for Development Economics Research, Helsinki, May.
Davis, Mike. 2006. *Planet of slums*. London: Verso.
"Deadly Massage." 2008. *Economist*. 19–25 July.
De Krivoy, Ruth. 2002. *Colapso del sistema bancaria venezolana de 1994*. Caracas: Ediciones IESA.
De la Cruz, Rafael. 1988. *Venezuela en busca de un nuevo pacto social*. Caracas: Alfadil Ediciones.
———, ed. 1997. *Federalismo fiscal y decentralización*. Caracas: Ediciones IESA.
de Schweinitz, Karl. 1964. *Industrialization and democracy*. New York: The Free Press.
Demsetz, Harold. 1997. *The economics of the business firm*. Cambridge: Cambridge University Press.
Deininger, Klaus, and Lynn Squire. 1996. A new data set for measuring income inequality. *World Bank Economic Review* 10, no. 3: 565–91.
———. 1998. New ways of looking at old issues: Inequality and growth. *Journal of Development Economics* 57, no. 2: 259–87.
Di John, Jonathan. 2004. The political economy of anti-politics and social polarisation in Venezuela, 1998–2004. Crisis States Programme Working Paper No. 76 (series 1), London School of Economics.
———. 2006. The political economy of taxation and tax reform in developing countries. WIDER Research Paper 2006/74, World Institute for Development Economics Research, Helsinki, July.
———. 2007. Albert Hirschman's exit-voice framework and its relevance to problems of public education performance in Latin America. *Oxford Development Studies* 35, no. 3: 295–27.

Di Tella, Torcuato. 1970. Populism and reform in Latin America. In *Obstacles to change in Latin America*, ed. Claudio Veliz. Cambridge: Cambridge University Press.

Díaz-Alejandro, Carlos. 1975. Trade policies and economic development. In *International trade and finance: Frontiers for research*, ed. Peter Kenen. New York: Cambridge University Press.

Dollar, David. 1992. Outward-oriented developing economies really do grow more rapidly: Evidence from 95 LDCs, 1976–85. *Economic Development and Cultural Change* 40 (April): 523–44.

Doner, Richard, and Ansil Ramsey. 1997. Competitive clientelism and economic governance. In *Business and the state in developing countries*, ed. Sylvia Maxfield and Ben Ross Schneider. Ithaca: Cornell University Press.

Dornbusch, Rudiger, and Sebastian Edwards, eds. 1990. *Economic populism in Latin America*. Chicago: University of Chicago Press.

Duno, Pedro. 1975. *Los doce apóstoles*. Valencia: Vadell Hermanos.

Easterly, William, and Luis Servén, eds. 2003. *The limits of stabilization: Infrastructure, public deficits and growth in Latin America*. Washington, D.C.: The World Bank; Palo Alto: Stanford University Press.

Edwards, Sebastian. 1995. Trade and industrial policy reform in Latin America. In *Policies for growth: The Latin American experience*, ed. Lara Resende. Washington, D.C.: International Monetary Fund.

Eggertsson, Thrainn. 1990. *Economic behaviour and institutions*. Cambridge: Cambridge University Press.

Ellner, Steven. 1985. Inter-party agreement and rivalry in Venezuela: A comparative perspective. *Studies in Comparative International Development* 19, no. 4 (Winter 84/85): 38–66.

———. 1996. Political party factionalism and democracy in Venezuela. *Latin American Perspectives* 23, no. 3: 87–109.

———. 1997. Recent Venezuelan political studies: A return to Third World realities. *Latin American Research Review* 32, no. 2: 201–18.

———. 1999. The heyday of radical populism in Venezuela and its aftermath. In *Populism in Latin America*, ed. Michael Conniff. Tuscaloosa: University of Alabama Press.

———. 2005. The emergence of a new trade unionism in Venezuela with vestiges of the past. *Latin American Perspectives* 32 (March): 51–71.

———. 2008. *Rethinking Venezuelan politics: Class, conflict, and the Chávez phenomenon*. Boulder, Colo.: Lynne Rienner Publishers.

Elster, Jon. 1985. *Making sense of Marx*. Cambridge: Cambridge University Press; Paris: Editions de la Maison des Sciences de l'Homme.

———. 1989. *The cement of society*. Cambridge: Cambridge University Press.

Engerman, Stanley, and Kenneth Sokoloff. 1997. Factor endowments, institutions, and differential paths to growth among New World economies: A view from economic historians in the United States. In *How Latin America fell behind: Essays on the economic histories of Brazil and Mexico, 1800–1914*, ed. Stephen Haber. Stanford: Stanford University Press.

Enright, Michael, Antonio Francés, and Edith Scott Saavedra. 1994. *Venezuela: El reto de la competitividad*. Caracas: Ediciones IESA.

Espinasa, Ramón, and Bernard Mommer. 1992. Venezuelan oil policy in the long run. In *International issues in energy policy, development, and economics*, ed. James Dorian and Fereidun Fesharaki. Boulder, Colo.: Westview Press.

Evans, Peter. 1995. *Embedded autonomy: States and industrial transformation.* Princeton: Princeton University Press.
Fearon, James, and David Laitin. 2003. Ethnicity, insurgency, and civil war. *American Political Science Review* 97, no. 1: 75–90.
Field, Alexander James. 1981. The problem with neoclassical institutional economics: A critique with special reference to the North/Thomas model of pre-1500 Europe. *Explorations in Economic History* 18, no. 2: 174–98.
Fields, Karl. 1997. Strong states and business organization in Korea and Taiwan. In *Business and the state in developing countries,* ed. Sylvia Maxfield and Ben Ross Schneider. Ithaca: Cornell University Press.
Findlay, Ronald, and Mats Lundhal. 1999. Resource-led growth—A long-term perspective. WIDER Working Paper 162, United Nations University / World Institute for Development Economics Research, Helsinki.
Fine, Ben. 2003. New growth theory. In *Rethinking development economics,* ed. Ha-Joon Chang. London: Anthem Press.
Fishlow, Albert. 1991. Some reflections on comparative Latin American economic performance and policy. In *Economic liberalization: No panacea,* ed. Tariq Banuri. Oxford: Clarendon Press.
Fitzgerald, Edmund Valpy Knox. 2000. ECLA and the theory of import-substituting industrialization in Latin America. In *An economic history of twentieth-century Latin America,* vol. 3, *Industrialization and the state in Latin America: The postwar years,* ed. Enrique Cárdenas, José Antonio Ocampo, and Rosemary Thorp. London: Palgrave; Oxford: St. Anthony's College.
Francés, Antonio. 1993. *Aló Venezuela: Apertura y privatización de las telecomunicaciones.* Caracas: Ediciones IESA.
———. 2001. Qué le paso a la empresa venezolana en los noventa? *Debates IESA* 6, no. 3 (January–March): 6–8.
Frank, Andre Gunder. 1969. *Latin America: Underdevelopment or revolution. Essays on the development of underdevelopment and the immediate enemy.* New York: Monthly Review Press.
Frieden, Jeffrey. 1991. *Modern political economy and Latin America, 1965–1985: Debt, development, and democracy.* Princeton: Princeton University Press.
Furtado, Celso. 1957. El desarollo reciente de la economía venezolana. Mimeograph, El Centro de Estudios del Desarrollo (CENDES), Universidad Central de Venezuela, Caracas.
———. 1970. *Economic development in Latin America.* Cambridge: Cambridge University Press.
García Araujo, Mauricio. 1975. *El gasto público consolidado en Venezuela.* Caracas: Artegrafía.
Garrido, Alberto. 2000. *La historia secreta de la revolución Bolivariana.* Mérida: Editorial Venezolana.
———. 2002. *Documentos de la revolución Bolivariana.* Caracas: Ediciones del Autor.
———. 2003. *Guerilla y revolución Bolivariana.* Caracas: Ediciones del Autor.
Gelb, Alan, and Associates. 1988. *Oil windfalls: Blessing or curse?* Oxford: Oxford University Press.
Gerschenkron, Alexander. 1962. *Economic backwardness in historical perspective.* Cambridge: Harvard University Press.

Giacopini Zárraga, Rafael. 1993. Historia de los gobiernos venezolanos en el siglo XX. *El Universal*, May 12. Caracas.

Gil Yepes, José Antonio. 1981. *The challenge of Venezuelan democracy*. New Brunswick, N.J.: Transaction Books.

———. 1992. De 1976 hasta nuestros dias. In *Politica y economia en Venezuela, 1810–1991*. Caracas: Fundación John Boulton.

Gilmore, Robert. 1964. *Caudillism and militarism in Venezuela, 1810–1910*. Athens: Ohio University Press.

Godio, Julio. 1980. *El movimiento obrero venezolano, 1850–1944*. Vol. 1. Caracas: Editorial Ateneo de Caracas.

Gomez, Terrence, and Kwame Sundaram Jomo. 1997. *Malaysia's political economy*. Cambridge: Cambridge University Press.

Goodman, Louis W., Johanna Mendelson Forman, Moisés Náim, Joseph S. Tulchin, and Gary Bland, eds. 1995. *Lessons of the Venezuelan experience*. Baltimore: Johns Hopkins University Press.

Gott, Richard. 2000. *In the shadow of the Liberator: Hugo Chávez and the transformation of Venezuela*. London: Verso.

Gramsci, Antonio. 1971. *Selection from the prison notebooks*. London: Lawrence and Wishart.

Grindle, Merilee. 2000. *Audacious reforms*. Baltimore: Johns Hopkins University Press.

Grupo Roraima. 1984. *Proposición al país*. Caracas: Grupo Roraima.

Gutiérrez Sanín, Francisco. 2003. Fragile democracy and schizophrenic liberalism: Exit, voice, and loyalty in the Andes. Paper presented to the Crisis States Programme (London School of Economics) Annual Workshop, Johannesburg, July 15.

Haber, Stephen. 1997. Introduction: Economic growth and Latin American economic historiography. In *How Latin America fell behind: Essays on the economic histories of Brazil and Mexico, 1800–1914*, ed. Stephen Haber. Stanford: Stanford University Press.

Haber, Stephen, Armando Razo, and Noel Maurer, eds. 2003. *The politics of property rights: Political instability, credible commitments, and economic growth in Mexico, 1876–1929*. Cambridge: Cambridge University Press.

Haggard, Stephen, Sylvia Maxfield, and Ben Ross Schneider. 1997. Theories of business and business-state relations. In *Business and the state in developing countries*, ed. Sylvia Maxfield and Ben Ross Schneider. Ithaca: Cornell University Press.

Hall, Peter. 1986. *Governing the economy: The politics of state intervention in Britain and France*. Oxford: Oxford University Press.

Harcourt, Geoffrey. 1972. *Some Cambridge controversies in the theory of capital*. Cambridge: Cambridge University Press.

Hausmann, Ricardo. 1981. State landed property, oil rent, and accumulation in the Venezuelan economy. Ph.D. diss., Cornell University.

———. 1989. Development planning in Venezuela. Paper prepared for World Bank Public Sector Investment Review, 1989–90, Washington, D.C.

———. 1990. *Shocks externos y ajuste macroeconómica*. Caracas: Banco Central de Venezuela.

———. 1995. Quitting populism cold turkey. In *Lessons of the Venezuelan experience*, ed. Louis W. Goodman, Johanna Mendelson Forman, Moisés Náim, Joseph S. Tulchin, and Gary Bland. Baltimore: Johns Hopkins University Press.

———. 2001. Venezuela's growth implosion: A neo-classical story? Paper presented at seminar "Analytic Country Studies on Growth," Harvard University, Cambridge, August.

———. 2003. Venezuela's growth implosion: A neo-classical story? In *In search of prosperity: Analytical narratives on economic growth*, ed. Dani Rodrik. Princeton: Princeton University Press.

Hausmann, Ricardo, and Gustavo Márquez. 1990. La crisis económica de Venezuela. In *Ensayos Escojidos*, ed. Hector Vallecillos and Omar Bello Rodríguez, vol. 3. Caracas: Banco Central de Venezuela.

Hausmann, Ricardo, Lant Pritchett, and Dani Rodrik. 2004. Growth accelerations. Mimeograph, John F. Kennedy School of Government, Harvard University.

Hausmann, Ricardo, and Francisco Rodríguez. 2006. Why did Venezuelan growth collapse? Paper presented at the conference "Venezuelan Economic Growth, 1970–2005," Kennedy School of Government, Harvard University, April 28–29.

Hausmann, Ricardo, and Dani Rodrik, 2006. Doomed to choose: Industrial policy as predicament. Paper prepared for the Blue Sky seminar organized by the Center for International Development, Harvard University, September 9.

Hausmann, Ricardo, Dani Rodrik, and Andrés Velasco. 2006. Getting the diagnosis right: A new approach to economic reform. *Finance and Development* 43, no. 1: 12–15.

Hayek, Friedrich A. 1935a. Socialist calculation I: The nature and history of the problem. In *Collectivist economic planning*, ed. F. A. Hayek. London: George Routledge & Sons.

———. 1935b. Socialist calculation II: The state of the debate. In *Collectivist economic planning*, ed. F. A. Hayek. London: George Routledge & Sons.

———. 1945. The use of knowledge in society. *American Economic Review* 35, no. 4: 519–30.

Hellinger, Daniel. 2005. When "no" means "yes to revolution": Electoral politics in Bolivarian Venezuela. *Latin American Perspectives* 32 (May): 8–32.

Hidalgo, César, Bailey Klinger, Albert-László Barabási, and Ricardo Hausmann. 2007. The product space conditions the development of nations. *Science*, July 27, 482–87.

Higley, John, and Richard Gunther, eds. 1992. *Elites and democratic consolidation in Latin America and Southern Europe*. Cambridge: Cambridge University Press.

Hillman, Richard. 1994. *Democracy for the privileged: Crisis and transition in Venezuela*. Boulder, Colo.: Lynne Rienner Publishers.

Hirsch, Fred. 1976. *Social limits of growth*. London: Routledge & Kegan Paul.

Hirschman, Albert, O. 1958. *The strategy of economic development*. New Haven: Yale University Press.

———. 1967. *Development projects observed*. Washington, D.C.: The Brookings Institution.

———. 1971. The political economy of import-substituting industrialization in Latin America. In *A bias for hope*. New Haven: Yale University Press. Originally published in the *Quarterly Journal of Economics* 82, no. 1 (February 1968): 1–32.

———. 1981a. A generalized linkage approach to development. In *Essays in trespassing: Economics to politics and beyond*. Cambridge: Cambridge University Press.

———. 1981b. Policymaking in Latin America. In *Essays in trespassing: Economics to politics and beyond*. Cambridge: Cambridge University Press.

———. 1995. Social conflicts as pillars of democratic market societies. In *A propensity to self-subversion*. Cambridge: Harvard University Press.

Hirshleifer, Jack. 1994. The dark side of the force. *Economic Inquiry* 32 (January): 1–10.

Hutchcroft, Paul. 2000. Obstructive corruption: The politics of privilege in the Philippines. In *Rents, rent-seeking, and economic development*, ed. Mushtaq Khan and K. S. Jomo. Cambridge: Cambridge University Press.
IADB. *See* Inter-American Development Bank.
IMF. *See* International Monetary Fund.
Innis, Harold. 1930. *The fur trade in Canada*. New Haven: Yale University Press.
Instituto Nacional de Estadística (INE). Various years. *Información estadísticas*. Caracas: INE.
———. Various years. *Principales Indicatores de la Industria Manufacturera*. Caracas: INE.
Inter-American Development Bank. 1998. *Facing up to inequality in Latin America*. Washington, D.C.: Inter-American Development Bank.
———. 2000. *Development beyond economics*. Washington, D.C.: Inter-American Development Bank.
International Monetary Fund. Various years. *International financial statistics*. Washington, D.C.: International Monetary Fund.
Janicke, Kiraz. 2008. Chávez announces major cabinet reshuffle. http://www.Venezuelan analysis.com. January 4.
Jesudasen, James. 1989. *Ethnicity and the economy: The state, Chinese business, and multinationals in Malaysia*. Singapore: Oxford University Press.
Johnson, Chalmers. 1982. *MITI and the Japanese miracle: The growth of industrial policy, 1925–1975*. Stanford: Stanford University Press.
Jomo, Kwame Sundaram. 1986. *A question of class: Capital, the state, and uneven development in Malaysia*. Singapore: Oxford University Press.
Jomo, Kwame Sundaram, and Terrence Gomez. 1997. Rents and development in multiethnic Malaysia. In *The role of government in East Asian economic development: Comparative institutional analysis*, ed. Masahiko Aoki, Hyung-Ki Kim, and Masahiro Okuno-Fujiwara. Oxford: Clarendon Press.
———. 2000. Malaysia's development dilemma. In *Rents, rent-seeking, and economic development*, ed. Mushtaq Khan and K. S. Jomo. Cambridge: Cambridge University Press.
Jones, Leroy, and Il Sakong. 1980. *Government, business and entrepreneurship in economic development*. Cambridge: Harvard University Press.
Kaldor, Nicholas. 1966. *Causes of the slow rate of economic growth of the United Kingdom*. Cambridge: Cambridge University Press.
———. 1967. *Strategic factors in economic development*. Ithaca: New York State School of Industrial and Labor Relations, Cornell University.
———. 1980. Public or private enterprise—the issues to be considered. In *Public and private enterprises in a mixed economy*, ed. William Baumol. London: Macmillan.
Karl, Terry Lynn. 1982. *The political economy of petrodollars: Oil and democracy in Venezuela*. Ph.D. diss., Department of Political Science, Stanford University.
———. 1986. Petroleum and political pacts: The transition to democracy in Venezuela. In *Transitions from authoritarian rule*, ed. Guillermo O'Donnell, Philippe Schmitter, and Laurence Whitehead. Baltimore: Johns Hopkins University Press.
———. 1997. *The paradox of plenty: Oil booms and petro states*. Berkeley and Los Angeles: University of California Press.
Karlsson, Weine. 1975. *Manufacturing in Venezuela*. Stockholm: Almquist and Wiksell International.

Karst, Kenneth, Murray Schwartz, and Audrey Schwartz. 1973. *The evolution of law in the barrios*. Los Angeles: University of California Latin American Center.

Katz, Jorge. 2000. Structural change and labour productivity growth in Latin American manufacturing industries, 1970–1996. *World Development* 28, no. 9: 1583–96.

Katzenstein, Peter. 1985. *Small states in world markets*. Ithaca: Cornell University Press.

Kaufman, Robert. 1977. Corporatism, clientelism, and partisan conflict: A study of seven Latin American countries. In *Authoritarianism and corporatism in Latin America*, ed. James Malloy. Pittsburgh: University of Pittsburgh Press.

Kaufmann, Daniel, Aart Kraay, and Massimo Mastruzzi. 2005. Governance matters IV: Governance indicators for 1996–2004. http://www.worldbank.org/wbi/governance/pubs/govmatters4.html.

Kelly de Escobar, Janet. 1984. Las empresas del estado. In *El caso Venezuela: Una ilusión de armonía*, ed. Moisés Naím and Ramón Piñango. Caracas: Ediciones IESA.

Kenny, Charles, and David Williams. 2000. What do we know about economic growth? Or, why don't we know very much? *World Development* 29, no. 1: 1–22.

Keynes, John Maynard. 1930. *A treatise on money*. Vol. 2. London: Macmillan.

Khan, Mushtaq. 1995. State failure in weak states: A critique of new institutionalist explanations. In *The new institutional economics and Third World development*, ed. John Harris, Janet Hunter, and Colin Lewis. London: Routledge.

———. 2000a. Rents, efficiency, and growth. In *Rents, rent-seeking, and economic development*, ed. Mushtaq Khan and K. S. Jomo. Cambridge: Cambridge University Press.

———. 2000b. Rent-seeking as a process: Inputs, rent-outcomes and net effects. In *Rents, rent-seeking, and economic development*, ed. Mushtaq Khan and K. S. Jomo. Cambridge: Cambridge University Press.

———. 2002a. State failure in developing countries and strategies of institutional reform. Paper presented at World Bank Conference on Development Economics, Oslo. June.

———. 2002b. Corruption and governance in early capitalism: World Bank strategies and their limitation. In *Reinventing the World Bank*, ed. Jonathan Pincus and Jeffrey Winters. Ithaca: Cornell University Press.

———. 2005. Markets, states and democracy: Patron-client networks and the case for democracy in developing countries. *Democratization* 12, no. 5: 705–25.

———. 2006. Governance, economic growth, and development since the 1960s. Background paper for World Economic and Social Survey, 2006. New York: United Nations.

Khan, Mushtaq, and K. S. Jomo. 2000. Introduction. In *Rents, rent-seeking, and economic development*, ed. Mushtaq Khan and K. S. Jomo. Cambridge: Cambridge University Press.

Kim, Hyung-Ki, and Jun Ma. 1997. The role of government in acquiring technological capability: The case of the petrochemical industry in East Asia. In *The role of government in East Asian economic development: Comparative institutional analysis*, ed. Masahiko Aoki, Hyung-Ki Kim, and Masahiro Okuno-Fujiwara. Oxford: Clarendon Press.

Klein, Benjamin, Robert Crawford, and Armen Alchain. 1978. Vertical integration, appropriable rents, and the competitive contracting process. *Journal of Law and Economics* 21, no. 2: 297–326.

Knight, Jack. 1992. *Institutions and social conflict*. Cambridge: Cambridge University Press.

Kohli, Atul. 1999. Where do high growth political economies come from? The Japanese lineage of Korea's developmental state. In *The developmental state*, ed. Meredith Woo-Cumings. Ithaca: Cornell University Press.

———. 2004. *State-directed development: Political power and industrialization in the global periphery.* Cambridge: Cambridge University Press.

Kornai, János. 1980. *Economics of shortage.* Amsterdam: North Holland.

Kornblith, Miriam. 1995. Public sector and private sector: New rules of the game. In *Venezuelan democracy under stress*, ed. Jennifer L. McCoy, Andrés Serbin, William C. Smith, and Andrés Stambouli. Boulder, Colo.: Lynne Rienner Publishers.

Kornblith, Miriam, and Daniel Levine. 1995. Venezuela: The life and times of the party system. In *Building democratic institutions*, ed. Scott Mainwaring and Timothy Scully. Stanford: Stanford University Press.

Kornblith, Miriam, and Thais Maignon. 1985. *Estado y gasto público en Venezuela, 1936–1980.* Caracas: Universidad Central de Venezuela.

Krueger, Anne. 1974. The political economy of the rent-seeking society. *American Economic Review* 64, no. 3: 291–303.

Lattes, Alfredo. 2000. Population distribution in Latin America. In *UN population distribution and migration.* New York: United Nations.

Lederman, Daniel, and William F. Maloney. 2007. Trade structure and growth. In *Natural Resources: Neither curse nor destiny*, ed. Daniel Lederman and William F. Maloney. Washington, D.C.: The World Bank; Palo Alto: Stanford University Press.

Levi, Margaret. 1988. *Of revenue and rule.* Berkeley and Los Angeles: University of California Press.

Levine, Daniel. 1973. *Conflict and political change in Venezuela.* Princeton: Princeton University Press.

———. 1978. Venezuela since 1978: The consolidation of democratic politics. In *The breakdown of democratic regimes*, ed. Juan Linz and Alfred Stepan. Baltimore: Johns Hopkins University Press.

———. 1989. Venezuela: The nature, sources, and prospects for democracy. In *Democracy in developing countries*, ed. Larry Diamond, Juan Linz, and Seymour Martin Lipset, vol. 4, *Latin America.* Boulder, Colo.: Lynne Rienner Publishers.

Levine, Daniel, and Brian Crisp. 1995. Legitimacy, governability, and reform in Venezuela. In *Lessons of the Venezuelan experience*, ed. Louis W. Goodman, Johanna Mendelson Forman, Moisés Náim, Joseph S. Tulchin, and Gary Bland. Baltimore: Johns Hopkins University Press.

Lewis, Arthur. 1954. Economic development with unlimited supplies of labour. *Manchester School of Economics and Social Studies* 22, no. 2: 139–91.

Lieuwen, Edwin. 1954. *Petroleum in Venezuela.* Berkeley and Los Angeles: University of California Press.

Lindbeck, Assar. 1981. Industrial policy as an issue in the economic environment. *The World Economy* 4 (December): 391–405.

Lipsey, Richard, and Kelvin Lancaster. 1956. The general theory of second best. *Review of Economic Studies* 24 (December): 11–32.

Little, Walter. 1996. Corruption and democracy in Latin America. *IDS Bulletin* 27, no. 2: 64–70.

Loayza, Norman, Pablo Fajnzybler, and César Calderón. 2002. Economic growth in Latin America and the Caribbean: Stylized facts, explanations, and forecasts. Mimeograph, The World Bank, Washington, D.C.

López Alves, Fernando. 2001. The transatlantic bridge: Mirrors, Charles Tilly, and state formation in the River Plate. In *The other mirror: Grand theory through the lens of Latin America*, ed. Miguel Angel Centeno and Fernando López Alves. Princeton: Princeton University Press.

López Maya, Margarita. 2003. Hugo Chávez Frías: His movement and his presidency. In *Venezuelan politics in the Chávez era: Class, polarization and conflict*, ed. Stephen Ellner and Daniel Hellinger. Boulder, Colo.: Lynne Rienner Publishers.

López Maya, Margarita, and Luis Lander. 2005. Popular protest in Venezuela: Novelties and continuities. *Latin American Perspectives* 32, no. 2: 92–108.

Lucas, Robert. 1988. On the mechanics of economic development. *Journal of Monetary Economics* 22, no. 1: 3–42.

Machiavelli, Niccolò. [1571] 1992. *The prince*. London: Penguin Classics.

Maddison, Angus. 1995. *Monitoring the world economy, 1820–1992*. Paris: OECD.

Mahdavy, Hossein. 1970. The patterns and problems of economic development in rentier states. In *Studies in the economic history of the Middle East*, ed. Michael A. Cook. London: Oxford University Press.

Mahon, James. 1992. Was Latin America too rich to prosper? Structural and political obstacles to export-led industrial growth. *Journal of Development Studies* 28, no. 2 (January): 241–63.

Maier, Gerald, and James Rauch. 2000. *Leading issues in economic development*. 7th ed. Oxford: Oxford University Press.

Mainwaring, Scott, and Timothy Scully. 1995. Introduction: Party systems in Latin America. In *Building democratic institutions*, ed. Scott Mainwaring and Timothy Scully. Stanford: Stanford University Press.

Malloy, James, ed. 1977. *Authoritarianism and corporatism in Latin America*. Pittsburgh: University of Pittsburgh Press.

Mann, Michael. 1993. *The sources of power: A history of power from the beginning to A.D. 1760*. Vol. 1. Cambridge: Cambridge University Press.

Manzano, Osmel. 2006. Venezuela after a century of oil exploitation. Paper presented at the conference "Venezuelan Economic Growth, 1970–2005," Kennedy School of Government, Harvard University, April 28–29.

Manzano, Osmel, and Roberto Rigobón. 2007. Resource curse or debt overhang? In *Natural resources: Neither curse nor destiny*, ed. Daniel Lederman and William F. Maloney. Washington, D.C.: The World Bank; Palo Alto: Stanford University Press.

Márquez, Gustavo. 1987. *Intervención del estado, crecimiento, y mercado de trabajo*. Caracas: Fundación Friedrich Ebert.

Márquez, Gustavo, and Carola Alvarez. 1996. *Poverty and the labor market in Venezuela, 1982–1995*. Social Programs Division. Washington, D.C.: Inter-American Development Bank.

Márquez, Gustavo, and Carmen Portela. 1991. *La economía informal*. Caracas: Ediciones IESA.

Martz, John D. 1966. *Acción democrática*. Princeton: Princeton University Press.

Martz, John D., and David Myers, 1986. The politics of economic development. In *Venezuela: The democratic experience*, ed. John Martz and David Myers. New York: Praeger.

Marx, Karl. [1867] 1976. *Capital*. Vol. 1. London: Penguin Books.

Matsuyama, Kiminori. 1997. Economic development as a coordination problem. In *The role of government in East Asian economic development: Comparative institutional

analysis, ed. Masahiko Aoki, Hyung-Ki Kim, and Masahiro Okuno-Fujiwara. Oxford: Clarendon Press.
Mauro, Paulo. 1995. Corruption and growth. *Quarterly Journal of Economics* 110, no. 3: 681–713.
———. 1998. Corruption: Causes, consequences, and agenda for further research. *Finance and Development* 35, no. 1: 11–14.
Maxfield, Sylvia, and Ben Ross Schneider. 1997. Business, the state, and economic performance in developing countries. In *Business and the state in developing countries*, ed. Sylvia Maxfield and Ben Ross Schneider. Ithaca: Cornell University Press.
Mayorbe, José Antonio. [1944] 1990. La paridad del bolivar. In *Ensayos escojidos*, ed. Hector Vallecillos and Omar Bello Rodríguez, vol. 1. Caracas: Banco Central de Venezuela.
Maza Zavala, Domingo, F. 1974. *Venezuela: Crecimiento sin desarollo*. Caracas: Monte Avila.
McBeth, Brian. 1983. *Juan Vicente Gómez and the oil companies in Venezuela, 1908–1935*. Cambridge: Cambridge University Press.
McCoy, Jennifer. 1999. Chávez and the end of "partyarchy" in Venezuela. *Journal of Democracy* 10, no. 3: 64–77.
McCoy, Jennifer L., and David J. Myers, eds. 2004. *The unraveling of representative democracy in Venezuela*. Baltimore: Johns Hopkins University Press.
McCoy, Jennifer L., Andrés Serbin, William C. Smith, and Andrés Stambouli, eds. 1995. *Venezuelan democracy under stress*. Boulder, Colo.: Lynne Rienner Publishers.
Merhav, Meir. [1971] 1990. Crecimiento y perspectivas de la industria venezolana. In *Ensayos escojidos*, ed. Hector Vallecillos and Omar Bello Rodríguez, vol. 2. Caracas: Banco Central de Venezuela.
Milgrom, Paul, and John Roberts, 1990. Bargaining costs, influence costs, and the organization of economic activity. In *Perspectives on positive political economy*, ed. James Alt and Kenneth Shepsle. Cambridge: Cambridge University Press.
———. 1992. *Economics, organization, and management*. London: Prentice-Hall.
Mill, John Stuart. 1848. *Principles of political economy, with some of their applications to social philosophy*. London: Longmans, Green.
Millward, Robert. 1988. Measured source of inefficiency in the performance of private and public enterprises in LDCs. In *Privatisation in less developed countries*, ed. Paul Cook and Colin Kirkpatrick. London: Harvester Wheatsheaf.
Ministerio de Minas y Energía. Various years. *Informe anual*. Caracas: Ministerio de Energía y Minas.
———. Various years. *Petroleo y otros datos estadísticas*. Caracas: Ministerio de Minas y Energía.
Ministerio de Finanzas. Various years. *Base de datos*. Caracas: Superintendencia de Bancos.
Molina, José. 2004. The unraveling of Venezuela's party system: From party rule to personalistic politics and deinstitutionalization. In *The unraveling of representative democracy in Venezuela*, ed. Jennifer L. McCoy and David J. Myers. Baltimore: Johns Hopkins University Press.
Monaldi, Francisco, and Michael Penfold. 2006. The collapse of democratic governance: Political institutions and economic decline in Venezuela. Paper presented at the conference "Venezuelan Economic Growth, 1970–2005," Kennedy School of Government, Harvard University, April 28–29.

Monaldi, Francisco, Rosa Amelia González, Richard Obuchi, and Michael Penfold. 2004. Political institutions, policymaking processes, and policy outcomes in Venezuela. Paper commissioned for the project "Political Institutions, Policymaking Processes, and Policy Outcomes," Latin American Research Network, Inter-American Development Bank, Washington, D.C.

Moncada, Samuel. 1995. Entrepreneurs and governments in Venezuela, 1944–1958. Ph.D. diss., Oxford University.

Moore, Barrington. 1966. *The social origins of democracy and dictatorship: Lord and peasant in the making of the modern world.* Boston: Beacon Press.

Moreno, María Antonia, and Francisco Rodríguez. 2005. Plenty of room? Fiscal space in a resource abundant economy. Paper presented at the United Nations Development Programme (UNDP) Conference "Pro-Poor Domestic Resource Mobilization: Securing Fiscal Space," Dakar, Senegal, November 28–29.

Mörner, Magnus, and Harold Sims. 1985. *The story of migrants in Latin America.* Pittsburgh: University of Pittsburgh Press.

Mueller, Dennis C. 1989. *Public choice II.* Cambridge: Cambridge University Press.

Murphy, Kevin, Andrei Shleifer, and Robert Vishny. 1989. Industrialization and the big push. *Journal of Political Economy* 97, no. 5: 1003–26

Myers, David J. 1980. The elections and the evolution of Venezuela's party system. In *Venezuela at the polls: The national elections of 1978*, ed. Howard Penniman. Washington, D.C.: American Enterprise Institute for Public Policy Research.

———. 2004. The normalization of Punto Fijo democracy. In *The unraveling of representative democracy in Venezuela*, ed. Jennifer L. McCoy and David J. Myers. Baltimore: Johns Hopkins University Press.

Myrdal, Gunnar. 1957. *Economic theory and underdeveloped regions.* London: Duckworth.

———. 1968. *Asian drama: An inquiry into the poverty of nations.* New York: Pantheon.

Naím, Moisés. 1988. El crecimiento de las empresas privadas en Venezuela: Mucha diversificacíon, poca organización. In *Las empresas venezolanas: Su gerencia*, ed. Moisés Naím. Caracas: Ediciones IESA.

———. 1993. *Paper tigers and minotaurs: The politics of Venezuela's economic reforms.* Washington, D.C.: Carnegie Endowment Book.

———. 2001. High anxiety in the Andes: The real story behind Venezuela's woes. *Journal of Democracy* 12, no. 2: 17–31.

Naím, Moisés, and Antonio Francés. 1995. The Venezuelan private sector: From courting the state to courting the market. In *Lessons of the Venezuelan experience*, ed. Louis W. Goodman, Johanna Mendelson Forman, Moisés Náim, Joseph S. Tulchin, and Gary Bland. Baltimore: Johns Hopkins University Press.

Naím, Moisés, and Ramón Piñango. 1984. El caso Venezuela: Una ilusión de armonía. In *El caso Venezuela: Una illisión del armonía*, ed. Moisés Náim and Ramon Piñango. Caracas: Ediciones IESA.

Nankani, Gobind. 1979. Development problems of mineral exporting countries. Staff Working Paper 354, The World Bank, Washington, D.C.

Navarro, Juan Carlos. 1995. Venezuela's new political actors. In *Lessons of the Venezuelan experience*, ed. Louis W. Goodman, Johanna Mendelson Forman, Moisés Náim, Joseph S. Tulchin, and Gary Bland. Baltimore: Johns Hopkins University Press.

Neary, Peter J., and Sweder van Wijnbergen. 1986. *Natural resources and the macro economy.* Cambridge: MIT Press.

Neuhoser, Kevin. 1992. Democratic stability in Venezuela: Elite consensus or class compromise? *American Sociological Review* 57, no. 1: 117–35.

Nolan, Peter, Jin Zhang, and Chunhang Liu. 2007. The global business revolution, the cascade effect, and the challenge for firms from developing countries. *Cambridge Journal of Economics*, forthcoming. Advanced access publication (August 13, 2007) available at http://cje.oxfordjournals.org/cgi/content/abstract/bem016v1.

North, Douglass. 1981. *Structure and change in economic history*. New York: W. W. Norton.

———. 1990. *Institutions, institutional change and economic performance*. Cambridge: Cambridge University Press.

North, Douglass, John Joseph Wallis, Stephen Webb, and Barry Weingast. 2007. Limited access orders in the developing world: A new approach to problems of development. World Bank Policy Research Working Paper No. 4359, The World Bank, Washington, D.C.

Nurkse, Ragnar. 1953. *Problems of capital formation in underdeveloped countries*. New York: Oxford University Press.

Nye, Joseph. [1967] 1989. Corruption and political development: A cost-benefit analysis. In *Political corruption: A handbook*, ed. Arnold Heidenheimer, Michael Johnston, and Victor T. LeVine. New Brunswick, N.J.: Transaction Publishers.

Obregón, Clara López, and Francisco Rodríguez. 2001. La politica fiscal venezolana, 1943–2001. *Reporte de coyuntura annual, 2001*. Caracas: Oficina de Asesoría Económica y Financiera.

Ocampo, José Antonio, and Camilo Tovar. 2000. Colombia in the classic era of "inward-looking development." In *An economic history of twentieth-century Latin America*, vol. 3, *Industrialization and the state in Latin America: The postwar years*, ed. Enrique Cárdenas, José Antonio Ocampo, and Rosemary Thorp. London: Palgrave; Oxford: St. Anthony's College.

OCEI. *See* Oficina Central de Estadísticas e Información.

OCEPRE. *See* Oficina Central de Presupuesto.

O'Donnell, Guillermo. 1973. *Modernization and bureaucratic authoritarianism*. Berkeley, Calif.: Institute of International Studies.

O'Donnell, Guillermo, and Philippe Schmitter. 1986. *Transitions from authoritarian rule: Tentative conclusion about uncertain democracies*. Baltimore: Johns Hopkins University Press.

Office of Public Finance Statistics. Various years. Caracas: Ministerio del Poder Popular para las Finanzas.

Oficina Central de Estadísticas e Información. Various years. *Encuesta de empleo* (Employment survey). Caracas: OCEI.

———. Various years. *Encuesta de hogares* (Household survey). Caracas: OCEI.

———. Various years. *Encuesta industrial* (Industrial survey). Caracas: OCEI.

Oficina Central de Presupuesto. Various years. *Presupuesto fiscal anual*. Caracas: OCEPRE.

Okuno-Fujiwara, Masahiro. 1997. Toward a comparative institutional analysis of the government-business relationship. In *The role of government in East Asian economic development: Comparative institutional analysis*, ed. Masahiko Aoki, Hyung-Ki Kim, and Masahiro Okuno-Fujiwara. Oxford: Clarendon Press.

Olson, Mancur. 1982. *The rise and decline of nations: Economic growth, stagflation, and social rigidities*. New Haven: Yale University Press.

———. 1993. Dictatorship, democracy, and development. *American Political Science Review* 87, no. 3: 567–76.

———. [1996] 2000. Big bills left on the sidewalk. In *A not-so-dismal science: A broader view of economies and societies*, ed. Mancur Olson and Satu Kähkönen. Oxford: Oxford University Press.

Olson, Mancur, and Satu Kähkönen. 2000. Introduction: The broader view. In *A not-so-dismal science: A broader view of economics and societies*, ed. Mancur Olson and Satu Kähkönen. Oxford: Oxford University Press.

Ortiz, Nelson. 2004. Entrepreneurs: Profits without power? In *The unraveling of representative democracy in Venezuela*, ed. Jennifer L. McCoy and David J. Myers. Baltimore: Johns Hopkins University Press.

Ortiz Ramirez, Eduardo. 1992. *La politica comerical de Venezuela*. Caracas: Banco Central de Venezuela, Colección de Estudios Económicos.

Ostrom, Elinor. 1990. *Governing the commons*. Cambridge: Cambridge University Press.

Padgett, Tim, and Jens Erik Gould. 2007. Petro slacker. *Time*, December 10, 51–53.

Pagano, Ugo. 1999. Is power an economic good? In *The politics and economics of power*, ed. Samuel Bowles, Maurizio Franzini, and Ugo Pagano. London: Routledge.

Palma, José Gabriel. 1978. Dependency: A formal theory of underdevelopment or a methodology for the analysis of concrete situations of underdevelopment? *World Development* 6, no. 7–8: 881–924.

———. 1987. Structuralism. In *The new Palgrave: A dictionary of economics*, ed. John Eatwell, Murray Milgate, and Peter Newman. London: Macmillan.

———. 1998. Three and a half cycles of "mania, panic, and [asymmetric] crash": East Asia and Latin America compared. *Cambridge Journal of Economics* 22, no. 6: 789–808.

———. 2000. Trying to "tax and spend" oneself out of the "Dutch Disease": The Chilean economy from the War of the Pacific to the Great Depression. In *An economic history of twentieth-century Latin America*, ed. Enrique Cárdenas, José Antonio Ocampo, and Rosemary Thorp, vol. 1, *Industrialization and the state in Latin America: The postwar years*. London: Palgrave; Oxford: St. Anthony's College.

———. 2002. Three sources of de-industrialisation and a new concept of the Dutch Disease. Mimeograph, International Labour Organisation, Geneva.

Papandreou, Andreas. 1994. *Externalities and institutions*. Oxford: Oxford University Press.

Penfold-Becerra, Michael. 2001. El colapso del sistema de partidos en Venezuela: Explicación de una muerte anunciada. In *Venezuela en transición: Elecciones y democracia, 1998–2000*, ed. José Vicente Carrasquero, Thais Maignon, and Friedrich Welsch. Caracas: CDB Publicaciones.

———. 2002. Federalism and institutional change in Venezuela. Paper presented at the annual meeting of the American Political Science Association, Boston, August 28.

Pérez Alfonzo, Juan Pablo. 1976. *Hundiéndonos en el excremento del diablo*. Caracas: Lisbona.

Pérez Perdomo, Rogelio. 1995. Corruption and political crisis. In *Lessons of the Venezuelan experience*, ed. Louis W. Goodman, Johanna Mendelson Forman, Moisés Náim, Joseph S. Tulchin, and Gary Bland. Baltimore: Johns Hopkins University Press.

Pérez Perdomo, Rogelio, and Teolinda Bolívar. 1998. Legal pluralism in Caracas, Venezuela. In *Illegal cities: Law and urban change in developing countries*, ed. Edésio Fernandes and Ann Varley. London: Zed Books.

Petras, James, Morris Morley, and Steven Smith. 1977. *The nationalization of Venezuelan oil*. New York: Praeger.

Posner, Robert. 1975. The social costs of monopoly and regulation. *Journal of Political Economy* 83 (August): 807–27.

Powell, John. 1971. *Political mobilization of the Venezuelan peasant*. Cambridge: Harvard University Press.

Prebisch, Raul. 1950. *The economic development of Latin America and its principal problems*. New York: United Nations.

Przeworski, Adam. 1985. *Capitalism and social democracy*. Cambridge: Cambridge University Press; Paris: Editions de la Maison des Sciences de l'Homme.

———. 1991. *Democracy and the market*. Cambridge: Cambridge University Press.

Pritchett, Lant. 1996. Measuring outward orientation in LDCs: Can it be done? *Journal of Development Economics* 49, no. 2: 307–35.

———. 2001. Where has all the education gone? *The World Bank Economic Review* 15, no. 3: 367–91.

———. 2003. A toy collection, a socialist star, and a democratic dud? Growth theory, Vietnam, and the Philippines. In *In search of prosperity: Analytical narratives on economic growth*, ed. Dani Rodrik. Princeton: Princeton University Press.

Purroy, Miguel Ignacio. 1982. *Estado e industrialización en Venezuela*. Valencia: Vadell Hermanos.

Putzel, James. 1992. *A captive land: The politics of agrarian reform in the Philippines*. London: Catholic Institute for International Relations; New York: Monthly Review Press; Quezon City: Ateneo de Manila University Press.

———. 1995. Why has democratization been a weaker impulse in Indonesia and Malaysia than in the Philippines? In *Democratization*, ed. David Potter, David Goldblatt, Margaret Kiloh, and Paul Lewis. Cambridge: Polity Press.

———. 1997. Survival of an imperfect democracy in the Philippines. *Democratization* 6, no. 1: 198–223.

Rangel, Domingo Alberto. [1968] 1990. Una económia parasitaria. In *Ensayos escojidos*, ed. Hector Vallecillos and Omar Bello Rodríguez, vol. 1. Caracas: Banco Central de Venezuela.

———. 1972. *La oligarquía del dinero*. Caracas: Ediciones Fuentes.

———. 1980. *Los Andinos en el poder*. Valencia: Vadell Hermanos.

Rasiah, Rajah. 1996. State intervention, rents, and Malaysian industrialization. In *Capital, the state, and industrialization*, ed. John Borrego, Alejandro Alvarez Bejar, and K. S. Jomo. Boulder, Colo.: Westview Press.

Rey, Juan Carlos. 1986. Los veinticinco años de la constitución y la reforma del estado. *Venezuela 86* 2: 26–34.

———. 1991. La democracia Venezolana y la crisis del sistema populista de conciliación. *Estudios Políticos* 74 (October–December): 533–78.

Rigobón, Roberto. 1993. El subsidio indirecto a la gasolina. In *Gasto publico y distribucion del ingreso en America Latina*, ed. Ricardo Hausmann and Roberto Rigobón. Caracas: Ediciones IESA; Washington, D.C.: Inter-American Development Bank.

Roberts, Kenneth. 1996. Neoliberalism and the transformation of populism in Latin America: The Peruvian case. *World Politics* 48, no. 1: 82–116.

———. 2003. Social polarization and populist resurgence. In *Venezuelan politics in the Chávez era: Class, polarization and conflict*, ed. Stephen Ellner and Daniel Hellinger. Boulder, Colo.: Lynne Rienner Publishers.

Rodríguez, Francisco. 2000. Factor shares and resource booms: Accounting for the evolution of Venezuelan Inequality. WIDER Working Paper No. 205, World Institute for Development Economics Research, Helsinki.

———. 2003. Las consecuencias económicas de la Revolucíon bolivariana. *Nueva Economia*, no. 2 (April): 85–142.

———. 2005. Growth empirics when the world is not simple. Economics Department, Wesleyan University, Middletown, Conn.

———. 2006. The anarchy of numbers: Understanding the evidence on Venezuelan economic growth. *Canadian Journal of Development Studies* 27, no. 4: 503–29.

———. 2007. Policymakers beware: The use and misuse of regressions in explaining economic growth. IPC Policy Research Brief No. 5 (November), International Poverty Centre, United Nations Development Programme, Brasilia.

———. 2008. An empty revolution: The unfulfilled promises of Hugo Chávez. *Foreign Affairs* 87, no. 2: 49–62.

Rodríguez, Francisco, and Jeffrey Sachs. 1999. Why do resource-abundant economies grow more slowly? *Journal of Economic Growth* 4, no. 3: 277–303.

Rodríguez, Miguel A. 1991. Public sector behaviour in Venezuela. In *The public sector and the Latin American crisis*, ed. Felipe Larraín and Marcelo Selowsky. San Francisco: International Center for Economic Growth.

———. 2002. *El impacto de la política económica en el proceso de desarollo venezolano*. Caracas: Universidad Santa Maria.

Rodrik, Dani. 1995. Getting interventions right: How South Korea and Taiwan grew rich. *Economic Policy* 10, no. 20: 55–107.

———. 1996. Understanding economic policy reform. *Journal of Economic Literature* 34, no. 1: 9–41.

———. 2003. What do we learn from growth narratives? In *In search of prosperity: Analytical narratives on economic growth*, ed. Dani Rodrik. Princeton: Princeton University Press.

———. 2004a. Rethinking growth strategies. WIDER Annual Lecture 8, World Institute for Development Economics Research, Helsinki.

———. 2004b. Industrial policy for the twenty-first century. Paper prepared for the United Nations Industrial Development Organization (UNIDO), Vienna.

Romer, Paul. 1986. Increasing returns and long-run growth. *Journal of Political Economy* 94, no. 5: 1002–37.

———. 1994. The origins of endogenous growth. *Journal of Economic Perspectives* 8, no. 1: 3–22.

Rose-Ackerman, Susan. 1978. *Corruption: A study of political economy*. New York: Academic Press.

Rosenstein-Rodan, Paul. 1943. Problems of industrialization in Eastern and South-Eastern Europe. *Economic Journal* 53 (June–September): 202–11.

Ross, Michael. 1999. The political economy of the resource curse. *World Politics* 51, no. 2: 297–322.

Rousseau, Jean-Jacques. [1755] 1984. *A discourse on inequality*. London: Penguin Books.

Rowthorn, Bob. 1971. *Capitalism, conflict and inflation*. London: Lawrence and Wishart.

Rueschemeyer, Dietrich, Evelyn Huber Stephens, and John D. Stephens. 1992. *Capitalist development and democracy*. Cambridge: Polity Press.

Sachs, Jeffrey. 1985. External debt and macroeconomic performance in Latin America and East Asia. *Brookings Papers on Economic Activity* 2:523–73.

Sachs, Jeffrey, and P. Felipe Larrain. 1993. *Macroeconomics in the global economy.* London: Harvester Wheatsheaf.
Sachs, Jeffrey, and Andrew Warner. 1995. Natural resource abundance and economic growth. NBER Working Paper 5398, National Bureau of Economic Research, Cambridge, Mass.
———. 2001. The curse of natural resources. *European Economic Review* 45:827–38.
Sandbrook, Richard. [1972] 1998. Patrons, clients, and factions: New dimensions of conflict analysis in Africa. In *Africa,* ed. Peter Lewis. Boulder, Colo.: Westview Press.
———. 1985. *The politics of Africa's economic stagnation.* Cambridge: Cambridge University Press.
Schamis, Hector E. 1999. Distributional coalitions and the politics of economic reform in Latin America. *World Politics* 51, no. 2: 236–68.
Schumpeter, Joseph. [1918] 1954. The crisis of the tax state. *International Economic Papers,* no. 4.
Scott, Maurice Fitzgerald. 1989. *A new view of economic growth.* Oxford: Clarendon Press.
Segarra, Nelson. 1985. Como evaluar la gestión de la empresas públicas venezolanas. In *Empresas del estado en América Latina,* ed. Janet Kelly de Escobar. Caracas: Ediciones IESA.
Sen, Amartya. 1999. *Development as freedom.* Oxford: Oxford University Press.
Shapiro, Carl, and Joseph Stiglitz. 1984. Equilibrium unemployment as a worker discipline device. *American Economic Review* 74, no. 3: 433–44.
Shleifer, Andrei, and Robert Vishny. 1993. Corruption. *Quarterly Journal of Economics* 108, no. 3: 599–617.
———. 1998. *The grabbing hand.* Cambridge: Harvard University Press.
Short, Robert. 1984. The role of public enterprises: An international statistical comparison. In *Public enterprises in mixed economies,* ed. Robert Floyd, Clive Gary, and Robert Short. Washington, D.C.: International Monetary Fund.
Sidel, John. 1996. Siam and its twin? Democratization and bossism in contemporary Thailand and the Philippines. *IDS Bulletin* 27, no. 2: 56–63.
Simon, Herbert. 1991. Organizations and markets. *Journal of Economic Perspectives* 5, no. 2: 25–44.
Skitovsky, Tibor. 1954. Two concepts of external economies. *Journal of Political Economy* 62, no. 2: 143–51.
Solow, Robert. 1956. A contribution to the theory of economic growth. *Quarterly Journal of Economics* 70, no. 1: 65–94.
Stambouli, Andrés. 2002. *La política extraviada: Una historia de Medina a Chávez.* Caracas: Fundacion para la Cultura Urbana.
Stiglitz, Joseph. 1994. *Whither socialism?* Cambridge: MIT Press.
Stokes, Susan. 1999. What do policy switches tell us about democracy? In *Democracy, accountability, and representation,* ed. Adam Przeworski, Susan Stokes, and Bernard Manin. Cambridge: Cambridge University Press.
Sullivan, William. 1992. Situación económica durante el período de Juan Vicente Gómez, 1908–1935. In *Politica y economia en Venezuela, 1810–1991.* Caracas: Fundación John Boulton.
Svensson, Jakob. 2005. Eight questions about corruption. *Journal of Economic Perspectives* 19, no. 3: 19–42.
Swedberg, Richard. 1998. *Max Weber and the idea of economic sociology.* Princeton: Princeton University Press.

Székely, Miguel, and Marianne Hilgert. 1999. What's behind the inequality we measure: An investigation using Latin American data. Inter-American Development Bank, Research Department Working Paper 409, Inter-American Development Bank, Washington, D.C.

Tanzi, Vito. 2000. Taxation in Latin America in the last decade. Paper prepared for conference "Fiscal and Financial Reforms in Latin America," Stanford University, November 9–10.

Thorp, Rosemary. 1992. A reappraisal of the origins of import-substituting industrialisation, 1930–1950. *Journal of Latin American Studies* 24 (Quincentenary Supplement): 181–95.

———. 1998. *Progress, poverty and exclusion: An economic history of Latin America in the twentieth century.* Washington, D.C.: Inter-American Development Bank; Baltimore: Johns Hopkins University Press.

———. 2000. Has the Coffee Federation become redundant? Collective action and the market in Colombian development. WIDER Working Paper No. 183, World Institute for Development Economics Research, Helsinki.

Thorp, Rosemary, and Francisco Durand. 1997. A historical view of business-state relations: Colombia, Peru, and Venezuela compared. In *Business and the state in developing countries,* ed. Sylvia Maxfield and Ben Ross Schneider. Ithaca: Cornell University Press.

Tilly, Charles. 1990. *Coercion, capital, and European states: A.D. 990–1992.* Oxford: Blackwell.

———. 2003. *The politics of collective violence.* Cambridge: Cambridge University Press.

Tornell, Aaron, and Philip Lane. 1999. The voracity effect. *American Economic Review* 89, no. 1: 22–46.

Torres, Gerver. 1993. *Quienes Ganan? Quienes Pierden? La privatización en Venezuela.* Caracas: Banco Consolidado.

Transparency International. Various years. Subjective Corruption Index. http://www.transparency.org/.

Treisman, Daniel. 2000. The causes of corruption: A cross-national study. *Journal of Public Economics* 76, no. 3: 399–458.

Tugwell, Franklin. 1975. *The politics of oil in Venezuela.* Stanford: Stanford University Press.

Umbeck, John. 1981. Might makes right—A theory of the formation and initial distribution of property rights. *Economic Inquiry* 19, no. 1: 38–59.

UNCTAD. 2000. *World investment report.* New York: United Nations Committee on Trade and Development.

UNDP. 1997. *Human development report.* New York: United Nations Development Programme.

———. 2001. *Human development report.* New York: United Nations Development Programme.

UNESCO. Various years. *Statistical yearbook.* Paris: United Nations Educational, Scientific, and Cultural Organization.

UNIDO. Various years. *Handbook of industrial statistics.* Vienna: United Nations Industrial Development Organization.

———. Various years. *International yearbook of industrial statistics.* Vienna: United Nations Industrial Organization.

United Nations. Various years. *International trade statistics yearbook.* New York: United Nations.

———. Various years. *Survey on crime trends.* New York: United Nations.
Uslar Pietri, Arturo. 1936. Sembrar el petróleo. *Ahora* (Caracas), June 14.
———. 1984. *Venezeula en el petróleo.* Caracas: Doctrina.
Valencia Ramírez, Cristóbal. 2005. Venezuela's Bolivarian revolution: Who are the chavistas? *Latin American Perspectives* 32, no. 5: 79–97.
Vallenilla, Luis. 1975. *Oil: The making of a new economic order: Venezuelan oil and OPEC.* New York: McGraw Hill.
Vartianen, Juhana. 1999. The economics of successful state intervention in industrial transformation. In *The developmental state*, ed. Meredith Woo-Cumings. Ithaca: Cornell University Press.
Verdoorn, Petrus Johannes. 1949. Fattori che regolano lo sviluppo economico della produttiva de lavoro. *L'Industria* 1:45–53. English translation in Luigi Pasinetti, ed., *Italian economic papers*, vol. 2 (Oxford: Oxford University Press).
Viana, Horacio. 1994. *Estudio de la capacidad tecnologica de la industria manufacturera venezolana.* Caracas: Ediciones IESA.
von Mises, Ludwig. 1936. *Socialism: An economic and sociological analysis.* London: Jonathan Cape. Originally published in German under the title *Die Gemeinwirtschaft: Untersuchungen über den Sozialismus* (Jena: Gustav Fischer Verlag).
Wade, Robert. 1990. *Governing the market.* Princeton: Princeton University Press.
Watkins, Mel. 1963. A staple theory of growth. *Canadian Journal of Economics and Political Science* 29, no. 2: 141–58.
Weber, Max. 1978. *Economy and society*, ed. Guenther Roth and Claus Wittich. Berkeley and Los Angeles: University of California Press. Originally published in German under the title *Wirtschaft und Gesellschaft* (1921).
Weeks, John. 1985. *Limits to capitalist development: The industrialization of Peru, 1950–1980.* Boulder, Colo.: Westview Press.
Weisbrot, Mark, and Luis Sandoval. 2007. The Venezuelan economy in the Chávez years. *Center for Economic and Policy Research [CEPR]: Reports and Policy Papers*, July: 1–21.
Weisbrot, Mark, Luis Sandoval, and David Rosnick. 2007. Poverty rates in Venezuela: Getting the numbers right. *Center for Economic and Policy Research [CEPR]: Reports and Policy Papers*, June: 1–5.
Weyland, Kurt. 1999. Populism in the age of neoliberalism. In *Populism in Latin America*, ed. Michael Conniff. Tuscaloosa: University of Alabama Press.
Williamson, John. 1990. What Washington means by policy reform. In *Latin American adjustment: How much has happened?* ed. John Williamson. Washington, D.C.: Institute for International Economics.
Williamson, Oliver. 1985. *The economic institutions of capitalism.* New York: The Free Press.
———. 2000. Economic institutions and development: A view from the bottom. In *A not-so-dismal science: A broader view of economies and societies*, ed. Mancur Olson and Satu Kähkönen. Oxford: Oxford University Press.
Wilpert, Geoffrey. 2004. "The main obstacle is the administrative structure of the Venezuelan state." April 24. http://www.venezuelananlysis.com/.
Woo-Cumings, Meredith. 1997. The political economy of growth in East Asia: A perspective on the state, market, and ideology. In *The role of government in East Asian economic development: Comparative institutional analysis*, ed. Masahiko Aoki, Hyung-Ki Kim, and Masahiro Okuno-Fujiwara. Oxford: Clarendon Press.

———. 1999. Introduction. In *The developmental state*, ed. Meredith Woo-Cumings. Ithaca: Cornell University Press.
World Bank. 1970. Current economic position and prospects of Venezuela. World Bank Country Report. Washington, D.C.: The World Bank.
———. 1973. Current economic position and prospects of Venezuela. World Bank Country Report. Washington, D.C.: The World Bank.
———. 1989. *Malaysia: Matching risks and rewards in a mixed economy*. Washington, D.C.: The World Bank.
———. 1990. Venezuela: Industrial sector report. World Bank Country Report. Washington, D.C.: The World Bank.
———. 1993. *The East Asian miracle*. New York: Oxford University Press.
———. 1997a. *World development report: The state in a changing world*. Oxford: Oxford University Press.
———. 2000. *Helping countries to combat corruption: Progress at the World Bank since 1997*. Washington, D.C.: The World Bank.
———. 2002. *World development report: Building institutions for markets*. Oxford: Oxford University Press.
———. 2005. *Economic growth in the 1990's: Learning from a decade of reforms*. Washington, D.C.: The World Bank.
———. Various years. *World development indicators*. Washington, D.C.: The World Bank.
———. Various years. *World development report*. Washington, D.C.: The World Bank. The 1997 edition of this source is cited as World Bank 1997b.
———. Various years. *World tables*. Washington, D.C.: The World Bank.
Wright, Gavin, and Jesse Czelusta. 2003. Mineral resources and economic development. Paper prepared for the conference "Sector Reform in Latin America," Stanford Center for International Development, November 13–15.
———. 2007. Resource-based growth: Past and present. In *Natural resources: Neither curse nor destiny*, ed. Daniel Lederman and William F. Maloney. Washington, D.C.: The World Bank; Palo Alto: Stanford University Press.
Young, Allyn. 1928. Increasing returns and economic progress. *Economic Journal* 38, no. 152: 527–42.
Zettelmeyer, Jeromin. 2006. Growth and reforms in Latin America: A survey of facts and arguments. IMF Working Paper WP/06/210, The International Monetary Fund, Washington, D.C.:

INDEX

Page numbers in *italics* refer to tables.

Abdul Rahman, Tunku, 273, 278
absorptive capacity argument, 70
Acción Democrática (Venezuelan Social-Democratic Party, AD), 197–200, 203, 210; economic policies, 122–23, 209; factionalism within, 125, 207–8, 211–13, 215, 217, 224, 233–34, 246, 282; labor policies, 119, 251, 276; relationship to COPEI, 200, 201, 202, 207; weakening of, 195, 220, 262. *See also* Congress, AD-dominated
accumulation, 20, 211, 262–63. *See also* capital accumulation; primitive accumulation
Adelman, Jeremy, 141 n. 8
Adriani, Alberto, 49 n. 15
advanced economies, 93, 264, 287, 292
advanced import-substitution industrialization (ISI): challenges to, 151–59, 168, 266–67; early/easy-stage ISI compared to, 154, *154*; implementation of, 162, 165–66, 225, 230; patronage patterns, 265; policy failures, 252; politics of, 194–95, 287; R&D spending, 255; resource allocation for, 263; subsidies for, 250; transition to, 170, 184–85, 232, 291. *See also* big-push natural-resource-based heavy industrialization
aerospace sector, 75
Africa. *See* South Africa; sub-Saharan Africa
Agenda Venezuela, 122 n. 22, 218
Agrarian Reform Law of 1948, 203
agricultural sector, 31, 79 n. 1, 142 n. 11, 190, 196
Aitken, Brian, 74 n. 35
ALCASA (state aluminum firm), 182
Alianza Bravo Pueblo (political party), 216 n. 40
aluminum sector: big-push programs, 173; capacity utilization, 69, 252; exports, 175, 176, 182, 250; nationalization of, 182, 183; production statistics, 176 n. 7; public enterprises, 3, 70, 180
Alvarez, Carola, 20
Alvarez Paz, Oswaldo, 215
Amsden, Alice, 155 n. 25, 254 n. 20
Amuay Refinery Project, 181

Andean Pact of 1973, 246–47
Anderson, Benedict, 159 n. 27
Añez Fonseca, Ciro, 213
antipolitics, 127, 195–96, 217, 218–23
Argentina, 25, 29, 88, 97, 171
Arrow, Kenneth, 291
artisanal stage import-substitution industrialization. *See* early/easy stage import-substitution industrialization
Ascher, William, 71 n. 29
Asia. *See* East Asia; South/Southeast Asia; *and individual countries*
assets, 143; distribution of, 298; external public, 234, 235; privately-held, 45 n. 10, 146, 242; specificity of, 158 n. 26
Australia, 82, 85
authoritarianism: industrialization under, 32, 225, 264, 289–90; periods of, 21, 187–98, 200, 228
authority: centralized, 92, 94, 105, 107, 108, 133–34, 158–59, 291, 297; decision-making, 79, 82, 139, 145; jurisdiction *vs*., 80–81
automotive sector, 75 n. 36, 115, 170, 175, 182 n. 15, 248 n. 18
Auty, Richard, 79, 88

Baduel, Raùl, 220 n. 47
bailouts, 69–70, 120, 201 n. 20
balance-growth models, 153, 233 n. 2
balance of payments, 118; crises in, 42, 50, 112, 177 n. 8, 240, 289. *See also* capital flight
Banco Bilbao Vizcaya Argentaria (BBVA), 120 n. 19
Banco Industrial de Venezuela, 190
Banco Latino, 120 n. 18, 217
Banco Santander Central Hispano (BSCH), 120 n. 19
Bangladesh, 276
banking sector, 120 n. 19, 125, 151 n. 22, 241; crises in, 120, 121 n. 20, 201 n. 20, 217, 241, 251, 294 n. 8; GDP share, 117, 201 n. 20, 240–41; state-owned, 103, 190, 213, 228
Baptista, Asarúbal, 201 n. 20

Baran, Paul, 36
bargaining, 109, 139, 145, 289; Coasian approach, 141; labor, 119, 239; particularist, 157, 190, 248; state, 21, 22 n. 10, 80, 82, 158, 228
Baumol, William J., 258
bauxite production, 182, 183
Bello, Omar, 251
Bentham, Jeremy, 139 n. 4
Bermúdez, Adriana, 251
Betancourt, Rómulo, 50, 105, 197, 202, 212
Bhaskar, Venkataraman, 276
big-push natural-resource-based heavy industrialization: challenges to, 150, 151–59, 168, 266–67; coordination problems, 153, 155; early/easy-stage ISI compared to, 154; implementation of, 148, 162, 165–66, 177–83, 225, 232–33, 284; Pérez administration's, 69, 213; policy failures, 111, 242, 252, 265, 291; politics of, 194–95, 230, 239, 286, 287; R&D spending, 255; resource allocation for, 263; state-led, 3, 86, 184, 226–27, 237; transition to, 170, 181, 258. See also large-scale heavy industrialization
Bitar, Sergio, 175
Blomström, Magnus, 85
Bolívar, Simon, 187
bonded rationality, 145, 292–93
BO Petroleum, 181 n. 14
Botswana, 97, 299
bourgeoisie, 36, 188–89
Brazil, 22, 88, 126, 191 n. 8, 299, 300
Brewer-Carías, Allan, 104
bribery, 77, 92, 101, 145, 214 n. 37. See also corruption; cronyism
Buchanan, James, 89, 208 n. 32
budget constraints, 69–70, 71, 73, 150 n. 20, 165
bureaucracy, 289–90; corruption in, 105–6; expanding, 197, 261; industrialization and, 264; lack of, 80, 81, 243. See also state, the
business associations: Malaysian, 274, 276; mobilization of, 112–13, 210, 219 n. 44; weakening of, 121 n. 20, 122, 157, 189–90, 192. See also conglomerates; firms; multinational corporations; state-business relationships

cabinets, presidential, 113, 123, 215, 220 n. 48, 223. See also ministries, government
Caldera, Rafael, 119 n. 14, 120, 125 n. 27, 207 n. 28, 215–17; economic policies of, 122 n. 22, 180 n. 13, 183, 218, 246–47

campaign contributions, 113, 120, 125, 220, 248, 261, 281
Canada, 85
CANTV (national telephone company), 73, 244, 246
capability approach, 114, 121, 134; weaknesses in, 110, 124–25, 126, 128–29
capacity, 70, 109, 134–35, 178, 299–300; utilization levels, 36, 105, 250, 251–54, 253. See also state, the, capacity of
capital: cost of, 266; human, 24, 25–26, 31, 32, 70, 158 n. 26, 180; inefficient use of, 69, 91–92, 249, 251; measuring, 172 n. 3; physical, 24, 158 n. 26; public formation of, 19, 244; quality of, 260; sudden inflows of, 64
capital accumulation, 146, 150, 228, 276, 278–83, 292; state's role in, 142–44, 187–90
capital flight, 66 n. 25, 118, 120, 125, 241–42; cum debt crisis, 230, 232–34, 233, 235, 237–38. See also balance of payments
capital goods sector, 148 n. 18, 149, 185, 289
capital intensity, 101, 173; measurements of, 53, 54, 55, 56
capital-intensive sectors, 115, 151, 175, 250; growth rates, 55, 56; productivity levels, 62, 63; profitability levels, 59, 60, 61–62, 63; wages levels, 59, 60, 61–62. See also intermediate capital-intensive sectors
capitalism/capitalists: Chinese-Malay, 272–74, 278, 280–81; corruption's effects on, 189; creation of, 142–44, 262–63; labor unions vs., 238–39; rentier, 147
Capriles, Ruth, 206
Caracas stock exchange, 221 n. 54
Cardoso, Fernando Henrique, 197
Caribbean region, 17–18, 101
Castro, Cipriani, 151, 187
caudillos, 82–83, 85, 151, 187–90, 198. See also warlords
Central America, 296, 297
Chandler, Alfred, 52
Chang, Ha-Joon, 61, 71, 72, 92, 93, 134, 151, 264
Chávez, Hugo Frias, 119 n. 14; economic policies of, 118 n. 13, 183, 219–23, 239, 251; election of, 114, 125–26, 195, 217–18; military rebellion by, 120; opposition to, 219 n. 44
chavismo, 221, 222
chemicals sector, 47, 152; exports, 175, 176, 182, 183; growth rates, 61; productivity levels, 62–63, 176 n. 7; profitability levels, 61 n. 20
Chenery, Hollis, 148 n. 18

Chile: employment, 276; growth rates, 88, 97, 167; health indicators, 28; immigration to, 29; income inequalities, 126; ISI in, 171; operating surpluses, 60; productivity levels, 61, 62–63, 258; profitability levels, 57, 62; reforms, 111 n. 5, 127; resource inflow management, 107; strikes, 275; wage levels, 57, 59
Chinese, Malaysian, 272–74, 278, 280–81
Cisneros, Diego, 113 n. 8, 114 n. 9, 212 n. 36
civil wars, 82 n. 6, 189 n. 4, 227, 297
clientelism, 264–65; contention within, 149, 186; effects on growth, 292; in employment, 244; in Malaysia, 272, 273; political, 150, 202, 206, 261, 276; state-based, 133–34
clientelism, populist, 209; demands of, 202–3, 205, 291; economic effects, 230–60, 267, 276; factionalism in, 232, 239; Malaysian, 279, 281; origins of, 194–200; post-1968, 194–200, 210–11, 223–24
clothing sector, 56, 58, 59, 250
coalitions, 83, 145, 195, 201, 209 n. 33, 299
Coasian approach, 141
Cobb-Douglas production function, 259
cocoa sector, 49–50
coffee sector, 49–50, 164, 190
cognitive failure approach, 101–6, 136
cohesive-capitalist states, 161–62, 166 n. 41
collective action, 121–22, 157, 213–14, 223, 225
Collier, David, 199
Collier, Ruth Berins, 199
Colombia, 299; education indicators, 27; growth rates, 97, 164–65; health indicators, 26, 28; illiteracy rates, 26; immigrants from, 205 n. 26; ISI, 171; operating surpluses, 60; politics, 164, 210; productivity levels, 59, 60, 61, 62–63, 258; profitability levels, 57; skilled labor, 29; tariffs, 191 n. 8; technology, 165 n. 38; wage levels, 57, 59
Comité de Organización Política Electoral Independiente (COPEI), 125, 198, 207, 213, 1981; economic policies, 234–36; factionalism within, 216, 217, 233–34; labor policies, 211, 251; relationship to AD, 200, 201, 202, 207; weakening of, 195, 220, 262
Commission for the Integral Reform of Public Administration (CRIAP), 104
commodities, 35, 112. See also natural resources
Commons, John R., 141 n. 7
Communist Party of Venezuela (Partido Communista Venezolano), 199, 207
compatibility approach, 261, 297

competition: industrial, 73, 75, 142, 152, 157; political, 4–5, 264
computer sector, 156–57
concessions: oil, 188, 190; trade, 193–94. See also trade liberalization policies
Confederación de Trabajadores de Venezuela (CTV), 119
conflicts: avoidance of, 81, 202–3, 207, 227, 231; distributive, 239, 240, 242; effects on growth, 265–66; labor-related, 204, 211, 275, 275–76; managing, 119 n. 14, 124, 189
conglomerates, 75, 157, 211–12, 234, 247. See also family conglomerates
Congress, AD-dominated, 211, 234, 235, 236
consensus-building, 213–14, 220, 246. See also conflicts, avoidance of; democracy, pacted
consolidated states with centralized political organizations, 83, 151 n. 21, 159, 161–62, 167, 230, 283–84; Malaysia as, 272–76, 283–84; Venezuela as, 186, 187–94, 223–24, 225
consolidated states with fragmented political organizations, 163–66, 231–32; Venezuela as, 186, 224–25
construction sector, 192–93
consultative process, 195; weakening of, 122, 123–24, 126, 127. See also conflicts, avoidance of; consensus-building
contestations: electoral, 214–18; patronage-related, 213; political, 29, 125, 137, 159, 212, 224, 242, 262, 287; for resources, 163
Convergencia (Convergence, political party), 215–16
Coppedge, Michael, 261
Coronil, Fernando, 87 n. 12, 106, 111, 182, 196 n. 12, 200 n. 19, 248 n. 18
Corporación Venezuelona de Guyana (CVG), 178
corporatism. See politics, corporate
Corrales, Javier, 114 n. 9
corruption: growth and, 95, 96, 97–99, 98, 101, 106–7, 133, 144–45, 226; increases in, 77–80, 82, 106–7, 109, 206; liberalization's effects on, 124; oil abundance's effects on, 86, 95, 96, 98; political instability and, 121–26, 149, 263, 264–65; property rights and, 214 n. 37; reduction in, 108, 128–29, rent-seeking and, 88–95, 97–99, 101, 114, 188; state, 105–6, 136, 228. See also bribery; cronyism; scandals, corruption
Costa Rica, 126, 210
coup attempts: in 1899, 187; in 1992, 120, 121 n. 20, 123 n. 25, 124, 125; leftists', 209

crime rates, 221 n. 50
Crisp, Brian, 261–62
cronyism: industrialization and, 75, 133, 226, 228, 247; Malaysian, 281; under Pérez Jiménez, 105; rent-seeking and, 94, 188; in state-business relations, 87, 88, 120 n. 18, 157, 195
Cuba, 28 n. 17
currency: appreciation of, 191 n. 8, 236; devaluation of, 42–43, 48–50, 117, 237–38, 240. *See also* inflation
Customs Law of 1936, 191
Czelusta, Jesse, 85, 301

debt: crises of, 155 n. 25, 201 n. 20, 260 n. 25; external public, 45 n. 10, 70, 233, 233–34, 235, 238, 243; GDP share, 238; private, 235, 237; state, 65, 67, 201 n. 20. *See also* capital flight, cum debt crisis
decentralization, political, 110, 119, 128, 217, 261
decision-makers: authority of, 79, 92, 139, 145; corruption and, 79, 119, 125; ineffective, 8, 288–89; investment-related, 64, 77, 102–3, 107, 136; state, 37, 157, 158, 246, 247; technocratic, 123, 148. *See also* leaders
Decision 24, 246
deindustrialization, 39, 40 n. 1
DeKrivoy, Ruth, 23, 117, 120, 121, 125, 217, 251
democracy, 32; collapse of, 108–29, 261; importance of, 24, 301; industrialization and, 147, 290; late-developing, 4, 261–62; long-standing, 4–5; Malaysian, 273; pacted, 119 n. 14, 123, 124, 129, 195, 200–203, 205–10, 258, 261, 281; transition and consolidation, 3, 21, 103, 111, 194–96, 200–203, 205–10, 227, 230, 246, 287. *See also trieno*
dependency theory, 35–36
deregulation, 121, 122; financial, 86, 117, 251, 296 n. 10
developing countries, 30–31, 39–40, 99, 100
developmental state models, 134–38, 140–42, 144, 161–62, 189; weaknesses of, 146–47, 159, 290–91
development strategies, 32, 141–42, 265; inward-oriented, 152 n. 23; periodization of, 169–85, 184, 229, 258; political settlements' compatibility with, 137–38, 147, 160–67, 168, 227–30, 229, 232, 255, 256–57, 261, 266–67, 285–301. *See also* advanced import-substitution industrialization; early/easy-stage import-substitution industrialization; economic development

Diaz, Alejandro Carlos, 50
distribution: conflicts over, 265–66; income, 5, 29, 118, 238–39, 262, 278–79; Malaysian pattern of, 272–74, 279, 282; production vs., 142
diversification, 21, 47–48, 85, 103, 178, 301. *See also* overdiversification
Dollar, David, 50 n. 16
Duno, Pedro, 212 n. 36
durable goods sectors, 148 n. 18, 149, 151, 185
Dutch Disease models, 38–64, 77, 79; early, 38–45; recent, 45–48; wage rigidity argument, 48–64; weaknesses in, 68, 76

early/easy-stage import-substitution industrialization (ISI): big-push/advanced ISI compared to, 154, 155, 158, 289; capacity utilization, 252, 254; challenges to, 148–51, 155, 157, 167, 168; effects of oil windfalls on, 171 n. 2; growth during, 159, 185, 230; implementation of, 163–65, 190, 228, 258; periodization of, 170, 171–75; politics of, 266, 287; promotion of, 189; resource allocation for, 263
East Asia: debt crises, 155 n. 25; developmental states, 31; education indicators, 27; growth rates, 17–18, 99, 101, 167, 296; health indicators, 28; income distribution, 29; labor unions, 274, 276; manufacturing sectors, 26, 254 n. 20; national savings levels, 24–25; NICS, 152; taxes, 273 n. 3
Economic Commission of Latin America and the Caribbean (ECLA), 179–80, 247
economic decline. *See* growth collapse
economic development: changes in, 295; corruption's effects on, 124, 189; natural resource-related, 35–76; oil revenues related to, 64–68; political challenges, 264, 265, 286–87, 301; state hindrances to, 78–82, 84, 88–89; sustainable, 301. *See also* development strategies; growth, economic
economic liberalization policies, 78, 107, 108–29, 166 n. 40, 214, 239–40; during authoritarian rule, 191–93; of 1989, 20, 74, 86, 111–14, 115, 182–83, 243. *See also* trade liberalization policies
economics, politics related to, 49, 138, 261–67, 285–301
economy: crises in, 15 n. 1, 16–17, 19–21, 68, 201 n. 20, 217, 221; management of, 110, 223; reforms of, 217, 243; state's role in, 89–91, 102, 106, 118, 129, 140–43, 149, 214, 219,

263–64, 280, 292–94; trends and cycles in, 15–32, 229; world, 99, 101
education indicators, 25, 25, 26, 26, 27, 27–28, 70, 199
egalitarianism, 30–31, 102 n. 23, 263
Eggertsson, Thrainn, 93 n. 17
elections, 208 n. 31, 216 n. 42, 281; in 1978, 105; of Bentancourt, 207 n. 29; of Chávez, 114, 125–26, 195, 217–18; Malaysian, 274; presidential, 202, 206; state and local, 111 n. 3, 125 n. 28, 215–16, 217. *See also* campaign contributions; voters
electoral rivalries, 205, 282; economic effects of, 195, 237–38, 242; Malaysian, 273, 274, 279; multiparty, 202 n. 23, 224, 231, 232, 261; two-party, 210–18
electrical machinery sector, 60, 252, 254
electricity, 180, 182. *See also* hydroelectric power
electronics sector, 157
elites: closed, 31, 261; conciliation with, 195; economic/social, 79, 199, 203; local, 36; Malaysian, 274 n. 4, 278; military, 199 n. 17; rural, 188–89
Ellner, Steven, 127, 199 n. 17, 213–14
El Partido del Pueblo (People's Party), 198. *See also* Acción Democrática
Elster, Jon, 293 n. 7
Embraer (Empresa Brasileira de Aeronáutica, S. A.), 300
employment, 243; creation of, 116–17, 234 n. 5, 301; industrial sectors, 173–74; informal, 20, 119; Malaysian, 274; productivity and, 15, 39; rural, 190 n. 6; state, 197, 242–43, 244. *See also* unemployment
enclave industries, 21, 36, 79, 103
endogenous growth theory, 24–25
energy sectors, 181–82
Engerman, Stanley, 31
engineers, 29, 29
England, 38, 292
ENP rates (effective number of parties), 216
entrepreneurships, 292
ethnic issues, Malaysia, 272–74, 278–81
exceptionalism, 18, 88, 285, 286, 287
exchange rates, 112; appreciation of, 38, 48–49; depreciation of, 47, 201 n. 20; evolution of, 42, 42–43; overvalued, 50, 52, 183, 235. *See also* foreign exchange; multiple exchange rate regime (RECADI)
exports: diversification of, 3, 103, 166 n. 42, 175, 178; growth of, 35, 112, 175; large-scale industries, 176, 183, 250; Malaysian, 281 n. 15; natural-resource-based, 78, 104; non-oil, 46–48, 175–76, 176, 182, 183; oil, 43, 44, 47 n. 11, 48, 91, 112, 175; promotion of, 50–51, 105; revenues from, 22, 40; state management of, 221, 247
externality theory, 136

factionalism, 211, 214–18; avoidance of, 209; in business associations, 247–48; economic effects of, 230–31, 237–40, 242, 265–66; increases in, 124–25, 205, 261; interparty, 123, 195, 225, 231–32, 246, 282; in Malaysia, 279 n. 10. *See also* electoral rivalries; fragmentation, political
failure complex (*fracasomania*), 111–12
Faja de Orinoco (Oil Belt of the Orinoco), 183
Faletto, Enzo, 197
family conglomerates, 143, 189, 228, 236, 281; challenges to, 117, 192, 212–13; competition among, 122; rent deployment and, 87, 101, 190, 231
Fearon, James, 79
FEDECAMERAS (national business chamber), 206, 213, 221 n. 53
Federal War (1859-63), 82–83
Field, Alexander James, 291
Fifth Plan of the Nation, 152, 170, 177–78, 180 n. 13, 181–82
Findlay, Ronald, 35 n. 1, 85
Finland, 85
firms: neoclassical theory of, 83 n. 7; number and size of, 75 n. 36, 248–50, 249, 250, 254
fiscal deficits, 65, 67
fiscal linkages, 21–24, 31, 32
fiscal policies, 17 n. 3, 43, 217
Fishlow, Albert, 152 n. 23
Fondo de Exportaciones (export credit fund, FINEXPO), 175
Fondo de Inversiones de Venezuela (Venezuelan Investment Fund, FIV), 70, 175
food processing sector, 115, 116
footwear sector, 56, 58, 59, 115, 116
foreign direct investment (FDI), 74 n. 35, 115, 120 n. 19, 183, 234–35
foreign exchange, 40, 45, 87 n. 11, 177, 238. *See also* exchange rates
Fourth Plan of the Nation, 152, 180 n. 13
fragmentation, political, 194–96, 214–18, 223, 225, 284, 287; economic effects of, 230–60. *See also* factionalism
France, 103, 225

Francés, Antonio, 75 n. 36, 122, 244
Free Trade Zone Act of 1971, 282 n. 17
fuel sectors, 35–37, 78, 116 n. 12
Fujimori, Alberto, 127
Furtado, Celso, 36, 104, 105

Garcìa Araujo, Mauricio, 213
Garrido, Alberto, 218 n. 43
gas prices, 123, 124, 180
Gelb, Alan, 22, 88
Generation of 1928, 196, 197
Gerschenkron, Alexander, 128, 264
Gil Yepes, José Antonio, 81, 192, 208 n. 31, 282
Gini coefficient, 30
Gómez, Juan Vicente, 87 n. 11, 187–91, 198
Gomez, Terrence, 279 n. 11
good governance paradigm, 24, 32, 75, 98, 101, 110, 114, 264, 297–98
Gott, Richard, 218 n. 43
governance/government, 4–5, 134, 136–38, 140–42; crisis in, 108–9, 121–29; economic development role, 40, 45, 48–49, 78–82, 139–40, 143; political economy of, 147, 300. *See also* state, the
Gramsci, Antonio, 143
Great Turnaround (*El Gran Viraje*), 111–14, 115
gross domestic product (GDP): banking share, 117, 201 n. 20, 240–41; debt share, 238; declines, 219 n. 46; growth rates, 16–17, *18*, 88, 234 n. 6; interest rate share, *19*; investment share, 18, 64 n. 23, *67*; manufacturing share, 41–42, 76, 171, *171*; non-manufacturing industrial share, 41; non-oil share, 40–42, *41*, 64 n. 22, 76, 86–87, 91, 178; oil-exporting economies, 99; per capita rates, 112, 114, 294; private sector share, 237; public sector share, 72 n. 31; service sector share, 41; social spending share, 222; tax share, 57 n. 17, 91
growth, economic, 133–34; in consolidated state, 187–88; constraints to, 35, 89–91, 102, 106, 143; corruption's effects on, 95, *96*, 97–99, *98*, 101, 106–7, 144–45, 226; curse of, 235 n. 9; declines in, 23, 24–31, 71, 226, 238, 262, 285–86; developmental/political aspects, 160–67; enhancement activities, 27–28, 108, 117–18; income related to, 45; industrialization and, 35; neoliberal reforms related to, 114–21; oil abundance's effects on, 47 n. 11, 95–101, *96*, *98*, 228, 230, 285; political economy of, 23, 271–301; preconditions for, 32, 167, 264; rapid, 86, 167, 266, 301; rates of, 114 n. 11; state's role in, 89–91, 102, 106, 140–43, 199, 219, 292–94; sustainable, 85, 166, 288, 294; theories of, 4, 24–25, 32, 261–67, 295. *See also* stagnation, economic
growth accelerations, 137–38, 167, 287, 296–97
growth collapse, 3–5; causes of, 81, 136, 168, 219 n. 46, 226–27, 261–66, 296–97, 301; economic explanations of, 35–76; political economy explanations of, 77–107, 109, 232
growth diagnostic approach, 297
Grupo Roraima, 113
grupos económicos. *See* family conglomerates
Grupo Táchira, 187
guerilla movements, 122, 127, 165 n. 39, 208, 210, 246
Guri Dam, 182
Guyana, 70

Haber, Stephen, 146 n. 15
Harrison, Anne, 74 n. 35
Hausmann, Ricardo, 38–39, 46–48, 65, 66, 67, 69, 70, 90 n. 16, 167, 247, 265–66, 287, 294
Hayek, Friedrich A., 162 n. 30
hay pa' todo (expression), 81 n. 4
head-count ratios, 20 n. 7
health indicators, 25–26, *26*, 28, *28*, 70, 199
heavy industries. *See* big-push natural-resource-based heavy industrialization; large-scale, heavy industrial sectors
Herrera Campins, Luis, 234–36
high-income countries, 167
Hikino, Takashi, 52
Hillman, Richard, 262 n. 26
Hirschman, Albert O., 49 n. 14, 72 n. 30, 111, 147, 148 n. 17
Hirshleifer, Jack, 140, 142
HIV/AIDS, 299
Hobbes, Thomas, 208–9, 227
Hydrocarbons Law of 1943, 191
hydroelectric power, 3, 70, 180, 182, 183, 301

ICOR (incremental capital output ratio), 69, 70, 279–80
immigration, 29, 220 n. 49
imports: costs of, 177, 238 n. 11; declines in, 174–75, 185; demand for, 289; licenses for, 114 n. 9; restrictions on, 42, 192
import-substitution industrialization (ISI): exhaustion of, 73–74; implementation of, 48, 103, 289–90; inefficiencies of, 50, 76; opposition to, 49 n. 14; politics of, 246, 286;

Index 333

risks in, 104; stages of, 147, 160, 170–76, *184*; state-led, 3, 86. *See also* advanced import-substitution industrialization; big-push natural-resource-based heavy industrialization; early/easy-stage import-substitution industrialization
incentives, 31, 102, 133–34, 137, 139–40, 290–91, 297–98
inclusiveness, problem of, 265
income: declines in, 20–21, 48, 108; distribution of, 5, 29, 118, *118,* 238–39, 262, 278–79; growth rates related to, 45; inequalities in, 4, *30,* 30–31, 108, 119, 125, 126, 127, 219 n. 45, 222 n. 57; national, 118, *118,* 234 n. 2; per capita, 21, 287; unearned, 79. *See also* wages
Income Tax Law of 1942, 191
India, 97, 254 n. 20
Indonesia, 47, 97, 247
industrial concentration, 73–76, *74,* 101
industrial deepening, 289–90. *See also* capital goods sector; intermediate capital-intensive sectors
industrialization: capacity utilization in, 36, 251–54; capital-intensive, 105, 147; initial conditions for, 26; investment in, 65, 70–71, 151 n. 22, 181; late-developing countries, 39, 75, 93, 128, 133–68; natural-resource-based, 35–36, 152; oil revenues related to, 21–24; periodization of, 169–85; political economy of, 133–68; problems facing, 38, 136; stagnation in, 56–57; state's role in, 3, 40, 70, 143, 153, 177, 190, 231. *See also* advanced import-substitution industrialization; big-push natural-resource-based heavy industrialization; import-substitution industrialization
industrial policies: implementation of, 146–48, 287; ineffective, 252, 261, 289; post-1968, 242–60; productivity and, 58 n. 18; selective, 109, 248–49; state-led, 88, 174; successful, 134–37, 264
industrial sectors, 20, 104, 178. *See also* large-scale, heavy industrial sectors; manufacturing sectors; traditional industries; *and individual manufacturing sectors*
infant industries, 72, 158; Malaysian, 284; overdiversification in, 254; polices regarding, 58 n. 18; protection of, 105, 113, 146, 166, 228, 230; rents in, 281 n. 13; subsidies for, 72, 94, 145, 158, 165, 166, 230, 247
infant mortality rates, 28, *28*

inflation: increase in, 65, 112, 124, 127, 222–23, 236, 238–39, 289, 301; rates of, 42, 239, *239*. *See also* currency
infrastructure, investment in, 40 n. 6, 65, 150, 244, *245*
initial conditions, 5, 15, 24–31, 32, 126–29
institutions: economic role of, 102, 144, 145–46, 290–91; effective, 137–38; governance and, 4–5, 141, 299–300; politics of, 90, 148, 168, 263, 298; suboptimal, 288–89
Instituto de Estudios Superiores de Administración (Institute of Higher Administration Studies, IESA), 113
insurrections. *See* coup attempts
Inter-American Development Bank, 113
interest groups, 9 n. 1; competition among, 214; conservative, 201; currency devaluation and, 48–49; economic effects of, 90–91, 102, 191; pressure from, 112 n.6, 142 n. 9, 144, 200, 291–92; rent-seeking, 49, 84, 231; state bargaining with, 80, 82, 222, 228
interest rates, 117, 236, 251, 265–66
intermediate capital-intensive sectors: capital intensity measurements, *54,* 56; growth rates, 63, 151–52, 289; import coefficients, 175; ISI and, 148 n. 18, 250; productivity levels, 60, 287; profitability levels, 60
International Monetary Fund (IMF), 113
interventionism, state, 68–73, 112 n. 6, 191; corruption and, 88–89, 91–92, 93; economic development and, 76, 82, 108, 290, 292; growth effects of, 103, 106, 133–35; political economy of, 147, 300
investment: ceded, 241 n. 14; corruption's effects on, 95; declines in, 67, 77, 118, 201 n. 20, 227; effective, 63–64, 140 n. 5, 148 n. 17, 265; foreign, 74 n. 35, 115, 120 n. 19, 183, 234–35; of human capital, 24, 25–26; industrial sectors, 149, 158, 166, 167, 178–79, 181; oil booms' effects on, 45; private-sector, 214, 236, 240–41; public, 68–73, 76, 230, 238, 243–44; rates of, 15, 16–19, *19,* 67, 266; risks in, 123, 240–41; state-controlled, 143, 148–49, 153, 155, 156, 179, 184, 213. *See also* research and development; *and under specific industry sectors*
iron sector: capacity utilization in, 252, 254; exports, 47, 176; growth rates, 61–62; nationalization of, 177–78, 181–82; productivity rates, 62
Izaguirre, Maritza, 221 n. 53

Japan, 292
joint ventures, oil industry, 183
Jomo, Kwame Sundaram, 274 n. 4, 279 n. 11
Jones, Leroy, 155
jurisdiction, authority vs., 80–81

Kähkönen, Satu, 102
Kaldor, Nicholas, 35, 39, 72
Karl, Terry Lynn, 73, 79, 80–81, 82, 85, 86, 88, 98, 206 n. 27, 258
Katz, Jorge, 115
Katzenstein, Peter, 159 n. 27
Kelly de Escobar, Janet, 244
Kenny, Charles, 295
Keynes, John Maynard, 37
Khan, Mushtaq, 93–94, 98, 273, 276, 276 n. 8, 280, 282
Kohli, Atul, 161, 163, 166 n.41
Kokko, Ari, 85
Korea, 72, 189 n. 5
Kornblith, Miriam, 123 n. 24, 189, 205
Krueger, Anne, 89

labor: conflicts in, *204*, 211, *275*; costs of, 48, 251; demand for, 19, 39; legislation affecting, 251; skilled, 28–29, *29*
labor-intensive sectors: capital intensity measurements, *53*, 56–57; declines in, 250; growth rates, 63, 115, 159; productivity levels, 58; profitability levels, 56–57; wage levels, 56, 57
labor productivity growth, 15, 31, 255, 258; manufacturing sectors, 25, *51*, 51–52, 58, 60–61, 228, 258–60, *259*
labor unions: capitalists vs., 238–39; democracy in, 262; Malaysian, 274; mobilization of, 196, 199, 203; power of, 228, 251, *275*, 276; wage demands, 234 n. 5; weakening of, 119, 190, 220, 258. *See also* bargaining; strikes; trade unions
La Causa R (The Radical Cause, political party), 216, 217
La Conferderación de Trabajadores de Venezola (CTV), 203 n. 24
La Gran Venezuela, 177
Laitin, David, 79
land, ownership of, 188 n. 3, 192–93, 196
landlord class, 190, 199 n. 17
land reform, 30, 203
Lane, Philip, 90–91
large-scale, heavy industrial sectors, 172–74, 239; capacity utilization, 252–54;
employment, 276, *277*; exports, 176, 183; import coefficients, 175; investment in, 65, 75, 167; number of firms in, 248–49, 251; production strategies, 147, 287; productivity levels, 166; public enterprises, 177–78. *See also* big-push natural-resource-based heavy industrialization
Larrazábal, Wolfgang, 207 n. 29
late-developing countries: corruption in, 101, 263; growth rates, 72, *160*, 286; industrialization, 39, 75, 93, 128, 133–68; investment trends, 70, 92–93, 155; ISI in, 148–59; mineral-abundant, 95, 97; political economy of, 144, 261–64, 288–90; productivity levels, 58–59, 63; profitability levels, 58, 62; rent deployment, 144–46; state formation, 85; technology gaps, 40; wage levels, 52
Latin America: capital flight, 233, *233*; debt crises, 155 n. 25, 260 n. 25; deindustrialization, 40 n. 7; economic development, 36; economic liberalization, 121, 166 n. 40; education indicators, 28 n. 17; growth rates, 17–18, 88, 99, 101, 167, 258, 286–88, 294, 296; immigration rates, 29; income distribution, 29, 30; inflation rates, *239*; ISI in, 152 n. 23, 226 n. 1, 289–90; labor unions, 274; manufacturing competitiveness in, 254 n. 20; national savings levels, 24–25; productivity levels, 62; state consolidation, 83; tariffs in, 177, 191 n. 8; wealth concentration, 31. *See also individual countries*
laws. *See individual laws*
leaders, 80, 83–84, 144–45. *See also* decision-makers; *and individual political figures*
learning, 40; cost of, 152, 156, 254; by doing, 72, 134–35, 157. *See also* rents, learning
Lecuna, Vicente, 49 n. 15
Lederman, Daniel, 35 n. 2
leftists, 127, 207 n. 30, 209, 212
legitimacy: crises of, 4, 20–21, 123 n. 25, 125–26, 163 n. 33, 199 n. 17, 262; political, 30–31, 187 n. 2; state, 79, 83, 142–43, 265, 283, 287
Levi, Margaret, 90, 291 n. 6
Levine, Daniel, 189, 199, 200, 206 n. 27, 209
licenses, 73 n. 32, 177, 248–49, 265
life expectancy rates, 28
light-industry sectors, 252, 254
limited access orders, 144, 209
loans, 117, 240–41. *See also* interest rates
López Contreras, Eleazar, 188, 190

López Maya, Margarita, 122 n. 22, 218, 219 n. 46
Los Muchachos (political splinter group), 212
lower classes, 282
low-income countries, 4, 99, 101, 102, 128, 167, 296–97
low-technology sectors, 115, 151 n. 21
Lucas, Robert, 4, 24, 25, 50
Lundhal, Mats, 35 n. 1, 85
Lusinchi, Jaime, 112, 211, 237–38

Machiavelli, Nicolò, 84 n. 8
macroeconomic policies, 36 n. 3, 49–52, 236, 300; failures in, 230, 232–42, 266–67, 285; management, 190, 194, 239, 261; populist, 31, 48 n. 12, 112; stabilizing, 145–46, 209, 228, 289; sustainable, 164 n. 36, 165
Macroeconomic Stabilization Fund, 124
Maddison, Angus, 88
Mahdavy, Hossein, 81
Mahon, James, 48
Maignon, Thais, 205
Malay Chinese Association, 273–74
Malaysia, 85; capital intensity measurements, 52–56; comparisons with, 5, 65–68, 271–84, 290, 299; corruption rates, 97; education indicators, 26, 27; ethnic issues, 272–74, 278–81; growth rates, 64, 283–84; health indicators, 28; income distribution, 30, 262; industrialization, 70, 157, 166 n. 41, 225, 239; national savings rates, 66, 66, 67; operating surpluses, 60, 62; productivity levels, 59, 61, 62, 63, 258, 282–83; profitability levels, 57; skilled labor, 29; strikes, 275–76; taxes, 273 n. 3; wage levels, 57, 59
Malaysian Indian Congress, 273–74
Maloney, William F., 35 n. 2
manufacturing sectors: capacity utilization in, 252–54, 253; capital intensity measurements, 5, 52–56, 53, 54, 55; competition in, 39, 166 n. 42; declines in growth, 3–5, 24–25, 31, 36, 40, 48–49, 52, 68–76, 115–17, 230, 232, 251, 255; demand for, 35; development of, 194, 293; economic liberalization policies, 191–92; export credits, 175; GDP share, 41–42, 76, 171, 171; growth rates, 15, 16–19, 18, 29, 31, 39, 43, 51–52, 56–57, 57, 87, 104–5, 116, 151, 166, 227–30, 259, 260; investment trends, 15, 16–19, 43, 44, 63–64, 69, 76, 178, 181, 182; number and scale of firms, 75 n. 36, 248–50, 249, 250, 254; private-public sector comparisons, 70–71; productivity levels, 105, 174, 174, 183, 184, 238, 255, 258, 260, 266, 267; profitability levels 52-64, 53, 54, 55; public enterprises, 178; shares of, 172; structure of, 117; wage levels, 48–49, 52; weaknesses in, 104–5. *See also individual product sectors*
Manzano, Osmel, 235 n. 9
Márquez, Gustavo, 20, 48
Martz, John, 197 n. 14, 203
Marx, Karl, 90 n. 16, 139 n. 4, 143
Marxism, 36, 138, 208 n. 31
Mauro, Paulo, 95
Mayorbe, José Antonio, 38
Maza Zavala, Domingo, 36
McCoy, Jennifer, 108, 261
media sector: conflicts within, 121 n. 20, 122; coverage of scandals, 125, 217; opposition to Chávez, 219 n. 44, 220
Medina Angarita, Isaías, 103, 188
metals sector, 152, 170, 173, 183, 252, 254
Mexico: exports, 47; growth rates, 88, 247; income inequalities, 126; neoliberal reforms, 127; oil windfalls, 91; pact-making, 210; property rights, 140 n. 6
microeconomic policies, 50, 68–73, 242–60
middle class: clientelism in, 205–6, 243, 276; concessions to, 236; Malaysian, 272–74, 276; mobilization of, 196, 202, 203; opposition to Chávez, 219 n. 44; urban, 196–97, 199
middle-income countries, 128, 262, 287; growth rates, 17–18, 99, 101, 167, 267, 296–97. *See also* Malaysia
military, the, 127, 187–90, 198, 289
Mill, John Stuart, 140, 142
mineral-abundant economies, 78–79, 85, 95–101, 96, 98, 151, 209–10, 271 n. 2, 300–301. *See also* natural resources
mineral resources, 35–37, 78, 300–301
minerals sector, 116 n. 12, 183, 235 n. 9
Minimum Program of Government, 203
ministries, government, 104, 106, 128, 193 n. 10; Planning, 111; weakening of, 206, 217, 221, 223, 232. *See also* cabinets, presidential
Miquilena, Luis, 220 n. 47
mobilization, political, 141, 195–97, 202 n. 21, 231, 276, 278
Mohamad, Mahathir bin, 280, 281 n. 15
Molina, José, 216 n. 39
Monaldi, Francisco, 214, 216 n. 42
monetary policies, 43, 240–41. *See also* currency; exchange rates
monopolies, 73 n. 32, 89, 129
Moore, Barrington, 147, 293
Moreno, María Antonia, 91, 198

Movimient Bolivariano Revolucionario 200 (Bolivarian Revolutionary Movement, MBR-200), 218–19
Movimiento al Socialismo (Movement Toward Socialism, MAS), 215–16, 217, 220 n. 47
Movimiento Quinta República (Fifth Republican Movement, MVR), 217, 218
mudslides, 205 n. 26
multinational corporations, 17 n. 3, 36, 74 n. 35, 95, 162; in Malaysia, 281 n. 15, 282; oil-related, 87, 188, 191, 196
multiple exchange rate regime (RECADI), 175, 177, 237, 240
murder rates, 221 n. 50
Myers, David J., 206 n. 27
Myrdal, Gunnar, 92, 295

Naím, Moisés, 73, 75 n. 36, 81, 109, 119 n. 15, 121, 122
National Civic Crusade, 208 n. 31
National Coffee Federation, 164
nationalism, 159 n. 27
nationalization. *See* oil industry, nationalization of
natural-resource-based industrialization, 75, 115, 180, 182–83, 301. *See also* big-push natural-resource-based heavy industrialization
natural resources: economic development related to, 35–76; Malaysian, 280, 282; rents for, 35, 37, 46, 79, 88. *See also* resources; *and specific natural resources*
Navarro, Juan Carlos, 113
Neary, Peter J., 40
neoliberalism: reforms based on, 111 n. 5, 113, 114–26, 215, 218; views of state interventions, 68–76, 92, 136; weaknesses of, 134, 159, 329
Netherlands, the, 38, 85
Neuhoser, Kevin, 210
New Economic Policy (NEP, Malaysia), 278, 279–80
new institutional economics (NIE), 139, 140 n. 5, 141, 146
newly industrialized countries (NICS), 152 n. 23
Nigeria, 91, 97
nondurable goods sectors, 148–49
nonelectrical machinery sector, 60, 252, 254
non-ferrous metals sector, 47, 61–62, 181–82, 254
non-manufacturing industrial sector, 41
non-mineral sectors, 36

non-mining sectors, 22
non-oil economy: declines in, 3–5, 20, 24–25, 31, 39, 43, 48, 69, 114–15, 230, 232, 255; development of, 293; exports, 46–48, 175–76, 176, 182, 183; GDP share, 40–42, 41, 64 n. 22, 76, 86–87, 91, 178; growth rates, 15, 16–19, 17, 19, 31, 39, 46–48, 64 n. 22, 86–87, 227–30, 251; investment rates, 15, 16–19, 39, 64 n. 22, 67, 70–71, 118, 178, 240–41; oil prices related to, 23, 23, 31; productivity levels, 23–24, 46–48, 227–30, 260, 266, 267; state management of, 40
non-tariff barriers, 113–14, 177, 191, 247, 251
North, Douglass, 94, 102, 143–44, 208–9, 209 n. 33
North America, 31
Norway, 85; capital intensity measurements, 52–56; oil discoveries, 82; operating surpluses, 60, 62; productivity levels, 58, 60, 61, 62–63; profitability levels, 57, 58, 61 n. 20, 62; wage levels, 56, 57, 58, 59

Ocampo, José Antonio, 164, 165
O'Donnell, Guillermo, 147, 289–90
oil: abundance of, 77, 86, 89, 106, 133, 228, 230, 285; "dance of concessions" 188; dependence on, 79, 80–81, 301; discoveries of, 3, 21–22, 38, 82, 87 n. 11, 181 n. 14, 188; exploration for, 17 n. 3, 178; exports of, 43, 44, 47 n. 11, 48, 91, 112, 175; rents related to, 21, 37, 80, 85, 88–89, 112, 181
oil booms: corruption and, 86, 95, 96, 97–99, 98, 101, 106; employment related to, 20; exchange rates and, 4, 22–24, 42; growth related to, 166 n. 42, 214; investments related to, 43, 45, 68, 76; premature tertiarization and, 40–41; problems related to, 39, 64; state formation and, 82, 85; voracity effect of, 91; wage increases during, 51
oil-exporting countries, growth rates, 98–99, 100, 101
oil industry: enclave nature of, 103; growth rates, 38, 190, 192, 196; investment in, 181, 182, 183, 201 n. 20; nationalization of, 17 n. 3, 22, 71, 177–79, 182, 191 n. 7, 214; strikes against, 219, 220 n. 49, 222 n. 58, 241
oil prices: booms in, 19, 22, 65 n. 24, 107, 234–35; declines in, 201 n. 20; fluctuations in, 16 n. 2; non-oil growth related to, 23, 23, 31
oil production/refining, 3, 17, 36–37, 194; growth rates, 178, 180–81, 182, 191, 222 n. 58

Index 337

oil revenues: declines in, 32, 47; growth generated by, 22–23, 47 n. 11, 64–68, 166 n. 41; increases in, 235, 266, 301; industrialization related to, 21–24, 40–41, 43, 76, 171 n. 2; instability in, 64–68; management of, 5, 21–22; from multinational corporations, 191; per capita, 86–87, 87; political effects of, 196–97; social programs funded by, 183, 222, 301; state management of, 5, 24, 32, 43, 76, 91; windfalls, 4, 16, 65, 148 n. 17, 184, 219, 230, 301
oil stabilization fund, 65 n. 24
oligarchies, 73 n. 32, 75, 121, 188–89, 213
Olson, Mancur, 83–84, 102, 291–92
OPEC (Organization of Petroleum Exporting Countries), 17 n. 3
open economy models, 49–52
operating surpluses, 52, 56–57, 60. *See also* profitability levels
orimulsion (fuel), 181 n. 14
Ortiz, Nelson, 121 n. 20, 221
overdiversification, 75 n. 36, 248–49, 252 n. 19, 254. *See also* diversification
overshooting phenomenon, 45, 65 n. 24

Pacific region, 17–18, 101
pact-making, 123–24, 209–10, 214, 223, 230–31, 273. *See also* democracy, pacted
Pacto de Nueva York, 202 n. 23
Pact of Punto Fijo, 200, 202, 203, 206–7, 208
Palma, José Gabriel, 36, 40 n. 7, 107, 159 n. 27, 179 n. 12, 285
paradox of plenty, 133
path dependency theory, 129
patronage: clientelist, 243, 251; contestations over, 212–13; employment-related, 242–43, 244; in Malaysia, 276, 280; oil-related, 112, 190, 198, 209; political, 206–7, 251, 291; state-controlled, 142–44, 202, 231, 265
patron-client networks, 79, 189, 280
peasants, 196, 199, 200, 202, 210, 274
Penfold, Michael, 214, 216 n. 42
Penfold-Becerra, Michael, 217
Pérez, Carlos Andres: campaign contributions, 120, 125; economic policies of, 111–14, 122–23, 177–78, 180 n. 13, 212–13, 236; failures of, 125, 127; as political outsider, 215
Pérez Alfonso, Juan Pablo, 81–82
Pérez Jimémez, Marcos: authoritarianism of, 188, 200, 208 n. 31, 1805; economic policies, 87 n. 11, 193, 205; fall of, 201 n. 20
Peru, 88, 126

petrochemicals sector, 3, 36, 69, 180, 183, 192, 212
Petroleum of Venezuela (PDVSA), 71, 178, 181 n. 14, 222, 237
petro-states, 79, 80–82, 84, 88–89, 99
Philippines, the, 30, 97
Pigou, Arthur, 90
Piñango, Ramon, 81
plantations, 79 n. 1
PODEMOS (political party), 220 n. 47
polarization, political, 15 n. 1, 201, 218–23, 231, 232; under Chávez, 195–96; economic effects of, 242, 284, 286–87; income distribution and, 119–20, 124, 127; patronage and, 213. *See also* factionalism; fragmentation, political
policies: failures in, 49, 65–68, 101–6, 238, 249, 288–89; growth-related, 5, 77; political nature of, 168; resource-based, 86; switches in, 126–27, 215, 231–32. *See also* industrial policies
political organizations/parties, 208 n. 31, 261–63; centralized, 194, 200–203, 205–10, 258, 266; deinstitutionalization of, 220, 231; fragmentation in, 194–96, 214–18, 223, 230–60, 284, 287; fundamental, 202 n. 21; patronage through, 206; periodization of, 224; power of, 228; weakening of, 15 n. 1, 126, 128, 189–90, 222, 266, 286. *See also* consolidated states with centralized political organizations; consolidated states with fragmented political organizations; fragmentation, political
political settlements, 9 n. 1; development strategies' compatibility with, 137–38, 147, 160–67, 168, 227–30, 229, 232, 255, 256–57, 261, 266–67, 285–301; periodization of, 186–225, 261; populist-clientelist, 194–95; rent deployment related to, 271
politics: antiparty, 127, 195–96, 217, 218–23; cooperative, 159 n. 27, 205; economics related to, 49, 138, 261–67, 285–301; effects on state capacity, 110, 138, 300; electoral, 198; ethnic, 278; instability in, 4, 19, 21, 108–29; Malaysian, 273, 278, 281–82; patrimonial, 264–65. *See also* fragmentation, political; polarization, political; rents, politics of; rents, centralized deployment, politics of
politics of privilege, 122 n. 21, 133, 142, 143
Polo Patriótica (Patriotic Pole, PP), 218
poor countries, 102. *See also* low-income countries

population. *See* rural-urban migration; urbanization, rapid
populism, 119 n. 14, 186, 201, 286; macro-management policies related to, 31, 48 n. 12, 112; Malaysian, 272, 274; mobilization of, 195, 197–98. *See also* clientelism, populist
POSCO (South Korean steel company), 62 n. 21
poverty, 20, 205
power, 83–84, 142 n. 9, 228, 238–39, 274–76
Prebisch, Raul, 179
premature tertiarization, 40–41
prices, 17 n. 3, 58 n. 18, 238 n. 11. *See also* oil prices
Primera Justicia (political party), 216 n. 40
primitive accumulation: divisiveness in, 211, 263; lack of control over, 238; politics of, 142–44, 188, 231, 234–35; rents and, 149, 188
Pritchett, Lant, 167, 287
private sector: capacity, 200, 250; conflicts within, 122, 223; GDP share, 237; infant industries, 230; investment in, 18–19, 118, 213, 234–36, 240–41; national savings rates, 66; oil windfall management, 24; productivity levels, 110, 183, *184*; public enterprises compared to, 70–71, 143; stifling of, 73; weaknesses in, 189, 221, 249
privatization, 73, 86, 110, 113, 182–83, 280
production: capacity problems, 222; capital intensity measures, 69; costs of, 58; distribution *vs.*, 142; diversification of, 3, 103; energy requirements, 150; fragmentation of, 75 n. 36; implementation of, 146–48; investment in, 116, 179; Marxist theory of, 138; private sector, 110; public enterprise, 212; state's role in, 93, 178, 221
productivity growth: declines in, 32, 46, 63–64, 76, 230, 232, 238, 255, 267, 285–86; enhancement of, 39, 112, 117–18, 157; levels of, 15, 52–64, *53*, *54*, *55*; Malaysian, 279–80, 282–83; of rent deployment, 145; state hindrances to, 106. *See also* labor productivity growth; *and under specific industry sectors*
profitability levels, manufacturing sectors, 52–64, *53*, *54*, *55*
profits, 72, 118, 158 n. 26
property rights: appropriation of, 83, 84–85; confiscation of, 264; political economy of, 139–44; politics of, 139–44, 168, 298; security of, 77, 92, 93, 106, 146, 214 n. 31, 227; structures of, 136–37

Proposal to the Country *(Proposición al país)*, 113
protectionism, 291–92; blanket, 247, 281; industrial, 49, 105, 115, 150, 166, 250–51; manufacturing slowdown related to, 73–76; productivity and, 58 n. 18, 117; rents as, 177, 231; trade-related, 12–22, 50, 121–22, 193–94. *See also* non-tariff barriers; tariffs
Proyecto Venezuela (political party), 216 n. 40, 217
public enterprises: budgets for, 66; economic performance, 72–73, 244; employment, 276; exports, 182; investment rates, 18–19, 76, 234–36; ISI and, 150, 153, 155; manufacturing sectors, 68, 177–78, 180; private sector compared to, 70–71, 143; productivity levels, 110, 212; soft budget constraint, 69, 134 n. 1. *See also* state-owned enterprises
Public Property Protection Law of 1982, 106
public spending, 91, 193 n. 9, 193 n. 10, 243–44, 244, 301. *See also* social programs
Purroy, Miguel Ignacio, 73
Putzel, James, 187 n. 2, 291 n. 6

Rangel, Domingo Alberto, 30, 49, 143, 151
real estate sector, 192–93
real money balances, 240–41
RECADI. *See* multiple exchange rate regime
Reciprocal Trade Agreement, 193–94, 247
reforms: economic, 217; land, 30, 203; neoliberal, 111 n. 5, 113, 114–26, 215, 218; political, 110, 111 n. 3. *See also* economic liberalization policies; trade liberalization policies
regimes: evolution of, 229; exclusionary, 261; instability of, 21; legitimacy of, 79, 187 n. 2; Malaysian, 273; threats to, 209–10, 224, 231, 246, 271; types of, 133–34, *256–57*
regulations: anticompetitive, 291; industry, 75; labor, 121; state, 9 n. 1, 73 n. 32, 128–29, 242
rentier state models, 77–82, 95, 110, 133–34, 144; weaknesses of, 88–89, 97–99, 106–7, 146, 258
rents: costs of, 92; cronyism's effects on, 188; definitions of, 89–90; deployment of, 186, 194, 209, 267, 286; deregulation's effects on, 122; divisible, 150; learning, 75, 92–93, 94, 149, 280; Malaysian use of, 276, 278–83; mineral, 82–84, 99; oil, 21, 37, 80, 85, 88–89, 112, 181; point, 79; politics of, 83–84, 142, 168, 202, 207, 208, 230–31; resource, 35, 37, 46, 81,

Index 339

88, 133; state control of, 105, 107, 156, 158, 246, 249; state-created, 78, 93, 109, 139–44, 176–77; trade protection, 194
rents, centralized deployment: benefits of, 134–36; corruption and, 101, 188–89; growth related to, 151, 184; politics of, 95, 141–42, 144–46, 231, 255, 271–72, 280–85
rent-seeking: corruption and, 88–95, 97–99, 101, 106–7, 114, 226; costs of, 145, 280–81; growth-restricting, 49, 133–34, 144; net effects of, 94–95; politics of, 75, 202
rent-seeking models, 77–82, 108, 110, 112 n. 6, 114; weaknesses of, 102, 128–29, 146, 189, 260, 283
research and development (R&D), 150, 156–57, 254–55
resource allocation: for advanced ISI, 263; corruption and, 91–92, 105, 108; effects on governance, 137, 139–40; efficiency of, 258–59; opportunity costs to, 142; state's role in, 89, 110, 128–29, 231, 263; trade protections and, 194
resource curse models, 23, 35–36, 77, 80, 101, 300–301. *See also* Dutch Disease models
resources: abundance of, 40, 84–85, 86, 88–89, 150, 235 n. 9; competition over, 142, 161 n. 29; constraints on, 128; inefficient management of, 23, 114; mobilization of, 21, 24, 190; point, 79, 80; rents for, 46, 81, 133; revenues from, 85; scarcity of, 136–37, 142; waste of, 77. *See also* natural resources
Reversion Law of 1960 (Ley de Reversion), 17 n. 3
Rey, Juan Carlos, 195, 201
Rigobón, Roberto, 180, 235 n. 9
riots *(Caracazo)*, 122–23
risks: advanced-stage ISI, 152, 157; investment, 123, 240–41
Rodríguez, Francisco, 23–24, 45, 46–48, 91, 198, 247, 260
Rodríguez, Gumersindo, 103, 212
Rodríguez, Miguel A., 46, 111, 123, 233, 238
Rodríguez Araque, Ali, 221
Rodrik, Dani, 50 n. 16, 135, 153, 167, 287, 294
Romer, Paul, 4, 32
Rosenstein-Rodan, Paul, 179
Ross, Michael, 84
Rousseau, Jean-Jacques, 139 n. 4
Rowthorn, Bob, 238
Rueschemeyer, Dietrich, 147
rules of the game, 140, 213–14, 237–38
rural-urban migration, 196, 197, 205 n. 26

Sachs, Jeffrey, 35 n. 2, 45, 46
Sakong, Il, 155
savings, national, 24–25, 25, 66, 66–67, 238
scandals, corruption, 101, 123, 124, 214, 234; media coverage of, 125, 217
Scandinavia, developmental states in, 31
Schumpeter, Joseph, 92
scientists, 28–29, 29
Segarra, Nelson, 69
service sector, 20, 39 n. 5, 41, 197
Shleifer, Andrei, 101
Short, Robert, 72
short-termism, curse of, 64–68
Shweinitz, Karl de, 147
Sidel, John, 150 n. 20
SIDOR (Siderúrgica de Orinoco, S.A., state steel company), 71, 73, 170, 180, 181–82
Simon, Herbert, 291 n. 6
Singapore, relationship with Malaysia, 278
single-issue theory, 136
Sixth Plan of the Nation, 182
small-scale industrial sectors, 150, 156–59, 165, 170, 185, 254, 255
Smith, Adam, 90 n. 16
Social Christian Party, 207 n. 28
social exclusion, 140, 261–62, 265
social interaction, process of, 139
socialism, 28 n. 17, 208 n. 31, 215–16, 217, 218–19, 220 n. 47, 223 n. 59
Social Missions, 222
social programs, 65 n. 24, 183, 205, 220–22, 230, 301
society. *See* state-society relations
soft budget constraint, 69, 134 n. 1
Sokoloff, Kenneth, 31
Solow, Robert H., 4
South Africa, 97, 298–99
South Korea: capital intensity measurements, 52–56; comparisons with, 5, 276, 300; corruption rates, 97; education indicators, 27; government intervention, 109; health indicators, 26, 28; industrialization in, 225, 239; investment rates, 64, 70; manufacturing development, 155; operating surpluses, 60, 62; private-public sector comparisons, 71; productivity levels, 58–59, 61, 62–63, 258; profitability levels, 57; rent-seeking, 93; wage levels, 57, 59
South/Southeast Asia, 17–18, 101, 274, 276
Soviet bloc countries, 28 n. 17, 99 n. 21, 162, 179
sowing the oil, 3, 21–24, 103
stabilization programs, 65 n. 24, 289

stagnation, economic, 4, 74, 86, 106; causes of, 63–64, 68, 114, 129, 146; corruption during, 101; effects of, 19–21, 119, 127; industrial, 56–57; investment related to, 117, 118; oil abundance related to, 86, 285; political crises and, 262; prolonged, 32, 76, 102, 114, 133, 226. *See also* growth collapse

staples thesis, 35, 77, 79

state, the: capacity of, 82, 102, 108–29, 133–34, 145, 147, 265, 298–300; corruption in, 87, 101, 105–6, 136; economic role of, 89–91, 102, 106, 118, 129, 140–43, 149, 214, 219, 263–64, 280, 292–94; factionalism in, 151; failures of, 5, 76, 83, 264–65; formation of, 82, 85, 137, 191, 208; fragmentation of, 186, 221, 266; industrialization role of, 3, 40, 70, 143, 153, 177, 190, 231; investment monitoring by, 143, 148–49, 153, 155, 156, 179, 184, 213; ISI role of, 148–49; legitimacy of, 79, 83, 142–43, 265, 267, 283; motives of, 144–45; oil windfall management, 5, 24, 32, 43, 76, 91; patronage from, 142–44, 202, 231, 265; planning by, 103–4; productivity and, 106, 145; protectionist policies of, 73–76; rents created by, 78, 93, 109, 139–44, 176–77; resource allocation by, 89, 110, 128–29, 231, 263; weakening of, 80–82, 120–29, 160, 163 n. 33, 219, 227, 237, 281 n. 13, 297. *See also* bailouts; consolidated states with centralized political organizations; consolidated states with fragmented political organizations; interventionism, state; petro-states; public spending; rentier state models; subsidies, state deployment of

state-business relationships, 158–59, 292; challenges to, 156, 232; politics of, 247–48, 261–62; weakening of, 121–22, 195, 220

state-owned enterprises (SOEs): capacity utilization in, 252; control of, 104; corruption and, 101; exports, 182, 250; industrialization and, 3, 213; investment in, 155, 243; Malaysian, 279–80; soft budget constraint, 69, 134 n. 1. *See also* public enterprises

state-society relationships, 195–96, 230

stationary bandit model, 84

steel sector: big-push programas, 69, 173; capacity utilization, 252, 254; development of, 180, 181–83; exports, 175, 176, 182, 183, 250; growth rates, 47, 61; nationalization of, 3, 71; productivity levels, 62, 176 n. 7; profitability levels, 61 n. 20; public enterprises, 70

Stiglitz, Joseph, 51, 71, 140

strikes, 19, 236, 275–76; increases in, 199, 203, 205; oil-sector, 219, 220 n. 49, 222 n. 58, 241

structuralist theory, 35–36, 179–80

sub-Saharan Africa, 17–18, 101, 167, 296–97

subsidies: for disadvantaged groups, 31; industrial, 93, 145, 150, 159, 165, 166; rents as, 231; state deployment of, 81, 94, 109, 142–43, 156, 158, 213, 248–50, 265, 286; targeted, 134, 157

Svensson, Jakob, 97 n. 19

Sweden, 85, 292

Taiwan: corruption rates, 97; development in, 155, 189 n. 5; government intervention, 109; industrialization in, 157, 225; private-public sector comparisons, 71–72; rent-seeking, 93

Tanzania, 299

tariffs, 42, 58 n. 18, 177, 213, 247; concessions regarding, 191–94; Malaysian, 281; manufacturing, 149, 191–94; reductions in, 49 n. 14, 113, 251

taxation, 148 n. 17; evasion of, 237; excise, 17 n. 3; income, 57 n. 17, 124, 191, 273 n. 3, 281 n. 13; oil, 196; ratio to GDP, 91; reduced, 79, 80, 84

technology: acquisition of, 70, 134; high-value-added, 156–57, 182; industrial, 28, 148–49, 178, 265; lack of, 36, 40; state development of, 146–47, 156–57, 182; weaknesses in, 254–55

technology-intensive sectors, 115, 116, 167, 287, 301

telecommunications sector, 86, 182–83

textiles sector, 56–57, 58, 59, 183, 191 n. 8, 194, 248 n. 18, 250

Thailand: corruption rates, 97; development strategy, 164, 165; education indicators, 26, 27; growth rates, 97, 149, 165 n. 39; health indicators, 28; industrialization, 150 n. 20, 166 n. 41; national savings levels, 24–25; politics, 282; productivity levels, 258; skilled labor, 29

Theorem of Second Best, 294

Thorp, Rosemary, 81, 83, 164, 164 n. 37, 247

Tinoco, Pedro, 104, 113 n. 8, 120 n. 18

Tornell, Aaron, 90–91

Torres, Gerver, 69–70

total factor productivity (TFP), 258–60

trade-goods sector, 36 n. 3, 39, 43, 50

trade liberalization policies, 49 n. 14, 86, 117, 176–77, 193, 282 n. 17; effects of, 166 n. 40,

248–51; manufacturing sectors, 41, 50, 74, 196, 254. *See also* economic liberalization policies
trade unions, 200, 210, 274. *See also* labor unions
traditional industries, 171–72, 175, 185
Transparency International index of corruption, 95 n. 18, 99 n. 20, 114 n. 10
transport sector: capacity utilization, 252, 254; exports, 176, 183; growth rates, 60; investment rates, 182; productivity levels, 61, 62; promotion of, 170; wage levels, 59
Tribunal de Responsabilidad Administrativa, 105
trienio, 198–200, 201, 203, 220
Troncoso, Eduardo, 175
trust funds, 241 n. 14
Twelve Apostles, 212–13, 234

unemployment, 19, 234 n. 6, 242. *See also* employment
unionization, rate of, 119
Unión Republicana Democrática (Democratic Republican Union, URD), 198–99, 201, 207
United Malays National Organization (UMNO), 273–74, 278, 280, 281, 282
United Socialist Party of Venezuela (PSUV), 223 n. 59
United States: growth rates, 85, 292; influence of, 207 n. 30, 228; productivity levels in, 52, 53, 54, 55, 56, 58, 60, 62, 63; trade with, 193–94, 247; wage levels, 52, 53, 54, 55, 56, 58, 59, 62
unit of activity, 141 n. 7
Un Nuevo Tiempo (political party), 216 n. 40
upper class, 236
urban bias theory, 49
urbanization: Malaysian, 274; rapid, 48–50, 190 n. 6, 196–97, 205, 242
Uruguay, 29, 79 n. 1, 126
Uslar Pietri, Arturo, 103 n. 24

Velasco, Andrés, 294
VENALUM (state aluminum firm), 182
Venezuelan Disease, 238
Viana, Horacio, 254
VIASA (national airline), 73
violence: monopoly control of, 146, 187 n. 2, 189, 208, 227; political, 264–65
Vishny, Robert, 101
voracity effect, oil booms, 91
voters, 199; abstention rates, 125 n. 28; cost per, 212 n. 35

wages: declines in, 19, 61, 117–18, 125, 234 n. 6, 238, 262; increases in, 39, 51–52, 199, 234 n. 5; levels of, 53, 54, 55, 56; rigidity model of, 48–64. *See also* income
warlords, 85, 187 n. 2, 189 n. 4, 227, 228
Warner, Andrew, 35 n. 2
wars, 82–85. *See also* coup attempts
Washington Consensus, 117
water supply investment, 199
wealth concentrations, 31
Weber, Max, 143
Weisbrot, Mark, 19, 20, 169, 183, 221 n. 50, 222
West Germany, 87 n. 11, 292
whistle-blowing, 125, 214
white-collar workers, 276
Wijnbergen, Sweder van, 40
Williams, David, 295
Williamson, Oliver, 141
working class, 196, 202, 203, 205, 278, 282
World Bank, 50 n. 16, 75 n. 36, 102, 108 n. 1, 109, 113, 294
World War II, 292
Wright, Gavin, 85, 301

Zambia, 299
ZULIA (state steel company), 182

www.ingramcontent.com/pod-product-compliance
Lightning Source LLC
Chambersburg PA
CBHW032127010526
44111CB00033B/157